Herbicide Resistance in Plants

Herbicide Resistance in Plants

Special Issue Editor

Hugh J. Beckie

MDPI • Basel • Beijing • Wuhan • Barcelona • Belgrade • Manchester • Tokyo • Cluj • Tianjin

Special Issue Editor
Hugh J. Beckie
University of Western Australia
Australia

Editorial Office
MDPI
St. Alban-Anlage 66
4052 Basel, Switzerland

This is a reprint of articles from the Special Issue published online in the open access journal *Plants* (ISSN 2223-7747) (available at: https://www.mdpi.com/journal/plants/special_issues/herbicide_resist_plant).

For citation purposes, cite each article independently as indicated on the article page online and as indicated below:

LastName, A.A.; LastName, B.B.; LastName, C.C. Article Title. *Journal Name* **Year**, *Article Number*, Page Range.

ISBN 978-3-03936-008-6 (Hbk)
ISBN 978-3-03936-009-3 (PDF)

Contents

About the Special Issue Editor . . . . **vii**

Hugh J Beckie
Herbicide Resistance in Plants
Reprinted from: *Plants* **2020**, *9*, 435, doi:10.3390/plants9040435 . . . . **1**

Ricardo Alcántara-de la Cruz, Pablo Alfredo Domínguez-Martínez, Hellen Martins da Silveira, Hugo Enrique Cruz-Hipólito, Candelario Palma-Bautista, José Guadalupe Vázquez-García, José Alfredo Domínguez-Valenzuela and Rafael De Prado
Management of Glyphosate-Resistant Weeds in Mexican Citrus Groves: Chemical Alternatives and Economic Viability
Reprinted from: *Plants* **2019**, *8*, 325, doi:10.3390/plants8090325 . . . . **5**

Silvia Panozzo, Alberto Collavo and Maurizio Sattin
Sensitivity Analysis of Italian *Lolium* spp. to Glyphosate in Agricultural Environments
Reprinted from: *Plants* **2020**, *9*, 165, doi:10.3390/plants9020165 . . . . **19**

J António Tafoya-Razo, Ernesto Oregel-Zamudio, Sabina Velázquez-Márquez and Jesús R. Torres-García
10,000-Times Diluted Doses of ACCase-Inhibiting Herbicides Can Permanently Change the Metabolomic Fingerprint of Susceptible *Avena fatua* L. Plants
Reprinted from: *Plants* **2019**, *8*, 368, doi:10.3390/plants8100368 . . . . **33**

Brent P. Murphy and Patrick J. Tranel
Target-Site Mutations Conferring Herbicide Resistance
Reprinted from: *Plants* **2019**, *8*, 382, doi:10.3390/plants8100382 . . . . **45**

Mithila Jugulam and Chandrima Shyam
Non-Target-Site Resistance to Herbicides: Recent Developments
Reprinted from: *Plants* **2019**, *8*, 417, doi:10.3390/plants8100417 . . . . **61**

Sara L. Martin, Jean-Sebastien Parent, Martin Laforest, Eric Page, Julia M. Kreiner and Tracey James
Population Genomic Approaches for Weed Science
Reprinted from: *Plants* **2019**, *8*, 354, doi:10.3390/plants8090354 . . . . **77**

Eric L. Patterson, Christopher Saski, Anita Küpper, Roland Beffa and Todd A. Gaines
Omics Potential in Herbicide-Resistant Weed Management
Reprinted from: *Plants* **2019**, *8*, 607, doi:10.3390/plants8120607 . . . . **119**

Franck E. Dayan
Current Status and Future Prospects in Herbicide Discovery
Reprinted from: *Plants* **2019**, *8*, 341, doi:10.3390/plants8090341 . . . . **133**

Vijay K. Nandula
Herbicide Resistance Traits in Maize and Soybean: Current Status and Future Outlook
Reprinted from: *Plants* **2019**, *8*, 337, doi:10.3390/plants8090337 . . . . **151**

Hugh J. Beckie, Michael B. Ashworth and Ken C. Flower
Herbicide Resistance Management: Recent Developments and Trends
Reprinted from: *Plants* **2019**, *8*, 161, doi:10.3390/plants8060161 . . . . **161**

Martin M. Vila-Aiub
Fitness of Herbicide-Resistant Weeds: Current Knowledge and Implications for Management
Reprinted from: *Plants* **2019**, *8*, 469, doi:10.3390/plants8110469 . . . . **175**

About the Special Issue Editor

Hugh J. Beckie Professor, School of Agriculture and Environment, University of Western Australia; and Director, Australian Herbicide Resistance Initiative (AHRI). Hugh farmed in Saskatchewan, Canada, for 30 years. He spent 26 years as a weed scientist with Agriculture and Agri-Food Canada, and Adjunct Professor at the University of Alberta. His research focused on the surveillance, risk assessment, and management of herbicide-resistant weeds as well as impact assessments of GM crops. Hugh is a Fellow of the Canadian Weed Science Society and Weed Science Society of America, and received the QEII Diamond Jubilee medal. In 2018, he became Director of AHRI and Professor of Crop Weed Science at the University of Western Australia.

plants

MDPI

Editorial

Herbicide Resistance in Plants

Hugh J Beckie

Australian Herbicide Resistance Initiative (AHRI), School of Agriculture and Environment, The University of Western Australia, Perth, WA 6009, Australia; hugh.beckie@uwa.edu.au

Received: 11 March 2020; Accepted: 26 March 2020; Published: 1 April 2020

Abstract: Herbicide resistance in weeds is perhaps the most prominent research area within the discipline of weed science today. Incidence, management challenges, and the cost of multiple-resistant weed populations are continually increasing worldwide. Crop cultivars with multiple herbicide-resistance traits are being rapidly adopted by growers and land managers to keep ahead of the weed resistance tsunami. This Special Issue of *Plants* comprises papers that describe the current status and future outlook of herbicide resistance research and development in weedy and domestic plants, with topics covering the full spectrum from resistance mechanisms to resistance management. The unifying framework for this Special issue, is the challenge initially posed to all of the contributors: what are the (potential) implications for herbicide resistance management?

Keywords: herbicide resistance; non-target-site resistance; precision weed management; resistance management; weed biology; weed genomics

1. Introduction

Since the first global cases of herbicide resistance in weeds in the late 1950s, there are now over 500 unique cases reported in non-cropland and almost 100 different crops in 70 countries—over 260 species compromising the efficacy of over 160 herbicides or over 20 herbicide sites of action (SOA) [1]. The current rate of increase in the number of weed species resistant to glyphosate (e.g., see Alcántara-de la Cruz et al. [2] this issue) is second only to that of acetolactate synthase inhibitors. Since first introduced in the early 1980s, cultivars of major agronomic field crops possessing herbicide-resistance traits now occupy a significant proportion of the global crop production area [3,4]. This Special Issue presents a collection of papers that highlight the continuing breadth and depth of basic and applied herbicide resistance research and development in both weedy and crop species. As the privileged guest editor, I share my perspectives on key messages, and future directions gleaned from these volunteered or invited contributions.

2. Key Messages

An integral element of herbicide resistance surveillance is the periodic sensitivity analysis of populations of a weed species in an agroregion to commonly used herbicides. Such an analysis provides information on the inter- and intra-population variability in the effective dose (ED) required for 50 or 90%, etc., reduction in survival or biomass. Therefore, sensitivity analysis can determine if populations are becoming less sensitive to a herbicide over time, and if label rates need to be adjusted accordingly. These foundational studies are extremely important in mitigating quantitative (creeping) resistance evolution, particularly for key herbicides such as glyphosate and major problematic outcrossing weeds such as *Lolium* spp. [5,6]. Intentional or unintentional sublethal herbicide doses may even alter the metabolism, growth, and survival of susceptible plants of highly-selfing species, such as demonstrated for *Avena fatua* L. (wild oat) [7].

Target-site mutations conferring evolved herbicide resistance in weeds are known in nine different herbicide SOA. An emerging trend is increased cases of multiple mutations, including multiple amino acid changes at the glyphosate target site as well as mutations involving two nucleotide changes at a single amino acid codon [8]. Non-target-site resistance (NTSR) to herbicides in weeds, such as enhanced metabolism by P450 monooxygenases, is an increasingly serious threat to sustainable weed management as the efficacy of multiple SOA herbicides may be compromised. Although much more difficult to investigate than target-site resistance, steady advances are being made in the physiological, biochemical and molecular basis of NTSR mechanisms in weeds [9].

The fields of genomics, transcriptomics, proteomics, and metabolomics—collectively referred to as 'omics'—describe the component parts of the biological system that lead to the presentation of traits. Unravelling the genome of major global weedy species will greatly facilitate the identity and function of major and minor genes responsible for herbicide resistance [10]. Draft weed genomes can provide insights on the evolutionary origins of weeds, allowing identification of management practices that may mitigate resistance evolution. Moreover, genomics can identify strengths and weaknesses of weed populations that can be targeted for control, while providing fundamental information on how plants rapidly respond to herbicide selection. The weed omics era of today is enabling translational research to bridge from basic science to field applications, by linking systems-scale science to applied science for practitioners [11]. Weed science is still learning how to integrate omics technologies into the discipline; however, omics techniques are more frequently being implemented in novel ways to address basic questions in weed biology or practical questions of improving weed management; for the latter, the potential benefits of weed omics will be best realized for farms utilizing advanced data science approaches necessary for the implementation of digital farming [11].

After a 35-year hiatus in the commercialization of new SOA herbicides, there is now optimism in the agri-chemical industry as new SOA herbicides are being introduced for control of key economic weeds in major agronomic crops. A review in this issue of the current status and future prospects in herbicide discovery offer insights into novel potential target sites in plants and innovative approaches or processes to facilitate new herbicide SOA discovery [12]. Because of this hiatus in SOA discovery and commercialization, cultivars of the major agronomic crops, particularly maize (*Zea mays* L.) and soybean (*Glycine max* L. Merr.), are being conventionally bred or genetically engineered with combined (stacked) pesticide-resistance traits. A review in this issue summarizes their current status and future outlook [13]. Recent global developments and trends in herbicide resistance management also include the increasing reliance on pre-emergence vs. post-emergence herbicides because of weed resistance, breeding for weed-competitive cereal crop cultivars, expansion of harvest weed seed control practices, and advances in site-specific or precision weed management (via prescription maps or in real-time) [14].

3. Future Directions

Natural selection for herbicide-resistant weed genotypes may act on standing genetic variation or on a genetic and physiological background that is altered because of stress responses to sublethal herbicide exposure. Stress-induced changes include DNA mutations, epigenetic alterations, transcriptional remodeling, and protein modifications, all of which can lead to herbicide resistance and various pleiotropic effects [15]. Studies examining stress-induced evolution of herbicide resistance and related pleiotropic effects are needed to inform improved herbicide-resistant weed prevention and management strategies [7]. As both the incidence of weed populations with NTSR and the worldwide occurrence of environmental stress are expected to increase, expanded research on NTSR evolution and its potential for pleiotropic effects should be a high priority [15].

A primary goal driving the need to characterize herbicide resistance mechanisms is the management of herbicide-resistant weeds. Better understanding is needed of the relationship between target-site resistance mutations or mechanisms in troublesome weed species, their geographic distribution and prevalence across an agroregion, resulting in cross-resistance patterns, and associated fitness costs. Continuing advances or improvements are expected in the efficiency and accuracy of high throughput in vitro diagnostic techniques [16]. A uniform and replicable system for in planta functional validation, which is the gold standard for demonstrating resistance and susceptibility, is necessary to facilitate high-throughput screening initiatives [8]. Because the evolution of NTSR via herbicide metabolism is a serious threat to weed management, identification of the genes endowing resistance and their functional characterization are important future research goals for possible mitigation and management strategies. The increasing availability of sequenced genomes for different weed species will greatly accelerate research in this area [10]. Accurately assessing fitness costs of resistance and deriving practical management tactics to potentially exploit this phenomenon in resistant weed populations will continue to be an important research endeavour [17].

Omics research in weed science faces several challenges, including management of large and complex omics datasets, efficient and accurate annotation of reference genome assemblies and eventual pan genomes, and the large number of weed species with a diversity of weedy traits and variation in evolutionary strategies. Examining the diverse ways that researchers working in model systems use omics technologies in their respective fields can provide established tools and templates to address the future needs of the weed science community [11]. In particular, method standardization for utilizing next generation sequencing in weed science, improving herbicide resistance diagnostics with omics, and improved gene function validation for herbicide resistance mechanisms are attainable medium-term (5 to 10 year) goals. Can we alter weed populations to make them easier to control? Current and future omics tools to improve herbicide-resistant weed management, such as gene drive systems for sensitizing herbicide-resistant weed populations, requires proof of concept studies but has promising long-term potential [10,11,18]. A better understanding of weed species at the population, genomic, and genic levels using population genomic approaches will help begin to address that question.

Ultimately, basic or applied herbicide resistance research and development should inform resistance management by growers and land managers. Sustaining the utility of existing herbicides and effective stewardship guidelines for herbicide-resistant crops will continue to demand innovative research and development to address these challenges. Adoption of some recommended best management practices by end-users may require private or public sector financial incentives. Recent advances in precision or digital agriculture have largely been driven by significant private-sector investments. It offers the best route for optimizing crop production and crop protection across a field by varying input levels commensurate with site-specific soil or environmental conditions that govern yield potential. The ongoing challenge is the development of user-friendly and cost-effective technologies or systems that can be easily integrated into existing farming enterprises.

Acknowledgments: I sincerely thank the contributors for agreeing to participate in this writing project. On their behalf, I extend my appreciation to each of the reviewers who generously gave of their time and energy to ensure the expected high scientific standards. Finally, I thank Ms. Sylvia Guo, Managing Editor *Plants*, for her professional assistance from conception to completion of this Special Issue.

Conflicts of Interest: The author declares no conflict of interest.

References

1. Heap, I.M. International Survey of Herbicide Resistant Weeds. 2020. Available online: http://www.weedscience.org (accessed on 10 March 2020).
2. Alcántara-de la Cruz, R.; Domínguez-Martínez, P.A.; da Silveira, H.M.; Cruz-Hipólito, H.E.; Palma-Bautista, C.; Vázquez-García, J.G.; Domínguez-Valenzuela, J.A.; de Prado, R. Management of glyphosate-resistant weeds in Mexican citrus groves: Chemical alternatives and economic viability. *Plants* **2019**, *8*, 325. [CrossRef] [PubMed]
3. Beckie, H.J.; Harker, K.N.; Hall, L.M.; Warwick, S.I.; Légère, A.; Sikkema, P.H.; Clayton, G.W.; Thomas, A.G.; Leeson, J.Y.; Séguin-Swartz, G.; et al. A decade of herbicide-resistant crops in Canada. *Can. J. Plant Sci.* **2006**, *86*, 1243–1264. [CrossRef]
4. Green, J.M.; Owen, M.D.K. Herbicide-resistant crops: Utilities and limitations for herbicide-resistant weed management. *J. Agric. Food Chem.* **2011**, *59*, 5819–5829. [CrossRef] [PubMed]

5. Panozzo, S.; Collavo, A.; Sattin, M. Sensitivity analysis of Italian *Lolium* spp. to glyphosate in agricultural environments. *Plants* **2020**, *9*, 165. [CrossRef] [PubMed]
6. Busi, R.; Powles, S.B. Evolution of glyphosate resistance in a *Lolium rigidum* population by glyphosate selection at sublethal doses. *Heredity* **2009**, *103*, 318–325. [CrossRef] [PubMed]
7. Tafoya-Razo, J.A.; Oregel-Zamudio, E.; Velázquez-Márquez, S.; Torres-García, J.R. 10,000-times diluted doses of ACCase-inhibiting herbicides can permanently change the metabolomic fingerprint of susceptible *Avena fatua* L. plants. *Plants* **2019**, *8*, 368. [CrossRef] [PubMed]
8. Murphy, B.P.; Tranel, P.J. Target-site mutations conferring herbicide resistance. *Plants* **2019**, *8*, 382. [CrossRef] [PubMed]
9. Jugulam, M.; Shyam, C. Non-target-site resistance to herbicides: Recent developments. *Plants* **2019**, *8*, 417. [CrossRef] [PubMed]
10. Martin, S.L.; Parent, J.-S.; Laforest, M.; Page, E.; Kreiner, J.M.; James, T. Population genomic approaches for weed science. *Plants* **2019**, *8*, 354. [CrossRef] [PubMed]
11. Patterson, E.L.; Saski, C.; Küpper, A.; Beffa, R.; Gaines, T.A. Omics potential in herbicide-resistant weed management. *Plants* **2019**, *8*, 607. [CrossRef] [PubMed]
12. Dayan, F.E. Current status and future prospects in herbicide discovery. *Plants* **2019**, *8*, 341. [CrossRef] [PubMed]
13. Nandula, V.K. Herbicide resistance traits in maize and soybean: Current status and future outlook. *Plants* **2019**, *8*, 337. [CrossRef] [PubMed]
14. Beckie, H.J.; Ashworth, M.B.; Flower, K.C. Herbicide resistance management: Recent developments and trends. *Plants* **2019**, *8*, 161. [CrossRef] [PubMed]
15. Dyer, W.E. Stress-induced evolution of herbicide resistance and related pleiotropic effects. *Pest Manag. Sci.* **2018**, *74*, 1759–1768. [CrossRef] [PubMed]
16. Délye, C.; Michel, S.; Pernin, F.; Gautier, V.; Gislard, M.; Poncet, C.; Le Corre, V. Harnessing the power of next generation sequencing technologies to the purpose of high-throughput pesticide resistance diagnosis. *Pest Manag. Sci.* **2020**, *76*, 543–552. [CrossRef] [PubMed]
17. Vila-Aiub, M.M. Fitness of herbicide-resistant weeds: Current knowledge and implications for management. *Plants* **2019**, *8*, 469. [CrossRef] [PubMed]
18. Neve, P. Gene drive systems: Do they have a place in agricultural weed management? *Pest Manag. Sci.* **2018**, *74*, 2671–2679. [CrossRef] [PubMed]

plants

MDPI

Article

Management of Glyphosate-Resistant Weeds in Mexican Citrus Groves: Chemical Alternatives and Economic Viability

Ricardo Alcántara-de la Cruz [1,*], Pablo Alfredo Domínguez-Martínez [2], Hellen Martins da Silveira [3], Hugo Enrique Cruz-Hipólito [4], Candelario Palma-Bautista [5], José Guadalupe Vázquez-García [5] José Alfredo Domínguez-Valenzuela [6,*] and Rafael De Prado [5]

1 Departamento de Química, Universidade Federal de São Carlos, São Carlos 13565-905, Brazil
2 National Institute of Forestry, Agriculture and Livestock Research (INIFAP)-Valle del Guadiana Experimental Field, Durango 34170, Mexico
3 Departamento de Fitotecnia, Universidade Federal de Viçosa, Viçosa 36570-900, Brazil
4 Bayer Crop Science Mexico, Mexico 11520, Mexico
5 Department of Agricultural Chemistry and Edaphology, University of Cordoba, 14071 Cordoba, Spain
6 Department of Agricultural Parasitology, Chapingo Autonomous University, Texcoco 56230, Mexico
* Correspondence: ricardo.cruz@ufscar.br (R.A.-d.l.C.); jose_dv001@yahoo.com.mx (J.A.D.-V.)

Received: 1 August 2019; Accepted: 2 September 2019; Published: 4 September 2019

Abstract: Glyphosate is a cheap herbicide that has been used to control a wide range of weeds (4–6 times/year) in citrus groves of the Gulf of Mexico; however, its excessive use has selected for glyphosate-resistant weeds. We evaluated the efficacy and economic viability of 13 herbicide treatments (glyphosate combined with PRE- and/or POST-emergence herbicides and other alternative treatments), applied in tank-mixture or sequence, to control glyphosate-resistant weeds in two Persian lime groves (referred to as SM-I and SM-II) of the municipality of Acateno, Puebla, during two years (2014 and 2015). The SM-I and SM-II fields had 243 and 346 weeds/m^2, respectively, composed mainly of *Bidens pilosa* and *Leptochloa virgata*. *Echinochloa colona* was also frequent in SM-II. The glyphosate alone treatments (1080, 1440, or 1800 g ae ha^{-1}) presented control levels of the total weed population ranging from 64% to 85% at 15, 30, and 45 d after treatment (DAT) in both fields. Mixtures of glyphosate with grass herbicides such as fluazifop-p-butyl, sethoxydim, and clethodim efficiently controlled *E. colona* and *L. virgata*, but favored the regrowth of *B. pilosa*. The sequential applications of glyphosate + (bromacil + diuron) and glufosinate + oxyfluorfen controlled more than 85% the total weed community for more than 75 days. However, these treatments were between 360% and 390% more expensive (1.79 and 1.89 \$/day ha^{-1} of satisfactory weed control, respectively), compared to the representative treatment (glyphosate 1080 g ae ha^{-1} = USD \$29.0 ha^{-1}). In practical and economic terms, glufosinate alone was the best treatment controlling glyphosate resistant weeds maintaining control levels >80% for at least 60 DAT (\$1.35/day ha^{-1}). The rest of the treatments, applied in tank-mix or in sequence with glyphosate, had similar or lower control levels (~70%) than glyphosate at 1080 g ae ha^{-1}. The adoption of glufosiante alone, glufosinate + oxyfluorfen or glyphosate + (bromacil + diuron) must consider the cost of satisfactory weed control per day, the period of weed control, as well as other factors associated with production costs to obtain an integrated weed management in the short and long term.

Keywords: *Citrus latifolia*; hairy beggarticks; integrated weed management; junglerice; tropical sprangletop

1. Introduction

Citriculture is an important activity in Mexico occupying ~40% of the total area devoted to fruticulture [1]. The state of Veracruz is the biggest producer of citrus fruits with ~225,000 ha, and the orange (*Citrus sinensis*) occupies the largest crop area (55%); however, the Persian lime (*C. latifolia*), with ~47,000 ha [2], makes the highest economic contribution [3]. In this way, Persian lime plantations receive more care than other citrus crops from small and large growers [4], such as fertilization (foliar and soil), prunings, control of diseases (*Colletotrichum gloeosporioides*), pests (*Diaphorina citri* and *Phyllocnistis citrella*), and weeds (chemical and mechanical) [5–7].

Weed competition require special attention in young citrus trees and during blooming and fruit setting [8], but in older plantations, the impacts of weeds are also indirect, mainly by limiting crop management [9], in addition to the fact that weeds can be hosts for pests and diseases [6]. Thus, weed management in the citrus-producing region of the Gulf of Mexico, by combining chemical and non-chemical (manual or mechanical mowing) methods, is carried out four to six times a year [7,9,10], representing ~8–12% of production costs ha^{-1} [11].

Glyphosate is a systemic and non-residual and post-emergence (POST) herbicide that controls a wide range of weeds [12], making it preferred by the Mexican citrus growers [7,13]. The doses recommended by the manufacturers of this herbicide range from 700 to 2100 g ae ha^{-1} [14], according to the weed species, phenological stage, and infestation level. Due to the frequent glyphosate applications, most Mexican citrus growers have widely adopted and applied doses ranging from 720 to 1080 g ae ha^{-1} for up to 15 years [9]. The high dependence of glyphosate-based herbicides has led to the selection of resistant populations of *Bidens pilosa* [15], *Eleusine indica* [4], *Leptochloa virgata* [7], and *Parthenium hysterophorus* [13] between 2010 and 2016 in the citrus-producing region of the states of Puebla and Veracruz, Gulf of Mexico.

Despite the evident loss of glyphosate efficacy in controlling weeds, Mexican citrus growers continue using this herbicide for its low cost, which makes it necessary to look for weed management alternatives that help to extend the useful life of this herbicide [12]. Glufosinate and paraquat, POST, broad-spectrum, and non-residual herbicides like glyphosate are also used in citrus groves [7]. However, these herbicides have short control periods forcing growers to make more applications, which is more expensive than with glyphosate alone. In addition, when glufosinate or paraquat are applied in late POST in order to reduce the number of herbicide applications, weed control is poor (personal communication with growers). Management strategies with pre-emergence (PRE) and early POST herbicides could reduce the selection pressure exerted by glyphosate on weeds that are difficult to control [16], as well as production costs [11].

In this work, we evaluated the efficacy and economic viability of 13 herbicide treatments (glyphosate combined with PRE and/or POST herbicides and other alternative treatments), applied in tank-mixture or sequence (Table 1), to control weeds, including glyphosate-resistant species in the citrus-producing region of the Gulf of Mexico.

Table 1. Herbicides (Treatments), mechanism of action (MOA), field rates in g ai or ea ha^{-1} (Rate), liters of commercial herbicide ha^{-1} (liters), application time (Time) of pre- (PRE) and post-emergence (POST) herbicides for the weed control in two Persian lime groves of the "San Manuel" Farm, Puebla, Mexico, and cost of each treatment ha^{-1} (USD).

	Treatments [1]	MOA [2]	Rate	Liters	Time	USD [3]
-	Control	-	-	-	-	-
1	Gly 1080	EPSPS	1080	3	POST	29.0
2	Gly 1440	EPSPS	1440	4	POST	38.7
3	Gly 1800	EPSPS	1800	5	POST	48.4
4	Gly + Flua	EPSPS + ACCase	1080 + 250	3 + 2	POST	100.9
5	Gly + Ace †	EPSPS + Mitosis	1080 + 1678	3 + 2	POST+PRE	57.3
6	(Gly + Oxi) + Flua †	(EPSPS+PPO) + ACCase	(1080 + 480) + 250	(3 + 2) + 2	POST+PRE	97.4
7	Gly + Seth	EPSPS + ACCase	1080 + 368	3 + 2	POST	82.0
8	Gly + Cleth	EPSPS + ACCase	1080 + 236	3 + 2	POST	70.0
9	Glufos	GS	450	2	POST	54.1
10	Par + Diu	PSI + PSII	400 + 200	2	POST	26.8
11	Glufos + Oxi †	GS + PPO	420 + 480	2 + 2	POST + PRE	134.6
12	Gly + Oxa †	GS + PPO	1080 + 1000	4 + 3	POST + PRE	112.0
13	Gly + (Brom + Diu) †	EPSPS + (PSII + PSII)	1080 + (1200 + 1200)	3 + 3	POST + PRE	142.4

[1] Gly = Faena® Fuerte 360 (SC, 35.6% glyphosate w/v); Flua = Fusilade BIW® (EC, 12.5% fluazifop-p-butyl w/v); Ace = Harness® EC (EC, 60% acetochlor w/v); Oxi = Goal® 2XL (EC, 22.3% oxifluorfen w/v); Seth = Poast® (CL, 18.4% sethoxydim w/v); Cleth = Select® Ultra (CE, 12.5% clethodim w/v); Glufos = Finale® (CE, 15% glufosinate w/v); Par + Diu = Gramocil® (SC, 20 + 10% paraquat + diuron w/v); Oxa= Ronstar® 25CE (CE, 24.4% oxadiazon w/v); and Brom + Diu = Krovar® (WG, 40 + 40% bromacil + diuron w/w). *Mention of trade names in this publication is solely for providing specific information and does not imply their recommendation.* [2] Mechanism of action: Inhibitors of enolpyruvyl shikimate-3-phosphate synthase (EPSPS), acetyl-CoA carboxylase (ACCase), mitosis, protoporphyrinogen oxidase (PPO), glutamine synthetase (GS), photosystem I (PSI) and II (PSII). [3] Average exchange rate of the Mexican peso (MNX) to US dollar (USD) corresponding to January 2014 (13.20 = 1.0) and January 2015 (14.67 = 1.0), respectively. † Treatments applied in sequence 15 days after the first application.

2. Results

2.1. Initial Weed Density

The average density of weeds was 242.9 and 345.6 plants m^2 in the SM-I and SM-II fields, respectively. Weed community was composed mainly of *B. pilosa* and *L. virgata* in both fields. In addition, *Echinochloa colona* was frequent in SM-II. Density of weeds showed no differences between years and *B. pilosa* presented the highest density (Table 2). Species such as *Amaranthus viridis, Cynodon nlemfuensis, Digitaria sanguinalis, Eleusine indica,* and *Parthenium hysterophorus* were sporadic in SM-I, and *E. indica* and *Rottboellia cochinchinesis* in SM-II. Due to the low density of these weeds, they were not considered for the analysis of herbicide control per species.

Table 2. Initial weed density (plants m^2) in two Persian lime groves of the "San Manuel" Farm, Puebla, Mexico.

Species	San Manuel I		San Manuel II	
	2014	2015	2014	2015
B. pilosa	114.8 ± 4.6	117.3 ± 3.7	195.9 ± 8.6	182.5 ± 6.3
L. virgata	98.7 ± 4.7	104.8 ± 3.4	49.8 ± 3.0	58.7 ± 4.6
E. colona	7.3 ± 3.2	11.4 ± 2.08	96.7 ± 5.9	87.3 ± 4.1
Other weeds	19.6 ± 2.6	15.6 ± 1.7	8.0 ± 2.1	12.4 ± 3.8
Total	237.7	248.1	350.3	340.9

±Standard error of the mean (n = 28).

2.2. Total Control of Weeds

The glyphosate treatments of 1080, 1440, and 1800 g ae ha^{-1} presented similar control levels at 15, 30, and 45 days after treatment (DAT) in both SM-I and SM-II fields ranging from 64% to 85%. Weed control with 1080 g ae ha^{-1} of glyphosate was ~10–20% lower in relation to the other two-glyphosate treatments at the 60 and 75 DAT. Most of the herbicides, applied in tank-mix or in sequence with glyphosate, had similar control levels (~70%) than the lowest dose of glyphosate (1080 g ae ha^{-1}) at 15 DAT, except clethodim and oxadiazon in both fields, and fluazifop-p-butyl, oxyfluorfen + fluazifop-p-butyl (in sequence), and sethoxydim in SM-II that showed lower control level than glyphosate alone. Acetochlor in SM-II and fluazifop-p-butyl, sethoxydim, and clethodim in tank mixture, and fluazifop-p-butyl and oxadiazon in sequential application in SM-II, showed greater control at 30 DAT than at 15 DAT. As of this period, the control of these herbicides was similar or less than the control obtained with glyphosate at 1080 g ae ha^{-1}, except the sequential application of bromacil + diuron. The control level of the latter treatment increased from 30 DAT up to 90% until the end of the experiments. Glufosinate and glufosinate + oxyfluorfen (in sequence) showed the highest levels of control (>95%) at 15 and 30 DAT. The last treatment showed a control level above 90% at 75 DAT, while glufosinate alone decreased to 77%. Paraquat + diuron had control levels above 85% at 15 and 30 DAT but decreased to 48% at 75 DAT (Table 3).

Table 3. Total weed control percentage with pre- and post-emergence herbicides in two Persian lime groves of the "San Manuel" Farm, Puebla, Mexico from 15 to 75 days after treatment (DAT). Visual control was measured as 0 = no control and 100 = plant death.

Treatment [1]	15 DAT	30 DAT	45 DAT	60 DAT	75 DAT
		San Manuel I			
Control	-	-	-	-	-
Gly 1080	74.2 ± 2.4 c	64.2 ± 1.5 e	68.3 ± 1.7 d	48.3 ± 2.8 ef	41.7 ± 2.5 ef
Gly 1440	72.5 ± 3.1 c	78.3 ± 1.1 cd	73.3 ± 2.1 cd	64.2 ± 3.3 cd	49.2 ± 2.0 de
Gly 1800	84.2 ± 2.0 b	76.7 ± 2.1 d	78.3 ±1.1 c	62.5 ± 2.8 cd	55.0 ± 1.8 cd
Gly + Flua	68.3 ± 1.7 c	51.7 ± 1.7 f	53.3 ± 3.1 ef	41.7 ± 2.5 f	35.8 ± 2.0 f
Gly + Ace †	69.2 ± 2.4 c	84.2 ± 2.0 bcd	71.7 ± 1.7 cd	59.2 ± 3.0 de	63.3 ± 2.1 c
(Gly + Oxi) + Flua †	71.7 ± 2.5 c	64.2 ± 1.5 e	70.8 ± 1.5 cd	64.2 ± 2.4 cd	52.5 ± 1.7 d
Gly + Seth	73.3 ± 2.1 c	37.5 ± 1.7 g	58.3 ± 2.1 e	55.0 ± 2.2 de	45.0 ± 1.8 def
Gly + Cleth	66.7 ± 1.7 c	34.2 ± 2.7 g	46.7 ± 2.5 f	42.5 ± 2.1 f	20.0 ± 2.9 g
Glufos	100.0 ± 0 a	97.5 ± 1.1 a	88.3 ± 2.1 b	85.8 ± 1.5 b	77.5 ± 1.1 b
Par + Diu	86.7 ± 1.1 b	86.7 ± 2.1 bc	74.2 ± 0.8 cd	70.8 ± 0.8 cd	54.2 ± 2.4 cd
Glufos + Oxi †	100.0 ± 0 a	99.2 ± 0.8 a	97.5 ± 1.8 a	96.3 ± 1.7 a	91.7 ± 1.7 a
Gly + Oxa †	65.0 ± 1.3 c	68.3 ± 2.5 de	59.2 ± 1.5 e	38.3 ± 2.5 f	35.8 ± 2.4 f
Gly + (Brom + Diu) †	70.8 ± 2.7 c	88.3 ± 1.7 b	91.7 ± 2.5 ab	92.5 ± 2.5 ab	89.2 ± 2.7 a
		San Manuel II			
Control	-	-	-	-	-
Gly 1080	66.7 ± 2.5 cd	80.8 ± 4.0 de	69.2 ± 1.5 bc	61.7 ± 3.3 cd	43.3 ± 2.1 ef
Gly 1440	76.7 ± 2.1 b	80.8 ± 1.5 de	69.2 ± 2.0 bc	70.8 ± 2.0 bc	51.7 ± 2.5 de
Gly 1800	75.0 ± 2.6 bc	85.8 ± 2.0 cde	76.7 ± 2.1 b	74.2 ± 3.0 b	57.5 ± 2.1 cd
Gly + Flua	48.3 ± 1.1 fg	80.0 ± 2.9 de	31.7 ± 2.8 f	48.3 ± 2.5 e	39.2 ± 2.0 f
Gly + Ace †	75.8 ± 3.0 bc	85.0 ± 1.8 cde	88.3 ± 2.5 a	87.5 ± 2.5 a	65.8 ± 1.5 c
(Gly + Oxi) + Flua †	53.3 ± 2.1 ef	86.7 ± 2.5 bcd	42.5 ± 1.1 de	36.7 ± 2.1 f	27.5 ± 2.8 g
Gly + Seth	40.8 ± 2.4 g	77.5 ± 1.1 de	45.8 ± 2.7 de	52.5 ± 1.7 de	44.4 ± 1.5 ef
Gly + Cleth	51.7 ± 1.1 ef	75.8 ± 1.5 e	36.7 ± 3.1 ef	49.2 ± 2.4 e	40.8 ± 1.8 f
Glufos	99.2 ± 0.8 a	96.7 ± 1.7 ab	93.3 ± 1.7 a	88.3 ± 1.7 a	79.2 ± 2.0 b
Par + Diu	98.3 ± 1.1 a	85.8 ± 1.5 cde	63.3 ± 2.1 c	57.5 ± 1.7 de	48.3 ± 1.7 def
Glufos + Oxi †	100.0 ± 0 a	99.2 ± 0.8 a	97.5 ± 1.7 a	94.2 ± 2.4 a	90.8 ± 1.5 a
Gly + Oxa †	58.3 ± 2.8 de	76.7 ± 1.7 d	50.8 ± 3.0 d	54.2 ± 2.0 de	45.0 ± 1.8 ef
Gly + (Brom + Diu) †	74.2 ± 2.0 bc	94.2 ± 2.0 abc	95.8 ± 2.4 a	96.7 ± 2.1 a	90.8 ± 2.7 a

[1] Abbreviations of herbicides: Gly = Glyphosate, Flua = Fluazifop-p-butyl, Ace = Acetochlor, Oxi = Oxifluorfen, Seth = Sethoxydim, Cleth = Clethodim, Glufos = Glufosinate, Par = Paraquat, Diu = Diuron, Oxa = Oxadiazon, Bro = Bromacil. † Treatments applied in sequence 15 days after the first application. Same letter within a column showed no differences between treatments by the Tukey test ($P > 0.05$). ± Standard error of the mean of two field trials conducted in 2014 and 2015 ($n = 6$).

2.3. Control of Bidens Pilosa

The control of *B. pilosa* with the different treatments was more heterogeneous in SM-I at 30 DAT than in SM-II. None of the three-glyphosate treatments showed satisfactory levels of control. Glufosinate alone and the sequential applications of glufosinate + oxyfluorfen and glyphosate + (bromacil + diuron) presented the best control levels of *B. pilosa* in both fields at 30 and 75 DAT. Paraquat + diuron presented a control >80% at 30 DAT but decreased to 50% at 75 DAT. The other herbicides, applied in tank-mix or in sequence with glyphosate, had similar or lower control than any glyphosate treatment alone. As expected, graminicides such as fluazifop-p-butyl, sethoxydim, and clethodim did not contribute to the control of *B. pilosa* (Figure 1).

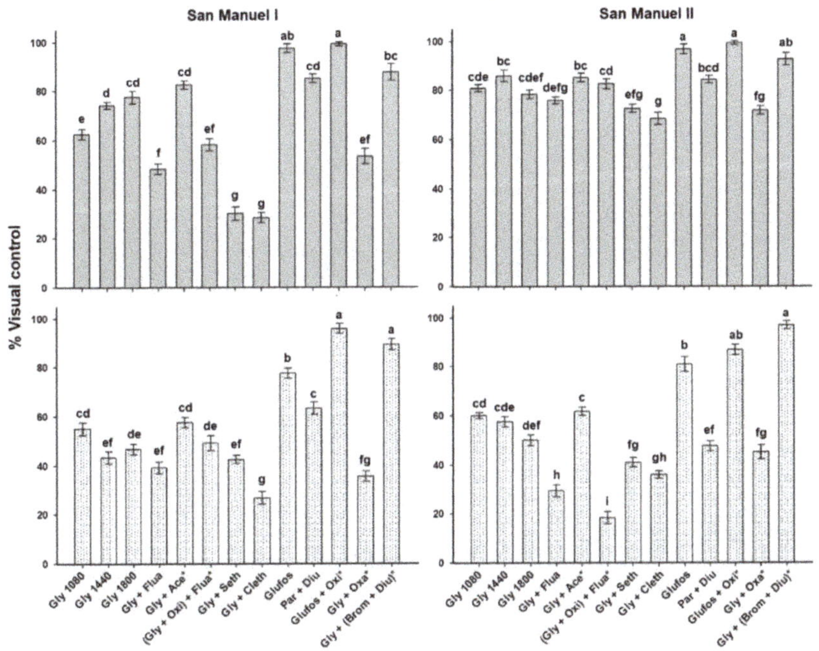

Figure 1. Control of *Bidens pilosa* in two Persian lime groves of the "San Manuel" Farm, Puebla, Mexico, at 30 (gray bars) and 75 (dotted bars) d after treatment. Same letter within a subfigure showed no differences between treatments by the Tukey test ($P > 0.05$). Vertical bars ± standard error from combined data of field trials carried out in 2014 and 2015 ($n = 6$).

2.4. Control of Leptochloa Virgata

Most of the treatments controlled *L. virgata* by 85% or higher in both fields at 30 DAT, except glyphosate alone (1080, 1440, and 1800 g ae ha^{-1}) and the sequential application with acetochlor in SM-II (control ~60%). At 75 DAT, tank-mixtures of glyphosate with fluazifop-p-butyl, oxyfluorfen + fluazifop-p-butyl (in sequence), and clethodim had control levels of ~90% in SM-I. Oxifluorfen, applied in sequence, extended the control of *L. virgata* (~80%) with glufosinate. Paraquat + diuron and the sequential applications of glyphosate with oxadiazon and bromacil + diuron maintained control levels of 76–82%. At SM-II, all herbicides, applied in tank-mixture or in sequence with glyphosate, showed greater control of *L. virgata* compared to the glyphosate alone treatments. Oxifluorfen + fluazifop-p-butyl, sethoxydim, oxadiazon, and bromacil + diuron contributed to maintaining control levels >85% at 75 DAT. Treatments that did not include glyphosate had similar control levels (glufosinate) or lower (glufosinate + oxyfluorfen and paraquat + diuron) than the previous treatments. However, these levels of control were similar to those observed in SM-I in the same period (Figure 2).

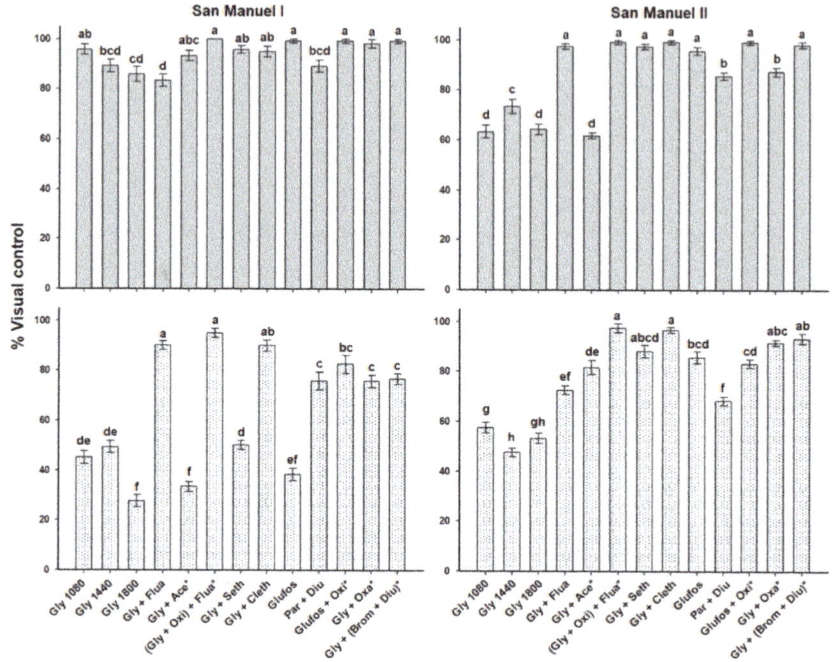

Figure 2. Control of *Leptoclhoa virgata* in two Persian lime groves of the "San Manuel" Farm, Puebla, Mexico, at 30 (gray bars) and 75 (dotted bars) d after treatment. Same letter within a subfigure showed no differences between treatments by the Tukey test ($P > 0.05$). Vertical bars ± standard error from combined data of field trials carried out in 2014 and 2015 ($n = 6$).

2.5. Control of Echinochloa Colona

Echinochoa colona occurred at high density in the field SM-II, but it was controlled by most of the treatments, including the glyphosate ones, in more than 90% at 75 DAT. Glufosinate and the mixture of paraquat + diuron showed the lowest control levels (~75%) due to low residuality (Figure 3).

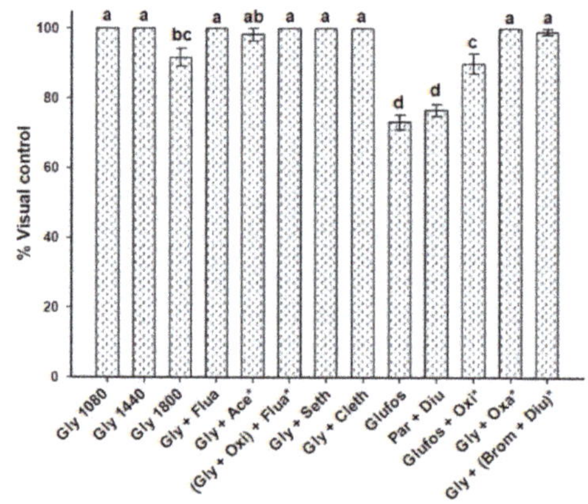

Figure 3. Control of *Echinochloa colona* in the Persian lime grove "San Manuel II" of the "San Manuel" Farm, Puebla, Mexico, at 75 days after treatment. Same letter shows no differences between treatments by the Tukey test ($P > 0.05$). Vertical bars ± standard error from combined data of field trials carried out in 2014 and 2015 ($n = 6$).

2.6. Cost of Glyphosate Resistance Management

Analyzing the cost of treatments compared to the representative treatment most used by producers (glyphosate at 1080 g ae ha-1 = USD \$29.0 ha^{-1}), with exception of the paraquat + diuron that was 8% cheaper, all treatments were between 33 and 391% more expensive. Taking into account the duration of the experiments (75 days), the cheapest daily cost of weed control was \$ 0.36/day for the cheapest treatment (duron + paraquat), and glufosinate + oxyfluorfen and glyphosate + (bromacil + diuron) were apparently the most expensive treatments (1.79 and \$ 1.89 \$/day ha^{-1}, respectively) (Table 4). The minimum satisfactoty and accdeptable level of weed control for an herbicide treatment is 80%, according to the European Weed Research Society [17]. Based on this criterion, several treatments did not reach this level of weed control since 15 DAT; mainly in the San Manuel I field (Table 2). Thus, the treatments that presented most expensive cost of satisfactory weed control per day were glyphosate (1800 g ae ha^{-1}, \$3.22/day ha^{-1}) in San Manuel I and glyphosate + fluazifop-p-butyl in San Manuel II (\$3.36/day ha^{-1}), since they only maintained control ≥ 80% for 15 and 30 DAT, respectively. Glufosinate + oxyfluorfen and glyphosate + (bromacil + diuron) presented control levels ≥ 80% for 75 DAT in both fields, i.e., they were almost half-cheaper than the previous two treatments. However, the best satisfactory cost of weed control per day (0.95 and 1.08 \$/day ha^{-1}) was achieved with glyphosate + acetochlor (only in San Manuel II) and glufosinate (in both fields), mainly by maintaining weed control ≥ 80% at least 60 DAT (Table 4).

Table 4. Cost of herbicide treatments (Cost), relative daily cost considering 75 days of weed control (R-\$/day ha^{-1}) *, percentage of cost increase (% Inc) * in relation to the representative treatment (RT; 1080 g ae ha^{-1} glyphosate = USD \$29.0 ha^{-1}), days of weed control <80% (DAT>80%), and cost of satisfactory weed control per day (SC \$/day ha^{-1}) * in two Persian lime groves of the "San Manuel" Farm, Puebla, Mexico. All costs are estimated in relation to one hectare (1 ha^{-1}).

Treatment [1]	Cost	R-\$/day	% Inc	San Manuel I		San Manuel II	
				DAT > 80%	SC \$/day	DAT > 80%	SC \$/day
Control	-	-	-	-	-	-	-
Gly 1080	29.0	0.38	*RT*	-	-	30	0.96
Gly 1440	38.7	0.51	33	-	-	30	1.29
Gly 1800	48.4	0.65	66	15	3.22	30	1.61
Gly + Flua	100.9	1.35	247	-	-	30	3.36
Gly + Ace †	57.3	0.76	97	30	1.91	60	0.95
(Gly + Oxi) + Flua †	97.4	1.30	183	-	-	30	-
Gly + Seth	82.0	1.09	236	-	-	-	-
Gly + Cleth	70.0	0.93	141	-	-	-	-
Glufos	81.2	1.08	86	60	1.35	75	1.08
Par + Diu	26.8	0.36	−8	30	0.89	30	0.89
Glufos + Oxi †	134.6	1.79	364	75	1.79	75	1.79
Gly + Oxa †	112.0	1.49	286	-	-	-	-
Gly + (Brom + Diu) †	142.4	1.89	391	75	1.89	75	1.89

* R-\$/day = cost of a determined treatment/75 days; % Inc = [cost of a determined treatment/\$29.0 ha^{-1} (cost of representative treatment)] − 100; and SC \$/day = cost/DAT > 80% of a determined treatment. [1] Gly = Glyphosate, Flua = Fluazifop-p-butyl, Ace = Acetochlor, Oxi = Oxifluorfen, Seth = Sethoxydim, Cleth = Clethodim, Glufos = Glufosinate, Par = Paraquat, Diu = Diuron, Oxa = Oxadiazon, Bro = Bromacil. † Treatments applied in sequence 15 days after the first application.

3. Discussion

Weed community was mainly composed of *B. pilosa*, *E. colona*, and *L. virgata* in both SM-I and SM-II fields. Weed control in citrus groves of the Gulf of Mexico has been based mainly on glyphosate-based herbicides [4,7,13,15,18]. This almost exclusive dependence exerted a strong selection pressure favoring the establishment of these species in the Persian lime groves of the San Manuel Farm. Natural weeds are megadiverse, composed of both dicot and monocotyledons [19,20]; however, most cropping practices reduce plant diversity favoring few weed species [21], as observed in the SM-II field that showed a lower weed diversity than SM-I field, but higher density of plants.

The glyphosate-based treatments with increasing doses did not increase the total control of weeds, mainly, of the dominant species *B. pilosa* and *L. virgata*. Glyphosate resistance of *L. virgata* in citrus groves from Martinez de la Torre and Cuitláhuac, state of Veracruz, was confirmed in 2010 [7,22], municipalities that are at least 450 km apart. Some of these populations selected glyphosate resistance independently, but the majority have a common resistance selection origin that was spread throughout the citrus-producing region of the Gulf of Mexico to both short and long distances mainly by dispersing its seeds by people, tractors, and machinery used for the cultural deals [22]. Considering that the municipality of Acateno is located ~25 km from Martinez de la Torre, it was expected that glyphosate-resistant populations of *L. virgata* found in the San Manuel Farm, as well as in other farms around the region. In addition, there could also be selection of glyphosate resistance of weeds in situ [22], because weed management strategies based on herbicides lead to an eventual loss of control [23], and consequently, the selection for weeds resistant to herbicides [24].

The continuous use of glyphosate in citrus groves of the San Manuel farm, in addition to consolidating the glyphosate resistance of *L. virgata*, also selected for resistance in *B. pilosa*, as later corroborated by characterizing representative populations (R1 and R2 collected in SM-II and SM-I fields, respectively [15]). Some grass-controlling herbicides (fluazifop-p-butyl, sethoxydim, and clethodim), which efficiently controlled *L. virgata* and *E. colona*, reduced the efficacy of glyphosate to control *B. pilosa*, i.e., were apparently antagonistic. However, the low levels in controlling this species were perhaps due to its resistance to glyphosate, because once grasses were controlled by the grass herbicides, uncontrolled *B. pilosa* plants regrowth reached high levels of coverage, which was reflected in the low control percentages of total weeds of these treatments. This shows that improper implementation of a weed management strategy may indirectly favor the selection of herbicide resistance in another weed(s) [25]. This situation is worrisome because infestations of *B. pilosa* may impede future manual activities of crop management such as pruning and harvesting (Figure 4), since its achenes have three barbed spines that easily attach to clothing or fur [26]. Bromacil + diuron applied 15 DAT following glyphosate application controlled *B. pilosa*, demonstrating its contribution of bromacil + diuron in the control of weeds. Thus, other mixtures of glyphosate with herbicides such as 2,4-D, dicamba, or picloram, with good efficacy on dicots [27], could also contribute to controlling *B. pilosa* in autumn–winter period when it reaches its highest population density, as well as other broadleaf weeds, when trees are not blooming or setting fruits.

Figure 4. Persian lime rows infested with *B. pilosa* in "San Manuel I", Acateno, Puebla, in 2014. (**A**) Untreated control plot 10 DAT, and (**B**) random rows at 30 DAT. Persian lime trees were 3 years old.

The poor control of *L. virgata* with the glyphosate-based treatments with acetochlor and oxadiazon in both the SM-I and SM-II fields was due to regrowth of plants. In situations where glyphosate poorly controls resistant weeds, the use of residual herbicides was proposed to control grasses that come from seeds [28]. The mixture of glyphosate with contact or residual herbicides can improve weed control in citrus plantations in Florida, USA, and at the same time reduce the selection of resistant populations and extend the weed control period [16]. Tank-mix of glufosinate + indaziflam controlled glyphosate-resistant *L. virgata* populations in Persian lime and orange groves from Veracruz, Mexico, up to 90 days due to the residually of the second herbicide [7]. However, most Mexican citrus growers have not adopted the practice of using PRE and residual herbicides for effective and timely control of weeds. In addition, growers are reluctant to adopt non-chemical weed control methods [29,30]. Thus, the misuse of herbicides (late application, subdose, or overdose) has selected for resistant weeds, which increased yield losses and production costs [30].

The good control of *E. colona*, a weed species in high plant density in the SM-II field, and the other sporadic weeds were also susceptible to glyphosate. However, the high occurrence rates of *B. pilosa* and *L. virgata* resistant to glyphosate make it necessary to seek weed management strategies, including non-chemical methods, without increasing production costs.

Without the occurrence of glyphosate resistant weeds, the cost of satisfactory weed control per day would be \$ 0.38/day ha^{-1} (1800 g ae ha^{-1} glyphosate). However, due to the occurrence of *B. pilosa* and *L. virgata* populations resistant to this herbicide, no glyphosate treatment alone showed control above 80%, classified as satisfactory or good [17], for more than 30 DAT. Glufosinate + oxifluorfen and glyphosate + (bromacil + diuron) maintained this control level up to 75 DAT in both fields, but in global terms, they were the more expensive treatments (1.79 and 1.89 \$/day ha^{-1}, respectively). The cost of an herbicide treatment, without considering the application cost, is a good indicator that can help in selecting a chemical treatment [29,31], but this cost is relevant when the choice considers the period of control. In this sense, mixtures of glyphosate with sethoxydim, cletodim or oxadiazon were not efficient to control weeds in practical and economic terms. However, the best treatment was glufosinate that maintained an acceptable weed control for 60 DAT in San Manuel I (\$1.35/day ha^{-1}) and 75 in San Manuel II (\$1.08/day ha^{-1}). This represents at least one application operation less in comparison to the other treatments that presented control > 80% for up to 30 DAT. Therefore, the choice of a relative more expensive herbicide treatments, but that provide an acceptable control of glyphosate resistant weeds for longer in Persian lime orchards could result in fewer herbicide applications, reducing the cost of this and other crop management tasks [32], such as pruning and harvest. Integrated herbicide-resistant weed management programs reduce profits in the first year of its implementation, but profits increase in the second and subsequent years [33,34], since once the seed bank decreases, less expensive and persistent active ingredients can be used [23,35]. This could improve the yield and quality of fruits, increasing the return, since weed management is a key component to sustaining productivity [33,34]. Therefore, the adoption of weed management strategies, either chemical, not chemical, or combined, must consider other factors associated with production costs rather than the individual cost of a determined control method, in order to obtain an integrated weed management in the short and long term [29,30].

4. Materials and Methods

4.1. Local Data and Experimental Design

Field trials was carried out in two Persian lime groves, of the "San Manuel" Farm (20.10° N, 97.16° W), in Acateno, Puebla, referred to as SM-I and SM-II (at 128 and 137 m above sea level, respectively), which had received glyphosate applications of 720–1080 g ae ha^{-1} between 3 and 5 times a year from 2006 to 2014. The groves were at least 800 m apart, and Persian lime trees were 3 years old and arranged at 3 × 5 m between trees and rows, respectively. Thirteen herbicide treatments plus an untreated control (Table 1) were distributed in each field in a randomized block design with three replications. Each plot consisted of a 6 × 5 m area including two trees. A representative soil sample from sub random samples (0–10 cm) taken in both fields was analyzed. The sandy clay loam (8% clay, 64% sand, and 28% silt) presented pH = 8.06, 3.3% organic matter; 10.4 and 15.1 mg kg^{-1} of nitrates and P; and 0.01, 24.8, 0.93, 0.03, and 0.39 Cmol kg^{-1} of Ca, Mg, Na, and P, respectively. Field trials were carried out in 2014 and repeated in 2015 and the climatic conditions, presented during the execution of the experiments that were recovered from the National Meteorological System of Mexico, there are outlined in the Figure 5.

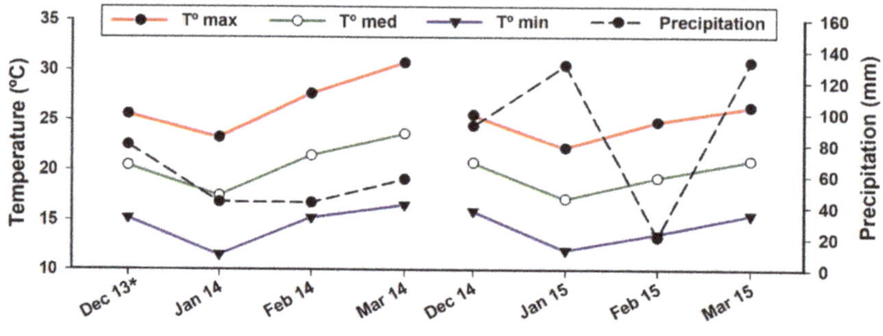

Figure 5. Accumulated precipitation (mm) and average monthly temperatures (°C) in Acateno, Puebla, Mexico from December 2013 to March 2014 and December 2014 to March 2015 (Source SMN, 2019 [36]). * Data are the historical average of the parameters presented since there is no data available for this month.

4.2. Herbicide Application

A mechanical mowing of weeds was carried out in the two experimental fields three weeks before application of herbicide treatments, following the crop management practices of the Mexican citrus growers. The applications were made on 10–15 cm tall weeds using a motorized manual spray backpack (Swissmex 425 SW), equipped with pressure regulator and an AI11002 nozzle, calibrated to spray 277 L ha^{-1} at 40 psi. The water pH of the glufosinate and glyphosate treatments was adjusted to 5.5 with the acidifying-buffing adjuvant pHase1® (Arysta LifeSciences México, Saltillo, México). The PRE herbicides of the treatments 5, 6, 11, 12, and 13 were applied 15 days after application of POST herbicides. Table 5 summarizes the action dates of field activities during the experiments.

Table 5. Summary of activities conducted for weed control with pre- (PRE) and post-emergence (POST) herbicides in two Persian lime groves of the "San Manuel" Farm, Puebla, Mexico.

Field Activity	Date	
	2014	2015
Weed mechanical mowing in both experimental plots	16 December 2013	19 December 2014
Initial counting of plants of each weed species	9 January2014	15 January 2015
Application of POST herbicides	10 January2014	16 January 2015
Application of PRE herbicides (treatments 5, 6, 11, 12, and 13), and evaluation at 15 DAT	25 January 2014	31 January 2015
Evaluation at 30 DAT	9 February 2014	15 February 2015
Evaluation at 45 DAT	23 February 2014	1 March 2015
Evaluation at 60 DAT	9 March 2014	15 March 2015
Evaluation at 75 DAT	24 March 2014	29 March 2015

DAT: days after herbicide treatment.

4.3. Evaluated Variables

One day before the herbicide applications, weeds were identified and counted using a square frame of 0.25 m^2, randomly placed twice in each experimental unit. These data were expressed in number of plants per m^2. The visual control of weeds total and by species was conducted at 15, 30, 45, 60, and 75 days after treatment (DAT). The evaluation at 75 DAT was included considering that PRE herbicides would have 60 days of activity. The control percentage of total weeds or a determined species was estimated with the equation $X = [(A - B)/A] \times 100$ [37], which compares the average coverage of weeds in plot (B) vs. the weed population in the control plot. The parameters of equation represent: X the control percentage of total weed or a determined species; A the coverage percentage of total weed or a determined species in the control plot, and B the coverage percentage of total weed or a determined species in the treatment evaluated.

Cost of herbicide treatments per ha was recorded to estimate the relative daily cost of weed control considering the evaluation period (75 days), as well as the percentage in cost increase of each treatment in relation to the representative treatment (1080 g ae ha^{-1} glyphosate = USD $ 29.0) most used by farmers. The herbicides treatments that presented weed control >80% as well as DAT that maintained that control level were identified to estimate the cost of satisfactory weed control per day.

4.4. Statistical Analysis

Percentage data were transformed to arcsine and, then, the model assumptions of normal distribution of errors and homogeneous variance were graphically inspected. The variance stability tests of control percentage data showed no differences for both cropping seasons, and data were pooled and subjected to ANOVA. Significant differences between means were analyzed using the Tukey's test at the 0.05 probability level. Statistical analysis was conducted using the Statistix 9.0 software (Analytical Software, Tallahassee, FL, USA).

5. Conclusions

Glyphosate alone, even at high doses, had difficulty controlling *B. pilosa* and *L. virgata*. Glyphosate applied in mixture with POST grass herbicides controlled *E. colona* and *L. virgata*, but favored the regrowth of glyphosate-resistant *B. pilosa* plants. Glufosinate alone, glufosiante + oxyfluorfen and glyphosate + (bromacil + diuron) controlled t weeds above 80% for at least 60 d, but they were also the most expensive herbicides. In practical and economic terms, glufosinate alone was the best treatment controlling glyphosate resistant weeds in Persian lime orchards. Therefore, the choice or not of these treatments must consider the cost of satisfactory weed control per day, the period of weed control, as well as other factors associated with production costs, and not only the relative cost of a certain treatment. Integrated weed management programs focused on reducing herbicide dependence that favor weed diversity in the short and long term, as well as facilitating other cultural tasks, should be implemented in the citrus-producing region of the Gulf of Mexico.

Author Contributions: R.A.-d.l.C.; J.A.D.-V., and H.E.C.-H. designed the experiments; P.A.D.-M.; C.P.-B.; J.G.V.-G., and J.A.D.-V. performed the field trials; R.A.-d.l.C., H.M.d.S., and R.D.P. analyzed the raw data; R.A.-d.l.C., J.A.D.-V., C.P.-B., and J.G.V.-G. wrote and reviewed the draft of the manuscript; J.A.D.-V. and H.E.C.-H., funding acquisition and project administration; J.A.D.-V. and R.D.P., supervision. All authors corrected and approved the submitted version of this manuscript.

Funding: This research was founded by Monsanto Comercial S. A. de C. V. (Mexico) in 2014 and Bayer de México S. A. de C. V.—CONACYT in 2015 under the research projects: "Manejo de *Leptochloa virgata* resistente a glifosato en cítricos en San José Acateno, Puebla, Mex." and "Investigación, desarrollo e innovación de nuevos herbicidas para el control y manejo de malezas susceptibles y resistentes en los cultivos de caña de azúcar, frutales y cereales", respectively, granted to JA Domínguez-Valenzuela.

Acknowledgments: The authors thanks to the agronomist Eduardo Sánchez-González for his technical assistance in the field trials.

Conflicts of Interest: The authors declare no conflict of interest.

References

1. FAOSTAT. Food and Agriculture Data. 2018. Available online: http://www.fao.org/faostat/en/#rankings/commodities_by_country (accessed on 27 December 2018).
2. SIAP-Servicio de Información Agropecuaria y Pesquera. Avance de Siembras y cosechas—Resumen Nacional por Estado. 2019. Available online: http://infosiap.siap.gob.mx:8080/agricola_siap_gobmx/AvanceNacionalCultivo.do (accessed on 1 August 2019).

3. USDA-U.S. Department of Agriculture. Citrus: World Markets and Trade. 2019. Available online: http://apps.fas.usda.gov/psdonline/circulars/citrus.pdf (accessed on 1 August 2019).
4. Gherekhloo, J.; Fernández-Moreno, P.T.; Alcántara-de la Cruz, R.; Sánchez-González, E.; Cruz-Hipolito, H.E.; Dominguez-Valenzuela, J.A.; De Prado, R. Pro-106-Ser mutation and EPSPS overexpression acting together simultaneously in glyphosate-resistant goosegrass (*Eleusine indica*). *Sci. Rep.* **2017**, *7*, 6702. [CrossRef] [PubMed]
5. Tiwari, S.; Mann, R.S.; Rogers, M.E.; Stelinski, L.L. Insecticide resistance in field populations of Asian citrus psyllid in Florida. *Pest Manag. Sci.* **2011**, *67*, 1258–1268. [CrossRef] [PubMed]
6. Torres-Pacheco, I.; López-Arroyo, J.I.; Aguirre-Gómez, J.A.; Guevara-Gonzalez, R.G.; Yáñez-López, R.; Hernández-Zul, M.I.; Quijano-Carraza, J.A. Potential distribution in Mexico of *Diaphorina citri* (Hemiptera: Psyllidae) vector of huanglongbing pathogen. *Fla. Entomol.* **2013**, *96*, 36–47. [CrossRef]
7. Pérez-López, M.; González-Torralva, F.; Cruz-Hipólito, H.E.; Santos, F.; Domínguez-Valenzuela, J.A.; De Prado, R. Characterization of glyphosate-resistance tropical sprangletop (*Leptochloa virgata*) and its alternative chemical control in Persian lime. *Weed Sci.* **2014**, *62*, 441–450. [CrossRef]
8. Futch, S.H.; Singh, M. Field evaluation of chemical weed control in Florida citrus. *Proc. Fla. State Hortic. Soc.* **2000**, *113*, 68–74.
9. Bracamonte, E.; Silveira, H.M.; Alcántara-de la Cruz, R.; Domínguez-Valenzuela, J.A.; Cruz-Hipólito, H.E.; De Prado, R. From tolerance to resistance: Mechanisms governing the differential response to glyphosate in *Chloris barbata*. *Pest Manag. Sci.* **2018**, *74*, 1118–1124. [CrossRef] [PubMed]
10. Fernández-Lambert, G.; Aguilar-Lasserre, A.; Azzaro-Pantel, C.; Miranda-Ackerman, M.A.; Purroy-Vázquez, R.; Pérez-Salazar, M.R. Behavior patterns related to the agricultural practices in the production of Persian lime (*Citrus latifolia tanaka*) in the seasonal grove. *Comput. Electron. Agric.* **2017**, *116*, 162–172. [CrossRef]
11. Almaguer-Vargas, G.; Ayala-Garay, A.V. Adoption of innovations in 'Persian' lemon (*Citrus latifolia* tan.) in Tlapacoyan, Veracruz. Use of logbook. *Rev. Chapingo Ser. Hortic.* **2014**, *20*, 89–100. [CrossRef]
12. Duke, S.O. The history and current status of glyphosate. *Pest Manag. Sci.* **2018**, *74*, 1027–1034. [CrossRef]
13. Palma-Bautista, C.; Gherekhloo, J.; Domínguez-Martínez, P.A. Characterization of three glyphosate resistant *Parthenium hysterophorus* populations collected in citrus groves from Mexico. *Pestic. Biochem. Physiol.* **2019**, *155*, 1–7. [CrossRef]
14. Syngenta. Trade Label of Touchdown (Glyphosate 62% m/v). Available online: https://www.syngenta.com.br/sites/g/files/zhg256/f/touchdown.pdf?token=1540408979 (accessed on 26 November 2018).
15. Alcántara-de la Cruz, R.; Fernández-Moreno, P.T.; Ozuna, C.V.; Rojano-Delgado, A.M.; Cruz-Hipolito, H.E.; Domínguez-Valenzuela, J.A.; Barro, F.; De Prado, R. Target and non-target site mechanisms developed by glyphosate-resistant Hairy beggarticks (*Bidens pilosa* L.) populations from Mexico. *Front. Plant. Sci.* **2016**, *7*, 1492.
16. Jhala, A.J.; Ramirez, A.H.M.; Knezevic, S.Z.; Van Damme, P.; Singh, M. Herbicide tank mixtures for broad-spectrum weed control in Florida citrus. *Weed Technol.* **2013**, *27*, 129–137. [CrossRef]
17. Vanhala, P.; Kurstjens, D.; Ascard, J.; Bertram, A.; Cloutier, D.C.; Mead, A.; Raffaelli, M.; Rasmussen, J. Guidelines for physical weed control research: Flame weeding, weed harrowing and intra-row cultivation. In Proceedings of the 6th EWRS—Workshop on Physical and Cultural Weed Control, Lillehammer, Norway, 8–10 March 2004; pp. 208–239.
18. Alcántara-de la Cruz, R.; Rojano-Delgado, A.M.; Giménez, M.J.; Cruz-Hipolito, H.E.; Domínguez-Valenzuela, J.A.; Barro, F.; De Prado, R. First resistance mechanisms characterization in glyphosate-resistant *Leptochloa virgata*. *Front. Plant. Sci.* **2016**, *7*, 1742. [CrossRef] [PubMed]
19. Marshall, E.J.P.; Brown, V.K.; Boatman, N.D.; Lutman, P.J.W.; Squire, G.R.; Ward, L.K. The role of weeds in supporting biological diversity within crop fields. *Weed Res.* **2003**, *43*, 77–89. [CrossRef]
20. Pollnac, F.W.; Maxwell, B.D.; Menalled, F.D. Weed community characteristics and crop performance: A neighbourhood approach. *Weed Res.* **2009**, *49*, 242–250. [CrossRef]
21. Alonso-Ayuso, M.; Gabriel, J.L.; García-González, I.; Del Monte, J.P.; Quemada, M. Weed density and diversity in a long-term cover crop experiment background. *Crop. Prot.* **2018**, *112*, 103–111. [CrossRef]
22. Alcántara-de la Cruz, R.; Romano, Y.; Osuna-Ruíz, M.D.; Domínguez-Valenzuela, J.A.; Menendez, J.; De Prado, R. Genetic relationships between tropical sprangletop (*Leptochloa virgata*) populations from Mexico: Understanding glyphosate resistance spread. *Weed Sci.* **2016**, *64*, 579–587. [CrossRef]

23. Bagavathiannan, M.V.; Davis, A.S. An ecological perspective on managing weeds during the great selection for herbicide resistance. *Pest Manag Sci.* **2018**, *74*, 2277–2286. [CrossRef] [PubMed]
24. Hicks, H.L.; Comont, D.; Coutts, S.R.; Crook, L.; Hull, R.; Norris, K.; Neve, P.; Childs, D.Z.; Freckleton, R.P. The factors driving evolved herbicide resistance at a national scale. *Nat. Ecol. Evol.* **2018**, *2*, 529–536. [CrossRef] [PubMed]
25. Shergill, L.S.; Bish, M.D.; Biggs, M.E.; Bradley, K.W. Monitoring the changes in weed populations in a continuous glyphosate- and dicamba-resistant soybean system: A five-year field-scale investigation. *Weed Technol.* **2018**, *32*, 166–173. [CrossRef]
26. Reddy, K.N.; Singh, M. Germination and emergence of hairy beggarticks (*Bidens pilosa*). *Weed Sci.* **1992**, *40*, 195–199. [CrossRef]
27. Ganie, Z.A.; Jhala, A.J. Interaction of 2,4-D or dicamba with glufosinate for control of glyphosate-resistant giant ragweed (*Ambrosia trifida* L.) in glufosinate-resistant maize (*Zea mays* L.). *Front. Plant. Sci.* **2017**, *8*, 1207. [CrossRef] [PubMed]
28. Bond, J.A.; Eubank, T.W.; Bond, R.C.; Golden, B.R.; Edwards, H.M. Glyphosate-resistant Italian ryegrass (*Lolium perenne* ssp. *multiflorum*) control with fall-applied residual herbicides. *Weed Technol.* **2014**, *28*, 361–370. [CrossRef]
29. Ervin, D.E.; Breshears, E.H.; Frisvold, G.B.; Hurley, T.; Dentzman, K.E.; Gunsolus, J.L.; Jussaume, R.A.; Owen, M.D.K.; Norswrothy, J.K.; Al Mamun, M.M.; et al. Farmer attitudes toward cooperative approaches to herbicide resistance management: A common pool ecosystem service challenge. *Ecol. Econ.* **2019**, *157*, 237–245. [CrossRef]
30. Moss, S. Integrated Weed Management (IWM): Why are farmers reluctant to adopt non-chemical alternatives to herbicides? *Pest Manag. Sci.* **2019**, *75*, 1205–1211. [CrossRef] [PubMed]
31. Fried, G.; Cordeau, S.; Metay, A.; Kazakou, E. Relative importance of environmental factors and farming practices in shaping weed communities structure and composition in French vineyards. *Agric. Ecosyst. Environ.* **2019**, *275*, 1–13. [CrossRef]
32. Singh, M.; Malik, M.S.; Ramirez, A.H.M.; Jhala, A.J. Tank mix of saflufenacil with glyphosate and pendimethalin for weed control in Florida citrus. *Hortic. Technol.* **2011**, *21*, 606–615.
33. Lamichhane, J.R.; Devos, Y.; Beckie, H.J.; Owen, M.D.K.; Tillie, P.; Messéan, A.; Kudsk, P. Integrated weed management systems with herbicide-tolerant crops in the European Union: Lessons learnt from home and abroad. *Crit. Rev. Biotechnol.* **2017**, *37*, 459–475. [CrossRef]
34. Livingston, M.; Fernandez-Cornejo, J.; Frisvold, G.B. Economic returns to herbicide resistance management in the short and long run: The role of neighbor effects. *Weed Sci.* **2016**, *64*, 595–608. [CrossRef]
35. Haring, S.C.; Flessner, M.L. Improving soil seed bank management. *Pest Manag. Sci.* **2018**, *74*, 2412–2418. [CrossRef]
36. SMN—Servicio Meteorológico Nacional. Información climatólogica por Estado, Acateno–21143. 2019. Available online: https://smn.conagua.gob.mx/tools/RESOURCES/Mensuales/pue/00021143.TXT (accessed on 20 August 2019).
37. Burril, L.C.; Cárdenas, L.; Locatelli, E. *Field Manual for Weed Control Research*, 1st ed.; Oregon State University Press: Corvallis, OR, USA, 1976; p. 63.

plants

MDPI

Article

Sensitivity Analysis of Italian *Lolium* spp. to Glyphosate in Agricultural Environments

Silvia Panozzo [1], Alberto Collavo [1,2] and Maurizio Sattin [1,*]

1 Institute for Sustainable Plant Protection (IPSP) – CNR, Viale dell'Università 16, 35020 Legnaro (PD), Italy; silvia.panozzo@ipsp.cnr.it (S.P.); alberto.collavo@bayer.com (A.C.)

2 Bayer AG, Industriepark Hoechst, H872, D-65926 Frankfurt am Main, Germany

* Correspondence: maurizio.sattin@cnr.it; Tel.: +39-049 8272820

Received: 31 December 2019; Accepted: 28 January 2020; Published: 30 January 2020

Abstract: Empirical observations generally indicate a shifting and decreased *Lolium* spp. susceptibility to glyphosate in Italy. This is likely due to the long history of glyphosate use and to the sub-lethal doses commonly used. There is, therefore, a need to determine the variability of response of *Lolium* spp. to glyphosate and identify the optimum field dose. To perform a sensitivity analysis on *Lolium* spp. populations in an agriculture area, collection sites were mainly chosen where glyphosate had not been applied intensely. Known glyphosate-resistant or in-shifting populations were included. Two outdoor dose-response pot experiments, including eleven doses of glyphosate, were conducted. The dose to control at least 93%–95% of susceptible *Lolium* spp. was around 450 g a.e. ha^{-1}. However, to preserve its efficacy in the long term, it would be desirable not to have survivors, and this was reached at a glyphosate dose of 560 ± 88 g a.e. ha^{-1}. Taking into account the variability of response among populations, it was established that the optimal dose of glyphosate to control *Lolium* spp. in Italy up to the stage BBCH 21 has to be at least 700 g a.e. ha^{-1}. As a consequence, it is recommended to increase the label recommended field rate for *Lolium* spp. control in Italy to a minimum of 720 g a.e. ha^{-1}.

Keywords: sensitivity line; ryegrass; herbicide dose; herbicide resistance; dose-response

1. Introduction

Lolium rigidum Gaud. (rigid ryegrass, LOLRI) and *Lolium multiflorum* Lam. (Italian ryegrass, LOLMU) are two self-incompatible species that have a global distribution [1]. According to reported cases [2], they are among the species most prone to evolve herbicide resistance. *L. rigidum* is one of the most troublesome weeds in grain crops as well as orchards, olive groves and vineyards, where it is also managed as a cover crop [3]. To date, *L. rigidum* populations resistant to 13 different herbicide Sites of Action (SoA) have been reported [4]. *L. multiflorum* occurs in several temperate countries and populations resistant to nine different herbicide SoA have been reported [4]. Both species originated from the Mediterranean, have a C_3 photosynthetic pathway and produce dense infestations [5]. The two species are often mixed in the field and not always easily identifiable, and in those cases, the population is defined as LOLSS (*Lolium* species).

The withdrawal of many herbicides from the EU market due to the strict regulation, the lack of herbicides with new SoA [6] and the propensity of *Lolium* spp. to evolve resistance to the most commonly used post-emergence herbicides have increased the importance of glyphosate for the management of these species in agricultural and in non-agricultural areas [1]. Glyphosate is the most successful herbicide in history [7,8], and its use is higher than any other herbicide SoA [6]. According to the HRAC (International Herbicide Resistance Action Committee) classification [9], which is based on the herbicides' SoA, it belongs to group G and exerts its action by inhibiting 5-enolpyruvylshikimate-3-phosphate (EPSP) synthase in plants, fungi and microorganisms, the only life forms that possess the shikimate pathway. Therefore, it has no measurable mammalian toxicity at the concentrations used [10]. It acts as a competitive inhibitor to the phosphoenol-pyruvate (PEP) binding site and a non-competitive inhibitor for the shikimate-3-phosphate (S3P) site, thus preventing the formation of EPSP [11].

Glyphosate is a non-selective, systemic, post-emergence herbicide that controls many dicotyledonous and monocotyledonous weeds [12]. It is neither active nor residual in the soil and, therefore, selection pressure for resistance is only exerted on emerged seedlings [13]. Glyphosate-resistant weeds were not found during the first 22 years of glyphosate use, whereas in the last 23 years (1996–2019), glyphosate resistance was documented in 45 weed species in 29 countries [4].

Glyphosate is also commonly used on a frequent basis between tree rows (i.e., olives, hazelnuts and vineyards) and for roadside weed control [14]. In these situations, glyphosate can be used for many years and applied several times per year. The continuous use of glyphosate in perennial crops, such as orchards, has imposed intense selection pressure for resistance evolution and has led to shifts in weed floras as well as towards glyphosate-resistant individuals [15,16]. In particular, in Europe, glyphosate resistance has evolved most often in two genetically diverse, but at the same time, resistance-prone genera, *Conyza* [17] and *Lolium* [18].

In Italy, the doses of glyphosate commonly used have become sub-lethal for *Lolium* spp. [19]. In the past, the rate of 360 g a.e. (acid equivalent) ha^{-1} gave satisfactory control, but it is likely that a few plants survived each treatment. Exposure to recurrent selection at sub-lethal glyphosate doses can result in a shift towards resistance within a few generations. It was clearly demonstrated by Busi and Powles [20] that in allogamous species, such as *Lolium*, minor resistance gene trait(s) may be additively enriched through cross-pollination among surviving plants. In Italy, empirical observations indicated a general decrease in susceptibility of *Lolium* spp. to glyphosate (i.e., relatively poor control at 360 g a.e. ha^{-1}) and the first resistance cases were reported in 2008 [19]. At the moment, 13 municipalities in five Italian regions and five different cropping systems (including orchards, olive groves, vineyards, wheat and no-tillage agriculture) are affected by glyphosate resistance [21].

As part of the herbicide resistance risk analysis and management, the availability of a robust baseline sensitivity for key-target species is critical to discriminate between susceptible (S) and resistant (R) populations and to identify early shifts in susceptibility. From a practical point of view, a population is ascribed as resistant (R) to a herbicide when more than 20% of treated plants survived the recommended herbicide field dose [22]. The identification of a first shift in susceptibility is particularly valuable when resistance evolution is rather slow, as in glyphosate resistance. Only a few herbicide sensitivity analyses are available in the literature [23–26]. The European and Mediterranean Plant Protection Organization defines the baseline as the mean of natural variability of a target species' sensitivity before the commercial introduction of an active ingredient and can be taken as a point of reference to be used in decision-making processes. Instead, glyphosate has been on the market for many years, and its selection pressure has been active for a long time. In such a case, the baseline term/approach is not correct, and a sensitivity analysis should instead be performed. The aim of a sensitivity analysis is to determine the average efficacy of an old herbicide on weed populations that may have been treated before with the same compound [27,28]. In other words, this is part of the monitoring procedure of herbicide efficacy. To our knowledge, only one paper in the literature has dealt with the glyphosate baseline sensitivity for *L. rigidum* in Spain using a quick Petri dishes test [29].

The establishment of a good sensitivity baseline should make it easier to identify any case of evolved herbicide resistance [30] and would have an added value if an effective monitoring program is initiated [29].

The aims of this research were (1) to determine the variability in glyphosate response of *Lolium* spp. populations collected from Italian agricultural environments and (2) to determine the glyphosate dose that is actually effective on the *Lolium* spp. populations in field conditions in order to preserve its efficacy in the long term.

2. Results

2.1. Dose-Response Experiments

Two outdoor pot dose-response experiments were performed during spring (March-May) and autumn (September-November) to test the effects of increasing glyphosate dose on plant survival and fresh weight for several *Lolium* spp. populations collected from Italian agricultural environments. The effective doses—EDs—and growth rates—GRs—causing 50% and 90% reduction in plant survival and fresh weight (ED_{50}-ED_{90} and GR_{50}-GR_{90}), respectively, were calculated using a regression analysis (see Section 4.3).

2.1.1. Spring Dose-Response Experiment

Twenty populations were included in the spring experiment (Table 1a). A variance test (F-test) was performed to compare the dose-response curves obtained for the different populations in the experiment. The lack-of-fit F-test on both plant survival and fresh weight indicated that it was not possible to simplify the glyphosate regressions to a model with a common slope for all populations: the slope tended to decrease when EDs (and GR_S) increased. The data of each population were, therefore, regressed as individual curves and treated separately.

Among the populations included in the spring experiment, eight were known resistant/shifting populations. A Box and Whisker analysis using the median and 25–75 percentiles was used to statistically exclude outliers. The analysis was performed including all populations, then repeated excluding the outliers until no further outliers were identified (Figure 1). The first analysis revealed three extreme value populations (403, 392 and 401) (Figure 1A), all previously confirmed as resistant (Table 1). The analysis was repeated excluding those populations and limiting the ED_{50} range to between 155 and 900 g a.e. ha^{-1}. In this second step, two other populations (343 and 384L) were found to be outliers (Figure 1B). The third analysis considered an ED_{50} range of 155–560 g a.e. ha^{-1} and highlighted three other outlier populations (384, 259 and 328) (Figure 1C), one included in the experiment as R check (384) and the two shifting populations (Table 1). The fourth analysis with 12 populations and an ED_{50} range of 155–260 g a.e. ha^{-1} did not reveal any outlier population (Figure 1D). The Box and Whisker analysis, also repeated for the ED_{90} and GRs values (data not shown), confirmed that all eight populations included in the experiment as resistant or partially resistant to glyphosate had a reduced sensitivity or resistance to this herbicide.

Considering the data of the other twelve populations resulted as being sensitive to glyphosate, it was highlighted that glyphosate ED_{50} ranged from 155 ± 5.9 to 260 ± 6.7 with a mean value of 206 g a.e. ha^{-1} and ED_{90} from 243 ± 20.8 to 506 ± 79.1 with a mean value of 342 g a.e. ha^{-1}. Concerning fresh weight, GR_{50} varied from 31 ± 8.8 to 98 ± 14.7 with a mean value of 64 g a.e. ha^{-1}, GR_{90} from 144 ± 15.7 to 272 ± 26.3 with a mean value of 198 g a.e. ha^{-1}.

Table 1. Details of the populations tested in the spring (a) and autumn (a in bold and b) dose-response experiments: species (LOL = *Lolium*, RI = *rigidum*, MU = *multiflorum*, SS = *multi-species*), sampling year, population code (progressive number, which, together with the sampling year, uniquely identifies a population), geographical origin and crop or collection site, where available. S = susceptible and R = resistant (i.e., plant survival >20% at the field dose); L = population reproduced in Legnaro greenhouse.

Species		Population Code		Origin (Municipality)	Crop or Collection Site	Notes
(a)						
LOL	RI	07	328	Santo Stefano Belbo	vineyard	Partially R [19]
LOL	**RI**	**08**	**204L**	**Legnaro**	**wheat**	**S check used by IPSP-CNR**
LOL	SS	08	259	Cortona	wheat	Partially R [31]
LOL	**SS**	**08**	**340**	**Collesalvetti**	**wheat**	
LOL	SS	08	343	Pomarance	wheat	R pop. [31]
LOL	**SS**	**10**	**381**	**Pontedera**	**wheat**	
LOL	SS	10	384	Cascina	wheat	R pop. [31]
LOL	SS	11	384L	Cascina		Reproduced from glyphosate-resistant plants of pop. 10–384
LOL	**SS**	**10**	**389**	**Castenaso**	**meadow**	
LOL	**MU**	**11**	**390**	**Legnaro**	**field margin**	
LOL	SS	11	392	Palo del Colle	olive grove	R pop. (unpublished data)
LOL	**RI**	**11**	**395**	**Acquaviva delle Fonti**	**roadside**	
LOL	**RI**	**11**	**400**	**Torchiarolo**	**set aside**	
LOL	SS	11	401	Lamezia Terme	olive grove	R pop. (unpublished data)
LOL	**RI**	**11**	**402**	**Lamezia Terme**	**meadow**	
LOL	SS	11	403	Cascina	sunflower	R pop. [31]
LOL	**RI**	**11**	**404**	**Lamezia Terme**	**olive grove**	
LOL	**SS**	**11**	**405**	**Livorno**	**lucerne**	
LOL	**MU**	**11**	**412**		**Commercial turf seed**	
LOL	**SS**	**11**	**425**	**Cascina**	**wheat**	
(b)						
LOL	MU	12	426	Ravenna	wheat	
LOL	SS	12	431	Duino Aurisina	meadow	
LOL	SS	12	432	Siena	wheat	
LOL	SS	12	434	Sovicille	wheat	
LOL	SS	12	444	San Casciano dei Bagni	wheat	
LOL	MU	12	449	Marsciano	vineyard	
LOL	SS	12	455	Gubbio	wheat	
LOL	MU	12	458	Montecchio Emilia	lucerne	
LOL	SS	12	461	Brisighella	roadside	
LOL	MU	12	462	Forlì	roadside	
LOL	MU	12	466	Coriano	wheat	
LOL	SS	12	472	Osimo	wheat	
LOL	SS	12	477	Pozzolengo	meadow	
LOL	SS	12	479	Cremona	field margin (maize)	
LOL	MU	12	483	Alessandria	meadow	
LOL	SS	12	487	Saluzzo	meadow	
LOL	SS	12	492	Cigliano	field margin	
LOL	SS	12	504	Pontoglio	roadside	

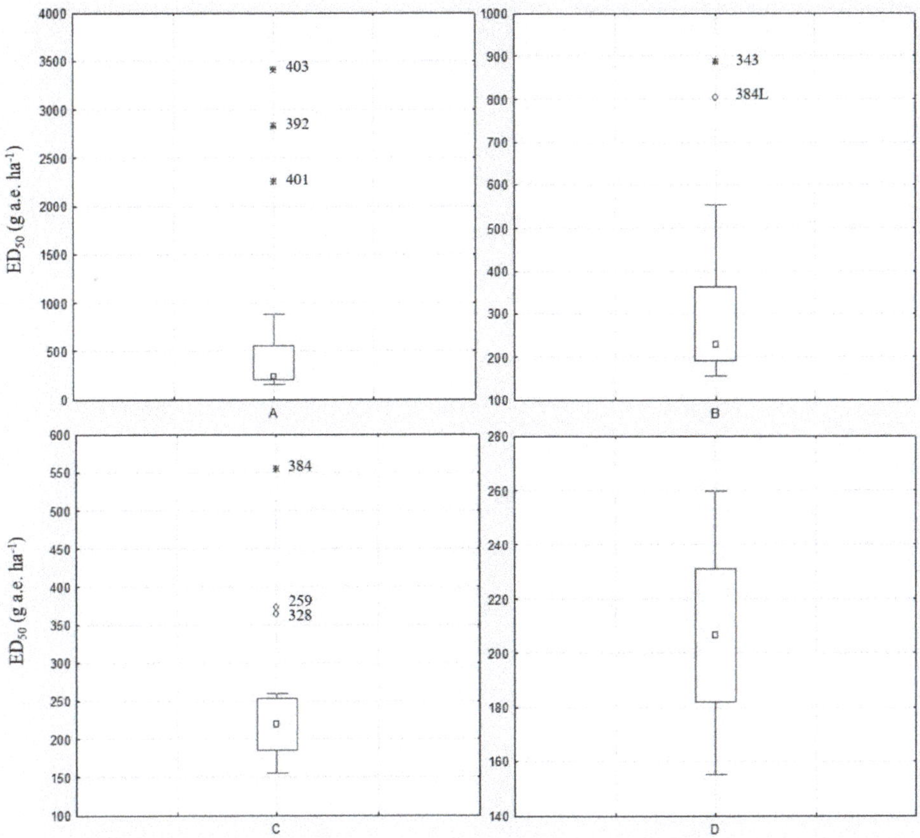

Figure 1. Box and Whisker plots illustrating the range of plant control (ED_{50}) for the twenty *Lolium* spp. populations included in the spring experiment: (**A**) all populations were included; (**B**) extreme values of Box Plot A were excluded; (**C**) extreme and outlier values of Box Plot B were excluded; (**D**) extreme and outlier values of Box Plot C were excluded. The central point is the median, the box represents the 25–75 percentiles and bars the non-outlier range, ○ and * represent outliers and extreme values, respectively. Population codes excluded during the analysis are reported.

2.1.2. Autumn Dose-Response Experiment

Twelve susceptible populations selected in the spring experiment were also tested in the autumn experiment, together with another eighteen populations reported in Table 1b. In this second experiment, the Box and Whisker analysis did not highlight any outliers among populations.

As for the spring experiment, the lack-of-fit F-test on both plant survival and fresh weight in the autumn experiment indicated that it is not possible to simplify the regressions to a model with a common slope for all populations, so a single-curve analysis was preferred.

Furthermore, an ad hoc lack-of-fit F-test performed on the data of each population included in both experiments comparing the dose-response curves obtained in the two experiments showed that most of the curves were significantly different at $p < 0.05$, so the two experiments cannot be merged (data not shown). Populations data could not be pooled considering both plant survival and fresh weight, so it was decided to consider the two experiments separately.

ED_{50} based on the autumn dose-response experiment ranged from 108 ± 10.1 to 282 ± 7.1 with a mean value of 186 g a.e. ha^{-1} (Figure 2a) and ED_{90} from 189 ± 16.7 to 561 ± 87.7 g a.e. ha^{-1} with a mean value of 317 g a.e. ha^{-1} (Figure 2b); concerning fresh weight, GR_{50} varied from 37 ± 4.8 to 148 ± 8.4 with a mean value of 78 g a.e. ha^{-1}, GR_{90} from 136 ± 10.3 to 295 ± 37.6 g a.e. ha^{-1} with a mean value of 199 g a.e. ha^{-1} (data not shown).

Cluster analyses were used to determine whether correlations were present between the calculated parameters (i.e., EDs and GRs) and the collection sites or geographical origin as well as species of the different populations. A cluster analysis based on the ED_{50} highlighted two clusters (Figure 2a), whereas the data were divided into four clusters when ED_{90} were considered (Figure 2b). In both cases, no correspondence was detected among these divisions and geographical origin of the populations, *Lolium* species or cropping system/collection site. Similar results were obtained considering GRs; therefore, data are not reported.

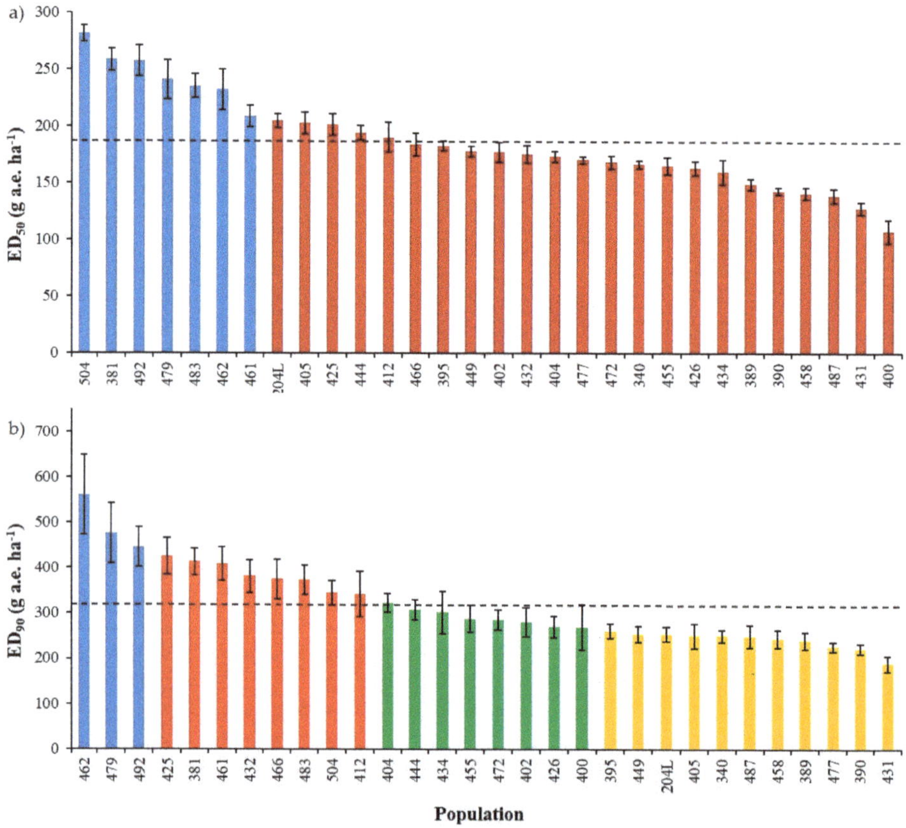

Figure 2. Response of thirty *Lolium* spp. populations included in the autumn experiment estimated by (**a**) the dose controlling 50% of plants (ED_{50}) and (**b**) 90% of plants (ED_{90}). Bars indicate standard errors (SE). Dashed horizontal lines represent the mean values of ED_{50} and ED_{90} in graphs (**a**) and (**b**), respectively. Different colours represent the subdivision of the populations obtained with the cluster analysis: (**a**) two clusters, (**b**) four clusters.

2.2. *Sensitivity Line Calculation*

Based on the results obtained through the dose-response experiments, the range of glyphosate susceptibility of *L. multiflorum* and *L. rigidum* sampled in Italian agricultural environments was established, a sensitivity line was calculated and the dose of glyphosate to fully control *Lolium* spp. in agronomic conditions was proposed.

The paired t-test proved that no significant differences were found between the mean EDs and mean GRs of the two experiments (considering only the common populations), so for calculation of the threshold value of the sensitivity analysis, only the data of the autumn experiment, having a higher number of populations, were considered.

Based on the mean value of ED_{90} as well as the variability across and within populations (Figure 2b), it was established that an agronomically suitable dose (i.e., at least 93%–95% of control) to adequately control susceptible *Lolium* species was around 450 g a.e. ha^{-1} of glyphosate. Therefore, the current dose indicated in Italy (480 g a.e. ha^{-1}) is enough to adequately control susceptible plants, confirming that the old dose (360 g a.e. ha^{-1}) was sub-lethal for many populations. As an anti-resistance measure, it is important to keep efficacy at or near 100% to avoid, or at least slow down, the selection and eventually the evolution of glyphosate resistance under tough climatic conditions or weed growth stages that can affect glyphosate efficacy level.

Figure 3 demonstrates how the sensitivity analyses data can be used to identify potentially resistant populations. The range of ED_{90} for the autumn experiment was 189-561 g a.e. ha^{-1}, with a mean sensitivity line of 317 g a.e. ha^{-1}. A population can, therefore, be considered as shifting (or partially resistant) if the difference between the threshold value and population is greater than 2x (634 g a.e. ha^{-1}) [26] (e.g., populations 259 and 328 included in the spring experiment). A population can be considered as resistant if the difference between threshold value and population is greater than 3x (951 g a.e. ha^{-1}) (e.g., populations 384, 384L, 343, 392, 403 and 401 included in the spring experiment, plus populations 332 and 336 [31]) (Figure 3).

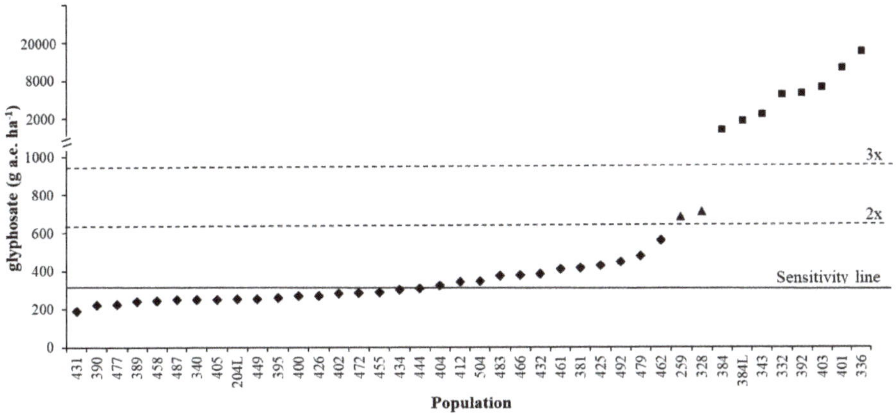

Figure 3. ED_{90} of different *Lolium* spp. populations: ♦ S populations tested in the autumn dose-response experiment, ▲ shifting populations tested in the spring dose-response experiment, ■ resistant populations tested in the spring dose-response experiment and/or discussed in [19] and [31]. Continuous line at 317 g a.e. ha^{-1} represents the sensitivity line calculated in this research, dashed lines represent 2x and 3x the sensitivity line value.

Two parameters that illustrate the variability of the response of populations to glyphosate were calculated (see Section 4.4). $ED_{50/90}$ variations do not fully explain the overall variability, and slope also has to be taken into account (Figure 4). When the ratio is close to one the slope tends to be vertical, i.e., small variations of glyphosate dose around EDs cause large variations in weed control. However, we did not observe any relation between collection site and slope. The Sensitivity Index (S.I.) proved to be three, demonstrating that there is a three-fold difference in sensitivity to glyphosate between *Lolium* spp. populations harvested across Italian agricultural environments.

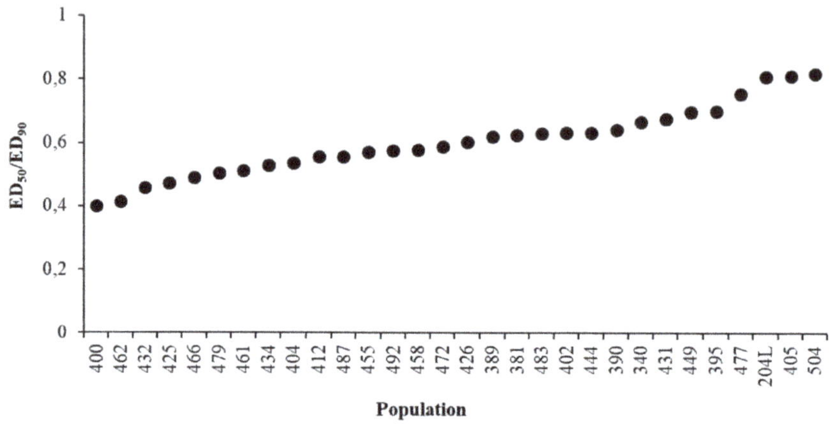

Figure 4. Variations of the ED_{50}/ED_{90} ratio among populations.

3. Discussion

Glyphosate is an efficient herbicide, and the evolution of resistant weeds is a big hindrance to efficient control in many circumstances [7]. Given that no herbicides with truly new molecular target sites have been marketed in the past 30 years and that there is no silver bullet chemistry ready to enter the marketplace [32], glyphosate efficacy should be preserved in the long term, especially in those cropping systems where there is a shortage of post-emergence herbicides (i.e., targeting grasses) or as a tool in weed resistance management. For these reasons, efficacy of herbicide treatments should be kept at or near 100% to avoid or at least slow down the selection and eventually the evolution of glyphosate resistance.

The two dose-response experiments performed to calculate the sensitivity line of glyphosate in Italian agricultural environments could not be pooled together for several reasons. First of all, in the second experiment, a larger number of populations coming from different parts of Italy and different agricultural systems were included in order to give a higher impact to the study. Secondly, it is very rare that two experiments conducted in outdoor conditions can be considered together because there are too many uncontrollable variables (e.g., variation in temperature, rainfall). In particular, this was expected using glyphosate because its performance is known to vary seasonally [18,33].

On the basis of ED_{50}, GR_{50}, ED_{90}, GR_{90} and slope, it was not possible to discriminate *L. multiflorum*, *L. rigidum* or intermediates, and neither a difference related to geographical areas nor collection site was found. Most probably, if a correlation is present among these values and the variables considered, a more specific study with a larger number of populations needs to be assessed. An example was reported for 80 accessions of *Echinochloa* spp. where *E. crus-galli* was found to be more sensitive than other *Echinochloa* species when sprayed with azimsulfuron or cyhalofop-butyl [23].

On average, GR_{50} and GR_{90} (78 and 199 g a.e. ha^{-1}, respectively) were lower than ED_{50} and ED_{90} (186 and 317 g a.e. ha^{-1}, respectively) indicating that a significant proportion of surviving plants had a low fitness and likely a low competitivity with the crop. However, it cannot be excluded that they could produce some seeds [34]. Therefore, in order to not underestimate this aspect, the ED values were used to calculate the sensitivity line.

The Box and Whisker analysis indicated that the selection of populations was adequate for the purpose of the study; in fact, only the populations included in the spring experiment as resistant or partially resistant checks were excluded through the analyses (Figure 2), whereas no outliers were found in the autumn experiment.

Data variability increased with EDs, which may indicate that less susceptible populations are also less homogeneous in terms of glyphosate susceptibility. The proposed field dose also considers this variability and was calculated excluding the eight outlier populations identified through the Box and

Whisker analysis in the spring experiment. The eight outliers correspond to the resistant and partially resistant (i.e., in-shifting) populations included for comparison (Table 1a) and for which the resistance mechanisms have been described elsewhere (Table 1a [19,31]). In this study, it was demonstrated that the optimal glyphosate dose to control *Lolium* spp. in Italy at the growth stage of first shoot visible (i.e., using the Extended BBCH scale at growth stage 21 [35]) should be 700 g a.e. ha^{-1} of glyphosate or higher. Indeed, in our experimental conditions no survivors were recorded for any susceptible population treated with 560 g a.e. ha^{-1} of glyphosate, whereas to completely control the shifting populations, 720 g a.e. ha^{-1} was necessary. This indicates that slightly higher doses, while remaining abundantly within the label recommendations, may be useful to control and hopefully reduce the evolution of resistance to this herbicide.

Guidelines for future herbicide-resistant weed management globally should focus on avoiding a general use of reduced herbicide, especially glyphosate [36]. Successful integrated weed management strategies should aim at decreasing weed seed banks and reducing herbicide use. This involves adjusting the herbicide doses applied to achieve both a reduction in the number of treatments as well as an increase in the number of weeds controlled by the treatments. In this context, this research provides useful information to avoid or slow down the selection of glyphosate resistance in *Lolium* spp. by establishing a threshold for identifying future shifts of susceptibility.

4. Materials and Methods

4.1. Plant Material

Seeds of *Lolium* spp. were collected in agricultural and non-agricultural sites including field margins, organic farms (winter cereals), conventional farms (winter cereals, sunflower and perennial crops) and roadsides (Table 1). When available, details of historical herbicide use on the sampled fields were recorded. Sampling sites covered all major Italian agricultural areas and were chosen according to the absence or moderate application of glyphosate during the last decade. Areas where glyphosate-resistant *Lolium* spp. had been already reported were excluded [21]. Preference was given to regions where *Lolium* spp. are widespread and potentially cause severe economic losses. In each site, seeds were randomly collected from at least 30 plants spatially distributed in a sampling area of about 400 m^2. Although morphological traits showed a high variability among and within populations, all of them were classified as *L. rigidum* or *L. multiflorum* or intermediates between the two species (LOLSS) (Table 1). The standard susceptible population S-204L, collected more than 15 years ago and reproduced in the greenhouse of the Institute for Sustainable Plant Protection (IPSP)- CNR (45°21′ N, 11°58′ E) was also tested.

After ripening, seeds were kept in paper bags and then stored in a cool chamber at 4 °C until use.

4.2. Dose-Response Experiments

4.2.1. Spring Dose-Response Experiment

Twenty populations were included in the spring experiment (Table 1a), twelve putative susceptible populations, six known resistant and two "shifting" populations [19,31]. To break dormancy, seeds were vernalized at 4 °C in Petri dishes on wet filter paper, in darkness for three days. They were then placed in transparent plastic dishes on 0.6% (wt/V) agar medium and placed in a germination cabinet at the following conditions: temperature (day/night) 25/15 °C, 12 h photoperiod with neon tubes providing a Photosynthetic Photon Flux Density (PPFD) of 15–30 μmol m^{-2} s^{-1}. Nine germinated seedlings at similar growth stage were transplanted into pots (15 × 15 × 20 cm) filled with a standard potting mix (60% silty loam soil, 15% sand, 15% perlite and 10% peat). To better mimic field conditions, pots were kept outside in a semi-controlled environment, and the soil water content was maintained at or near field capacity. Temperature ranged day/night from 18.8 °C to 7.9 °C. The experimental layout was a completely randomized design of three replicates per dose (a total of 27 plants per dose). Eleven doses (geometrically distributed) of glyphosate (MON 79351) 480 g a.e. L^{-1} were considered: 45, 90, 135, 180, 270, 360, 450, 540, 720, 1080 and 1440 g a.e. ha^{-1}. An untreated control was included for each population. Herbicide was sprayed when plants reached the stage BBCH 21 using a precision bench sprayer according to the following conditions: spray volume 200 L ha^{-1}, pressure 215 kPa, speed 0.75 m s^{-1} using TeeJet nozzles TP11001-VH. Plant survival and fresh weight were recorded four weeks after the treatment.

4.2.2. Autumn Dose-Response Experiment

Thirty populations sampled as described in Section 4.1 in different Italian agricultural environments (Figure 5) were included in the autumn experiment (Table 1, a in bold and b). All populations were putatively susceptible to glyphosate in order to calculate the threshold value of the sensitivity analysis. To compare the data of the two experiments, 12 susceptible populations (in bold in Table 1a) selected for the spring experiment were also included in the autumn experiment. Seeds preparation, seedlings transplanting and growth conditions, as well as treatment conditions, were as described in Section 4.2.1. Temperature ranged day/night from 20.3 °C to 11.2 °C. Plant survival and fresh weight were recorded four weeks after treatment.

Figure 5. Distribution and origin of *Lolium* spp. populations included in the autumn dose-response experiment (see also Table 1b and populations in bold in Table 1a): yellow = wheat, dark blue = Lucerne, light blue = field margin, dark green = meadow, light green = roadside, purple = perennial, pink = set aside. For IPSP S check 204L, the origin of the original population is reported (Civitella Paganico, GR). Population 412 is not included because it comes from a commercial seed stock.

4.3. Statistical Analyses

The mean survival and fresh weight per dose were expressed as a percentage of the untreated control. The ED_{50}, GR_{50}, ED_{90}, GR_{90} and relative standard errors for the mean percentage of plant survival and fresh weight were calculated by non-linear regression analysis performed using the macro BIOASSAY® developed by Onofri (2005) [37] and running in Windows Excel®. The macro is based on a log-logistic equation to fit the data: $Y = C + \{(D - C)/[1 + (x/I_{50})b]\}$ where Y is the fresh weight or survival, C and D are the lower and upper asymptotes at higher and zero doses, respectively, I_{50} (or I_{90}) is the herbicide dose resulting in a 50% (or 90%) reduction in plant biomass or survival, i.e., ED_{50} and GR_{50}, respectively (or ED_{90} and GR_{90}, respectively), b is the slope. The procedure estimates the standard error of the parameters and performs the Box-Cox power transformation family. For biological reasons and to improve the estimates of other parameters, the upper and lower asymptotes of survival data were forced to 100 and zero, respectively, whereas no parameters were constraints considering fresh weight data. Data of each population were first analyzed as a single curve to estimate the parameters and then all curves were regressed together. The data of the two experiments were analyzed separately. No parameters were fixed in the first analyses, and this complex model was then compared with progressively simplified models having common parameters among curves. The lack-of-fit F-test was performed at each step, and the simplification stopped when a significant lack of fit occurred.

The Box and Whisker plot analysis was used to identify possible outliers and extreme values described as the values greater, or lower, than 1.5 and 3 times the value of the Box, respectively [38]. The analysis was repeated excluding the outlier values at each step until no outliers were detected.

In order to determine if the data of the two experiments could be compared, an ad hoc lack-of-fit F-test was applied to the data of each population included in both experiments by comparing data singularly.

R software 3.2.5 and, in particular, the package *NbClust* [39] was applied to cluster ED_{50} and ED_{90} data. The package compares 30 different clustering methods and chooses as best the partition proposed by the majority of the methods.

4.4. Sensitivity Line Calculation

The mean values of EDs and GRs of the two experiments were compared using a paired t-test at $p < 0.05$ (excluding populations with extreme or outlier values in the Box and Whisker analyses).

The threshold value of the sensitivity analysis (sensitivity line) was calculated as the mean of the ED_{90} values with a slight modification compared to the method used by Paterson et al. [26] across biologically relevant populations and experiments. A population can be considered as shifting if the difference between the sensitivity line and population is greater than 2x and resistant if the difference between the sensitivity line and population is greater than 3x. To illustrate the variability of the response of populations to glyphosate, the $ED_{50/90}$ variation among populations and the S.I. were calculated as the ratio between ED_{50} and ED_{90} of each population and the ED_{90} of the most tolerant and most sensitive populations, respectively.

Author Contributions: The authors contributed equally to the research, in particular: conceptualization, A.C., M.S. and S.P.; methodology, A.C., M.S. and S.P.; formal analysis, S.P. and A.C.; investigation, S.P. and A.C.; data curation, S.P. and A.C.; writing—original draft preparation, S.P. and A.C.; writing—review and editing, S.P., M.S. and A.C.; supervision, M.S.; project administration, M.S.; funding acquisition, M.S. All authors have read and agreed to the published version of the manuscript.

Funding: The research was jointly funded by the National Research Council of Italy (CNR) and Monsanto International Srl.

Acknowledgments: The authors are grateful to Alison Garside for revising the English text, Ivo Brants (Monsanto Europe S.A.) for helpful discussions and Erica Manesso for the advice regarding statistical analyses.

Conflicts of Interest: The authors declare no conflict of interest. The funders had no role in the design of the study; in the collection, analyses, or interpretation of data; in the writing of the manuscript, or in the decision to publish the results.

References

1. Preston, C.; Wakelin, A.M.; Dolman, F.C.; Bostamam, Y.; Boutsalis, P. A Decade of Glyphosate-Resistant Lolium around the world: Mechanisms, genes, fitness, and agronomic management. *Weed Sci.* **2009**, *57*, 435–441. [CrossRef]
2. Heap, I.; Duke, S.O. Overview of glyphosate-resistant weeds worldwide. *Pest Manag. Sci.* **2018**, *74*, 1040–1049. [CrossRef]
3. Fernández-Moreno, P.T.; Bastida, F.; De Prado, R. Evidence, mechanism and alternative chemical seedbank-level control of glyphosate resistance of a rigid ryegrass (*Lolium rigidum*) biotype from Southern Spain. *Front. Plant Sci.* **2017**, *8*, 1–16. [CrossRef]
4. Heap, I.M. The International Survey of Herbicide Resistant Weeds. Available online: http://weedscience.org/ (accessed on 24 January 2020).
5. Barroso, A.A.; de S Costa, M.G.; Neto, N.J.; dos Santos, J.I.; Balbuena, T.S.; Carbonari, C.A.; Alves, P.L. Protein identification before and after glyphosate exposure in *Lolium multiflorum* genotypes. *Pest Manag. Sci.* **2018**, *74*, 1125–1133. [CrossRef]
6. Duke, S.O. Why have no new herbicide modes of action appeared in recent years? *Pest Manag. Sci.* **2012**, *68*, 505–512. [CrossRef]
7. Duke, S.O. The history and current status of glyphosate. *Pest Manag. Sci.* **2018**, *74*, 1027–1034. [CrossRef]
8. Duke, S.; Powles, S. Glyphosate: A once-in-a-century herbicide. *Pest Manag. Sci.* **2008**, *64*, 319–325. [CrossRef]
9. Mallory-Smith, C.; Retzinger, J. Revised classification of herbicides by site of action for weed resistance. *Weed Technol.* **2003**, *17*, 605–619. [CrossRef]
10. Solomon, K.R. Estimated exposure to glyphosate in humans via environmental, occupational, and dietary pathways: An updated review of the scientific literature. *Pest Manag. Sci.* **2019**. [CrossRef] [PubMed]
11. Schönbrunn, E.; Eschenburg, S.; Shuttleworth, W.A.; Schloss, J.V.; Amrhein, N.; Evans, J.N.; Kabsch, W. Interaction of the herbicide glyphosate with its target enzyme 5-enolpyruvylshikimate 3-phosphate synthase in atomic detail. *PNAS* **2001**, *98*, 1376–1380. [CrossRef] [PubMed]
12. Baylis, A.D. Why glyphosate is a global herbicide: Strengths, weaknesses and prospects. *Pest Manag. Sci.* **2000**, *56*, 299–308. [CrossRef]
13. Neve, P.; Diggle, A.J.; Smith, F.P.; Powles, S.B. Simulating evolution of glyphosate resistance in *Lolium rigidum* I: Population biology of a rare resistance trait. *Weed Res.* **2003**, *43*, 404–417. [CrossRef]
14. Powles, S.B. Evolved glyphosate-resistant weeds around the world: Lessons to be learnt. *Pest Manag. Sci.* **2008**, *64*, 360–365. [CrossRef]
15. Owen, M.D.K. Weed species shifts in glyphosate-resistant crops. *Pest Manag. Sci.* **2008**, *64*, 377–387. [CrossRef]
16. Gonzalez-Torralva, F.; Gil-Humanes, J.; Barro, F.; Dominguez-Valenzuela, J.; De Prado, R. First evidence for a target site mutation in the EPSPS2 gene in glyphosate-resistant Sumatran fleabane from citrus orchards. *Agron. Sustain. Dev.* **2014**, *34*, 553–560. [CrossRef]
17. Urbano, J.M.; Borrego, A.; Torres, V.; Leon, J.M.; Jimenez, C.; Dinelli, G.; Barnes, J. Glyphosate-resistant Hairy Fleabane (*Conyza bonariensis*) in Spain. *Weed Technol.* **2007**, *21*, 396–401. [CrossRef]
18. Powles, S.B.; Lorraine-colwill, D.F.; Dellow, J.J.; Preston, C. Evolved resistance to glyphosate in rigid ryegrass (*Lolium rigidum*) in Australia. *Weed Sci.* **1998**, *46*, 604–607. [CrossRef]
19. Collavo, A.; Sattin, M. Resistance to glyphosate in *Lolium rigidum* selected in Italian perennial crops: Bioevaluation, management and molecular bases of target-site resistance. *Weed Res.* **2012**, *52*, 16–24. [CrossRef]
20. Busi, R.; Powles, S.B. Evolution of glyphosate resistance in a *Lolium rigidum* population by glyphosate selection at sublethal doses. *Heredity* **2009**, *103*, 318–325. [CrossRef]
21. GIRE—Gruppo Italiano Resistenza Erbicidi. Available online: www.resistenzaerbicidi.it (accessed on 24 January 2020).
22. Panozzo, S.; Scarabel, L.; Collavo, A.; Sattin, M. Protocols for robust herbicide resistance testing in different weed species. *J. Vis. Exp.* **2015**, *101*, 1–10. [CrossRef]
23. Vidotto, F.; Tesio, F.; Tabacchi, M.; Ferrero, A. Herbicide sensitivity of *Echinochloa* spp. accessions in Italian rice fields. *Crop Prot.* **2007**, *26*, 285–293. [CrossRef]

24. Loddo, D.; Kudsk, P.; Costa, B.; Dalla Valle, N.; Sattin, M. Sensitivity analysis of *Alisma plantago-aquatica* L., *Cyperus difformis* L. and *Schoenoplectus mucronatus* (L.) Palla to Penoxsulam. *Agronomy* **2018**, *8*, 220. [CrossRef]
25. Davies, L.R.; Hull, R.; Moss, S.; Neve, P. The first cases of evolving glyphosate resistance in UK poverty brome (*Bromus sterilis*) populations. *Weed Sci.* **2018**, *67*, 41–47. [CrossRef]
26. Paterson, E.A.; Shenton, Z.L.; Straszewski, A.E. Establishment of the baseline sensitivity and monitoring response of *Papaver rhoeas* populations to florasulam. *Pest Manag. Sci.* **2002**, *58*, 964–966. [CrossRef] [PubMed]
27. Russell, P.E. *Sensitivity Baseline in Fungicide Resistance Research And Management*; Crop Life International: Brussels, Belgium, 2002; pp. 1–56.
28. Russell, P. Resistance management and the registration of pesticide products in Europe. *Pestic. Outlook* **2001**, *12*, 56–59. [CrossRef]
29. Barroso, J.; Loureiro, I.; Escorial, M.C.; Chueca, M.C. The response of *Bromus diandrus* and *Lolium rigidum* to dalapon and glyphosate I: Baseline sensitivity. *Weed Res.* **2010**, *50*, 312–319.
30. Moss, S.R. Baseline sensitivity to herbicides: A guideline to methodologies. In Proceedings of the British Crop Protection Conference—Weeds, Brighton, UK, 2001; pp. 769–774.
31. Collavo, A.; Sattin, M. First glyphosate-resistant *Lolium* spp. biotypes found in a European annual arable cropping system also affected by ACCase and ALS resistance. *Weed Res.* **2014**, *54*, 325–334. [CrossRef]
32. Dayan, F.E. Current status and future prospects in herbicide discovery. *Plants* **2019**, *8*, 341. [CrossRef]
33. Sammons, R.D.; Gaines, T.A. Glyphosate resistance: State of knowledge. *Pest Manag. Sci.* **2014**, *70*, 1367–1377. [CrossRef]
34. Pedersen, B.P.; Neve, P.; Andreasen, C.; Powles, S.B. Ecological fitness of a glyphosate-resistant *Lolium rigidum* population: Growth and seed production along a competition gradient. *Basic Appl. Ecol.* **2007**, *8*, 258–268. [CrossRef]
35. Hess, M.; Barralis, G.; Bleiholder, H.; Buhr, L.; Eggers, T.H.; Hack, H.; Stauss, R. Use of the extended BBCH scale-general for the descriptions of the growth stages of mono and dicotyledonous weed species. *Weed Res.* **1997**, *37*, 433–441. [CrossRef]
36. Beckie, H.J.; Ashworth, M.B.; Flower, K.C. Herbicide resistance management: Recent developments and trends. *Plants* **2019**, *8*, 161. [CrossRef] [PubMed]
37. Onofri, A. Bioassay97: A new Excel VBA macro to perform statistical analyses on herbicide dose-response data. *Riv. Ital. di Agrometeorol.* **2005**, *3*, 40–45.
38. Tukey, J.W. Exploratory data analysis. In *The Future of Data Analysis*; Addison-Wesley: New York, NY, USA, 1977; p. 688.
39. Charrad, M.; Ghazzali, N.; Boiteau, V.; Niknafs, A. NbClust: An R package for determining the relevant number of clusters in a data set. *J. Stat. Softw.* **2014**, *61*, 1–36. [CrossRef]

plants

MDPI

Article

10,000-Times Diluted Doses of ACCase-Inhibiting Herbicides Can Permanently Change the Metabolomic Fingerprint of Susceptible *Avena fatua* L. Plants

J António Tafoya-Razo [1,†], Ernesto Oregel-Zamudio [2,†], Sabina Velázquez-Márquez [3] and Jesús R. Torres-García [2,4,*]

1 Departamento de Parasitología Agrícola, Universidad Autónoma Chapingo, Texcoco 56230, Mexico; jtafoyar@chapingo.mx
2 Laboratorio de Ecología y Evolución Molecular, Centro Interdisciplinario de Investigación para el Desarrollo Integral Regional (CIIDIR) del Instituto Politécnico Nacional, Unidad Michoacán, Jiquilpan 59510, Mexico; eoregel@ipn.mx
3 Laboratorio de Genética Ecológica y Evolución, Departamento de Ecología Evolutiva, Instituto de Ecología, Universidad Nacional Autónoma de México, Ciudad de México 04510, Mexico; svelazquez@ecologia.unam.mx
4 Cátedras CONACyT, Ciudad de México 04500, Mexico
* Correspondence: jrtorresg@ipn.mx
† These authors contributed equally to this work.

Received: 30 July 2019; Accepted: 20 September 2019; Published: 24 September 2019

Abstract: Intentional use of low dosage of herbicides has been considered the cause of non-target resistance in weeds. However, herbicide drift could be a source of low dosage that could be detected by weeds and change their metabolism. Furthermore, the minimum dose that a plant can detect in the environment is unknown, and it is unclear whether low doses could modify the response of weeds when they are first exposed to herbicides (priming effects). In this study, we determined the metabolomic fingerprinting using GC-MS of susceptible *Avena fatua* L. plants exposed to a gradient of doses (1, 0.1, 0.001, 0.0001, and 0x) relative to the recommended dose of clodinafop-propargyl. Additionally, we evaluated the primed plants when they received a second herbicide application. The results showed that even a 10,000-fold dilution of the recommended dose could induce a significant change in the plants' metabolism and that this change is permanent over the biological cycle. There was no evidence that priming increased its resistance level. However, hormesis increased biomass accumulation and survival in *A. fatua* plants. Better application methods which prevent herbicide drift should be developed in order to avoid contact with weeds that grow around the crop fields.

Keywords: non-target metabolomics; GC-MS; non-target site resistance; priming; hormesis

1. Introduction

Evolution of herbicide resistance is the result of the strong selective pressure exerted by herbicides on weed populations [1]. This selection pressure is so strong that Harper predicted the evolution of herbicide-resistant weed populations even before the appearance of the first report [2]. Early studies of resistant populations attributed resistance to non-synonymous mutations in specific domains in herbicide target genes (Target-Site Resistance, TSR). These mutations change the protein conformation and limit herbicide effectiveness [1,3]. For this reason, some weed management models suggested the reduction of herbicide dosage to reduce the selection pressure in weed populations [4], a recommendation which was implemented in many countries. However, this practice had undesired effects [5], including an increase in the number of populations with metabolic adaptations to degrade herbicides (Non-Target site Resistance, NTSR) [6]. This type of resistance is due to the increase in the activity of specific enzymes such as Cytochrome oxidase P450 (hereafter P450) and Glutathione-S transferase [6]. NTSR has increased over time and has the potential to become a severe problem due to the capacity of degradation of multiple herbicides [1]. Moreover, experiments have shown that this type of resistance could evolve in less time than genetic resistance [7–9].

Other documented effects of low herbicide dosage on weeds are hormesis and priming [10–12]. Priming is defined as a physiological state caused when plants exposed to a low dose of a stressful agent (bacterial, fungus, herbicides, among others) develop an adaptative response, which increases their resistance to subsequent exposure [13]. This phenomenon has been documented in tumor cells, and it is possible that it is a mechanism of NTSR [14]. On the other hand, hormesis is growth stimulation at low doses of herbicides [12]. This phenomenon has been reported in many herbicide modes of action [15–17], and has even been proposed to use the hormetic effects to increase yield in some crops [18,19].

Due to the rapid evolution of NTSR based in P450, the current weed management practices suggest avoiding the application of herbicides at lower-than-recommended doses to prevent the risk of development of metabolic resistance [1,5,6]. However, several factors lead to this continuing in practice. On the one hand, farmers frequently reduce the dosage of herbicide to try to save money by purchasing less herbicide (Torres–Garcia, personal observation). Another cause of low herbicide dosage is late application; herbicide use is recommended when weeds are about 10 cm tall. However, farmers often apply chemical control when weeds are many times higher than the recommended size, thus diluting the effective dose (A Tafoya, Personal Observation).

Even when following all recommendations for their use, from manual application to airplane spraying, herbicides can be spread unintentionally, for example by leaf contact between treated and untreated plants, protection by taller plants, and drift of spray particles to nearby fields [20]. Clouds of vapor carrying nano-drops of herbicide could modify the metabolism of surrounding weeds. In highly advanced agricultural systems, this is not a problem, but only a small proportion of global agriculture has this level of mechanization.

With respect to herbicide spray particle drift, some basic questions that we asked are: (1) What is the minimum dose of herbicide that a plant could perceive? (2) Do weeds that received a non-lethal dose show differences in metabolism and survival when they receive a second application (i.e., are there priming effects)? and (3) Are the metabolic changes transgenerational, and therefore, a possible cause of metabolic resistance? The last question has been well responded by Neve and Powles [8]. However, the first two questions remain unclear. In this study, we are interested in generating information that could respond to the two initial questions.

Plant metabolism is complex, with a large number of chemical compounds and interactions among them, making the identification and quantification of all metabolic changes a complicated job [21–23]. The use of high-throughput metabolic methods can detect minimal changes in the metabolic state, giving a fingerprint of the metabolic state of the plant [24]. The metabolome is the final result of the plant's response and could provide us with a detailed "snapshot" of the changes caused by herbicides [25,26]. In this study, we applied a non-target metabolomic scope based on the identification of punctual metabolomic fingerprinting using GC-MS as an analytical approach. We expected that if plants could detect the presence of herbicides in the environment, then they would show changes in their global fingerprint compared to untreated plants, and this fingerprint would be different in plants that had previously been exposed to herbicides than those that had not been previously treated.

2. Results

2.1. Experiment 1. Biomass Accumulation and Survival

The application of clodinafop-propargyl showed a significant reduction in the dry matter accumulation of the studied biotype compared to unexposed plants, except at the 0.001x dose. Plants grew and accumulated significantly more dry matter even in the lowest dose (0.0001x the recommended dose). In plants treated with 0.001x, there was an increase in the dry matter accumulation caused by hormesis (Figure 1A). Dry matter accumulation showed a drastic reduction at the doses from 0.01x to 1x. Survival, on the other hand, was not affected in the two least concentrated doses (0.0001x and 0.001x) of clodinafop-propargyl, but there was a significant decrease in survival at the 0.01x dose (63% survival), and there was 100% mortality at both the 0.1x and 1x doses (Figure 1B).

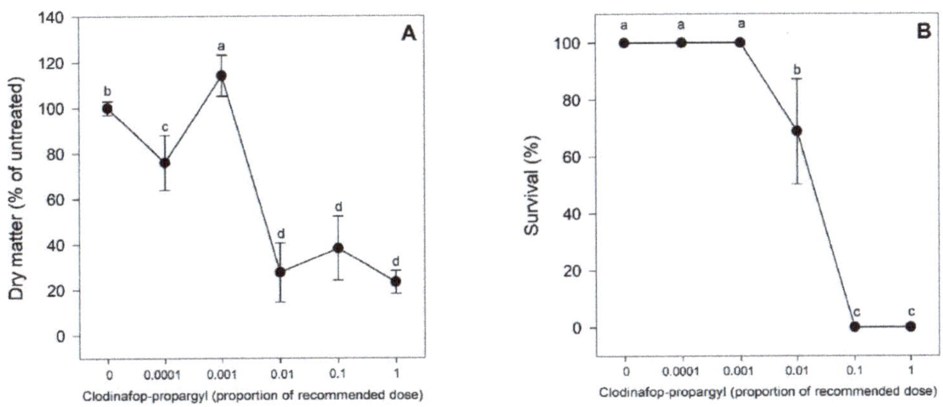

Figure 1. Dry matter accumulation (**A**) and survival (**B**) of susceptible *Avena fatua* L. plants treated with clodinafop-propargyl in proportional doses of the recommend rate (1x, 0.1x, 0.01x, 0.001x, 0.0001x, and 0x). Statistical significances are indicated with different letters. Datapoints represent mean values, and vertical bars represent standard errors. When they are absent, they are smaller than the symbol ($n = 4$).

Metabolomic Fingerprinting of the First Experiment

The metabolic fingerprint of the susceptible biotype of *A. fatua* with the application of clodinafop-propargyl was obtained by GC-MS recording a total of 67 metabolites. For the construction of the heatmap, we used only the 12 metabolites that had *p*- and *q*- values ≤ 0.05. The resulting heatmap shows significant changes in the metabolism caused by herbicide application, even with the most diluted concentration sprayed (Figure 2).

The dendrogram along the top Figure 2 shows the grouping among treatments. In this dendrogram, we found the formation of two main groups (branches). One of these branches (left side of the heatmap) includes the lower doses (including the control treatment; 0x, 0.0001x, and 0.001x; Figure 2) This grouping also corresponds with that observed in the dry matter accumulation and survival. Inside this branch, the control treatment (0x) comprised a different subgroup. This indicates that plants showed changes in the metabolic fingerprint even at a 10,000-fold reduction of the recommended herbicide dose. In the case of treatment with the 0.001x dose, there was an apparent hormetic effect on dry matter accumulation, but there was no evidence of significant changes in their expression pattern that would explain this effect, and the heatmap did not reveal metabolic differences between the 0.0001x and 0.001x treatments. The second main group (right side of the heat map) included the higher-dose treatments which caused the highest dry matter reduction and mortality. In this branch, the 0.01x treatment was divided as a subgroup from the 0.1x and 1x treatments.

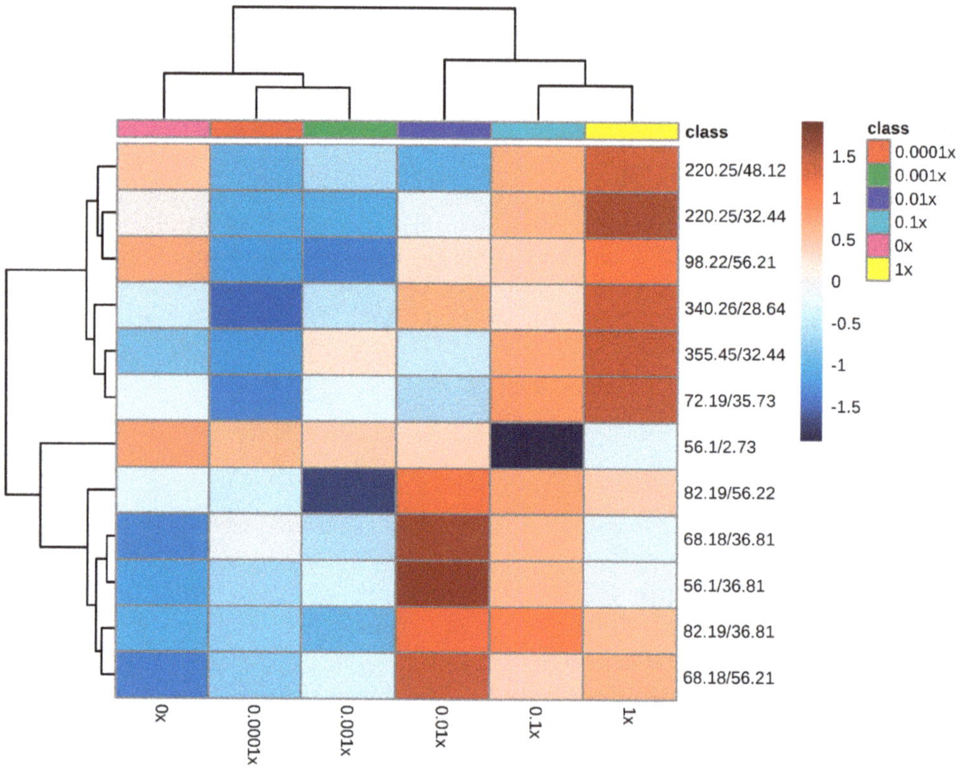

Figure 2. Metabolomic fingerprint of susceptible *Avena fatua* L. plants ($n = 4$) at 24 h after been sprayed with clodinafop-propargyl in proportional doses of the recommend rate (1x, 0.1x, 0.01x, 0.001x, 0.0001x, and 0x). The heatmap was constructed with the 12 metabolites that showed *p*- and *q*- values ≤ 0.05. Colors represent the abundance of metabolites; the blue color indicates down-expression and red color over-expression. The metabolites are clustered according to their Pearson correlation as a distance function, and the Ward clustering algorithm, the significance of the branches were of $p \leq 0.05$.

The dendrogram along the left side of the heatmap reveals the relationship among the metabolites detected. This dendrogram also has two main branches; in the upper section, there was a marked difference in the expression of the metabolites between the 0.0001x and 1x treatments. Metabolites of the 0.0001x treatment are shown in shades of blue, indicating that those metabolites were down-expressed. In contrast, the 1x treatment showed an over-expression of the same metabolites.

The second branch of the dendrogram (the lower half of the heatmap), according to random forest analysis, contains the five most important metabolites for classification (82.19/36.81; 68.18/36.81; 68.18/56.21; 56.1/36.81), such metabolites correspond to Hexane-2,6-di(isonitrile), 1-(formyloxymethyl)-Z-3, 17-Octadecadien-1-ol, and acetate(S)-2-methylbutanoic acid methyl ester 2-methylpropanoic.

In this zone, there was an evident change in expression. Treatments with the lowest doses (including 0x) showed a down-expression of those metabolites, while treatments with the most concentrated doses and with more biological changes (several reductions in the dry matter and high mortality) show overexpression of those metabolites.

2.2. Experiment 2. Biomass Accumulation and Survival

In the plants treated with 0.0001x and 0.01x doses, we did not find differences in the growth and survival among plants that had been previously sprayed with herbicides versus plants that were receiving their first application. In the case of the 0.001x treatment without previous herbicide application (0.001x-U), there was a significant increase in dry matter and survival (Figure 3). A significant increase of the dry matter was observed, even respect to control (60% of the rise). The survival was of the 100% in all pots sampled.

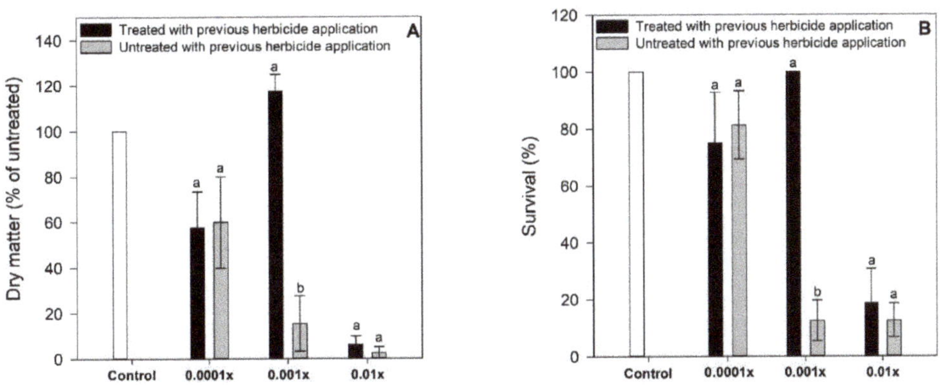

Figure 3. Dry matter accumulation (**A**) and survival (**B**) of susceptible *Avena fatua* L. plants that have been treated previously, and plants that received their first application of clodinafop propargyl at rates of 0.01, 0.001, 0.0001x of the recommended dose. Control plants only were sprayed with distillate water and adjuvant. Statistical significances are indicated with different letters. Datapoints represent mean values, and vertical bars represent standard errors. When they are absent, they are smaller than the symbol ($n = 4$).

Metabolomic Fingerprint the Second Experiment

In the second experiment, 51 metabolites were detected with q-values ≤ 0.05, and 46 metabolites had significant p-value ≤ 0.05, so these 46 were used to construct the heatmap. Four of those metabolites were shared with the first experiment: 220.25/48.11, 67.15/36.81, 72.12/35.72, and 220.25/32.45. According to the NIST (National Institute of Standards and Technology, Gaithersburg, MD, USA) library, such metabolites were 2-Methylamino-3-methylbutanoic acid, 2-methylpropanoic acid, Benzoic acid methyl ester, and 1-(p-Methoxycarbonylphenyl)-5-phenyl-3-(2-pyridyl)-2-pyrazoline, respectively.

The resulting heatmap shows that treatments were grouped into two main branches (Figure 4). One branch (Figure 4; right side of the heatmap) was conformed of the control, and the two most dilute herbicide applications (0.0001x-U, 0.0001x-T). Within this branch, the control and 0.0001x-U treatments had a very similar fingerprint, and for this reason, were grouped into the same sub-group. On the other main branch, all treatments were grouped in closed sub-branches. The only treatment that showed differentiation in this sub-branch was the treatment 0.001x-U. This same treatment also showed significant differences in dry matter and survival.

The dendrogram along the left side of the heatmap in Figure 4 shows the marked differences in metabolite expression of the treatments. The upper half of the heatmap shows that treatments with the lowest dose of herbicide (including the control) had increased expression of 17 metabolites. On the other branch, those metabolites were expressed less. In the lower branch, a set of tree metabolites had an inverse expression pattern compared to the other treatments. The metabolites 46.99/1.71, 81.98/1.71 and 47.95/1.71 (2-Nonen-1-ol, (S)-2-methylbutanoic acid methyl ester, and 3-Buten-1-ol, 2-methyl, respectively) were expressed less in the treatments with low doses of herbicide, while in the higher dose treatments, they were expressed more.

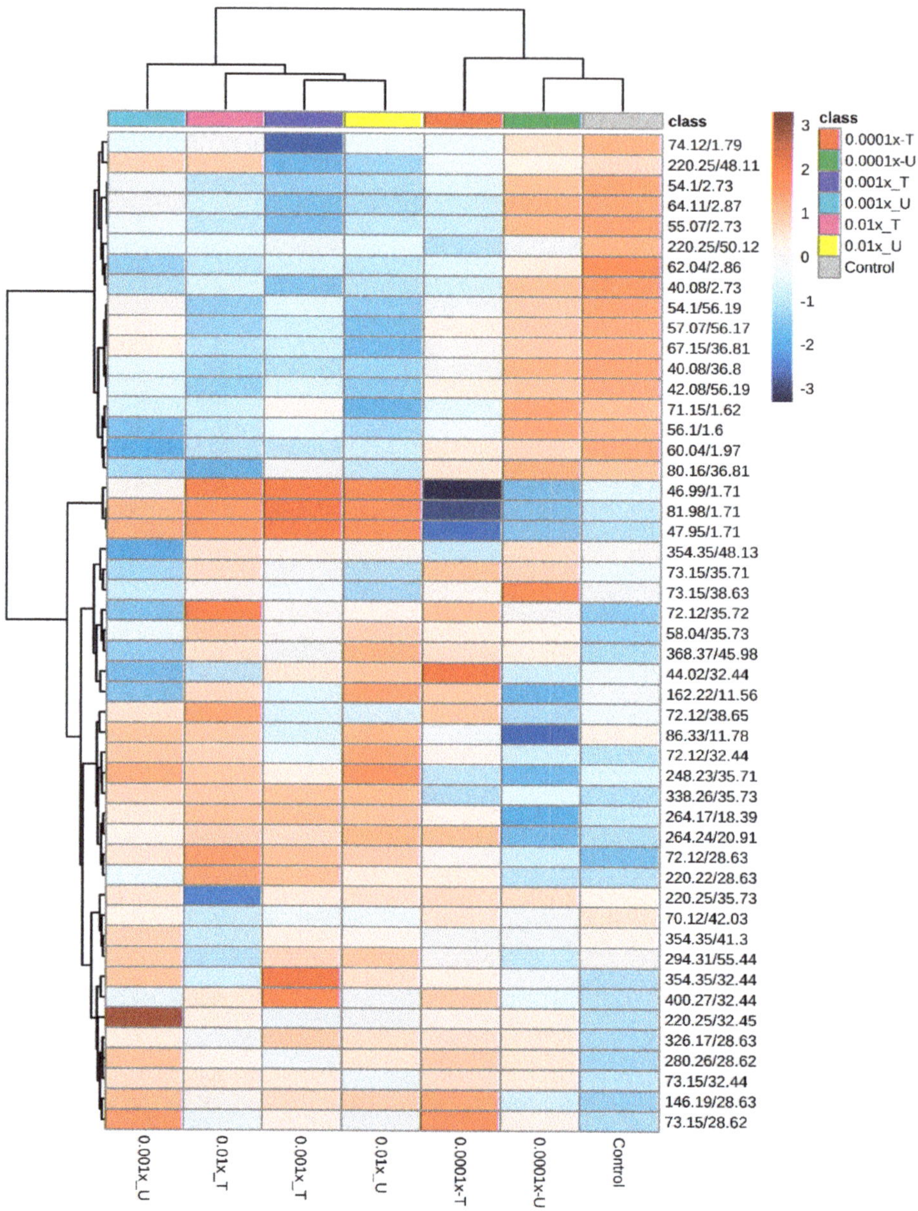

Figure 4. Metabolic fingerprint of susceptible *Avena fatua* L. plants that have been treated previously (T = Treated) and plants that received their first application (U = Untreated) of clodinafop propargyl at rates of 0.01, 0.001, and 0.0001x of the recommended dose. The heatmap was constructed with the 46 metabolites that showed *p*- and *q*- values ≤ 0.05. Colors represent the abundance of metabolites; the blue color indicates down-expression and red color over-expression. The metabolites are clustered according to their Pearson correlation as a distance function, and the Ward clustering algorithm, the significance of the branches were $p \leq 0.05$.

The 0.0001x-U treatment was grouped into a separate group in the heatmap, and it also had a significant change in dry matter and survival. In the fingerprinting of this treatment, a marked down-expression of eight metabolites (354.35/48.13, 73.15/35.71, 73.15/38.63, 72.12/35.72, 58.04/35.73, 368.37/45.98, 44.02/32.44, and 162.22/11.56) constitutes a notable difference compared to all of the other treatments. These compounds correspond to aromatic compounds as 1, 2-Hexadecanediol, trans 3-penten-1-ol, 3-methyl-2-buten-1-ol, Benzoic acid methyl ester, 1-pentenal, 1-(p-Methoxycarbonylphenyl) -5-phenyl-3-(2-pyridyl)-2-pyrazoline, and 2-heptanol. All treatments that received herbicide application showed increased expression of these metabolites; while the control treatment also showed decreased expression of the same metabolites, but this expression is not so evident as in treatment 0.0001x-U.

3. Discussion

In this study, we simulated the effect of herbicide drift (or any event that leads to exposure to a low herbicide dosage) on the metabolism of weeds surrounding crop fields. Our principal objectives were to determine: (1) What is the minimum dose of herbicide that a plant can perceive, and (2) if plants that received non-lethal doses show differences in their metabolism, growth, and survival when receiving a second application. The results showed that susceptible plants of *A. fatua* could detect the presence of clodinafop-propargyl in doses diluted by 10,000 fold with respect to the recommended dose. The changes observed in the metabolomic fingerprint were present at least 21 days after herbicide application. When the plants received a second application of herbicide, they showed differences in the metabolomic fingerprint compared to plants that were not treated previously, but this did not lead to changes in biomass accumulation and survival. Hormesis was observed with the application of a dose of 0.001x, but this did not increase their tolerance to a second application. In the second experiment, these hormetic effects also increased survival; this means that changes in plant metabolism can occur at extremely low doses and have effects on the plants' fitness, depending on plant age.

A. fatua plants showed reductions in their dry matter accumulation even at low doses (10,000 fold dilution of the recommended dose). This dry matter reduction was caused by the high effectiveness of herbicide in susceptible plants [27]. However, it is possible that other susceptible biotypes could display a different degree of susceptibility [28]. The metabolomic fingerprinting observed in the first experiment was congruent with the observations in dry matter and survival. Metabolic changes were observed 24 h after the herbicide application, with clear differentiation in the expression of some metabolites among treatments that showed high levels of damage and mortality. These differences in the fingerprint could be used as a predictive tool for determining susceptibility or resistance. Torres-García et al. [26] using Direct-Injection electrospray ionization mass spectrometry (DIESI) detected differences in the fingerprints of multiple herbicide-resistant biotypes of *A. fatua* (resistant to ACCase- and ALS-inhibiting herbicides) when sprayed with herbicides of each mode of action.

The analytical approach used in this study (GC-MS), has the disadvantage that only volatile compounds that can be detected, compared with other methods such as UPLC-MS, EI-MS, among others [29,30]. However, the 12 metabolites in the first experiment and 46 in the second fulfilled the requirements of q- and p- values ≤ 0.05, confirming their participation in the metabolic response to herbicide application. The non-target metabolic approach used also has the disadvantage that the identification of metabolites could be spurious due to the lack of standards for each metabolite found. However, one of the objectives of this study was to determine the minimum dose of herbicide that a plant can perceive, and it was accomplished. Besides, this could be an initial step for accurately determining the metabolites that participate in the plants' response to herbicides, i.e., the four metabolites that were shared in the two experiments and putatively were identified as 2-Methylamino-3-methylbutanoic acid, 2-methylpropanoic acid, Benzoic acid methyl ester, and 1-(p-Methoxycarbonylphenyl)-5-phenyl-3-(2-pyridyl)-2-pyrazoline.

The second experiment demonstrated that previous herbicide application (priming) did not have any effect on the biomass and survival during a second herbicide application. The expectation is that primed plants develop defense responses that are faster, stronger, and more sustained than in plants that were not primed [10]. However, in the phenotypic traits measured, this did not occur. On the other hand, there were significant changes in the metabolomic fingerprint. These changes caused by stimulation by low doses of herbicides were present throughout the biological cycle. The metabolomic fingerprint of plants that received a prior dose of herbicide was different from those that were receiving herbicide application for the first time. These effects have been called "metabolic memory" or "priming" and it has been demonstrated that this effect can be passed to the next generation, indicating an epigenetic component of transgenerational inheritance [11]. These transgenerational priming effects are likely one of the factors on the evolution of NTSR based in the overexpression of P450 genes. Neve and Powles [8,9] reported that recurrent selection at low doses for three generations caused a 55-fold increase in the resistance index. Later, Yu et al. [31] confirmed that the resistance mechanism of these biotypes was based on the over-expression of P450.

An exception to priming effects on metabolomic fingerprinting was the treatment 0.0001x-U. This treatment was grouped in the same branch as the control treatment. In this case, it is possible that the dose of 0.0001x (10, 000 diluted) was too low to provoke a metabolic change in plants with larger sizes (around 45 days after seedling). This elevated sensitivity is a factor to consider since the metabolic response of weeds to herbicide drift will depend on the plant size.

The treatment that showed hormetic effects was the only one that increased in survival. In addition, its metabolic fingerprint differed from those of the other herbicide treatments, particularly in eight metabolites (1, 2-Hexadecanediol, trans 3-penten-1-ol, 3-methyl-2-buten-1-ol, Benzoic acid methyl ester, 1-pentenal, 1-(p-Methoxycarbonylphenyl)-5-phenyl-3-(2-pyridyl)-2-pyrazoline, and 2-heptanol). However, the hormesis is a complex phenomenon that cannot be explained by the expression of only eight metabolites, and more research is needed [12].

The results presented in this study demonstrated the high sensitivity of susceptible biotypes to the presence of herbicides (clodinafop-propargyl) in the environment. This could have implications in the contamination caused by the application methods used. The unintentional low dosage caused by the drift of micro drops carrying ultra-diluted doses of herbicide can be detected by susceptible plants that grow around the crop fields and may cause metabolic resistance. Improvement in application methods is required to avoid drift. This study also documents the lack of priming effects in response to herbicides in *A. fatua* and that there is an increase in size and survival caused by hormesis.

4. Methods

4.1. Study System

Avena fatua L. is considered as the world's second-worst herbicide-resistant weed due to their worldwide presence in cereal-growing regions [32]. This weed has evolved resistance to at least seven modes of action (antimicrotubule mitotic disrupter and ACCase-, ALS-, PPO-, cell elongation-, long-chain fatty acids-, and lipid- inhibitors). A single biotype from Canada has even evolved multiple resistance to 5 modes of action (ACCase-, ALS-, PPO-, long-chain fatty acids-, and lipid- inhibitors) [33]. In recent years, the increase of cases of metabolic resistance also documented *A. fatua* biotypes [6].

The *A. fatua* biotype used in this study was collected in an alfalfa crop, in a zone where cereal has not been produced for the past 10 years (A Tafoya, personal observation). The susceptibility was confirmed in greenhouse conditions, and this biotype has been used in other studies as a susceptible biotype [26,34].

4.2. Experiment 1. Determination of the Minimum Dose of Herbicide That Produces Changes to Plants' Metabolism

We separated this study into two experiments to address each of the following questions separately: (1) What is the minimum dose that causes changes in the plants' metabolism? and (2) do weeds that received a non-lethal dose differ in metabolism and survival upon receiving a second dose compared to plants that are receiving their first dose (i.e., are there priming effects)?

Around 500 caryopses of the collected susceptible biotype of *A. fatua* of similar size and weight were selected. This selection was made to ensure the physiological maturity of the caryopses used. The florets (lemma and palea) were removed manually to synchronize the germination. Then, the caryopses were disinfected by immersing them in a solution of water and sodium hypochlorite at 5% for 10 min and washed three times with distilled water. Disinfected caryopses were placed in Petri dishes with wet filter-paper and maintained to 20 °C. Germination was considered to have occurred when the radicle measured 3 mm in length.

Plastic pots with 500 mL capacity were filled with a mix of peat moss and agrolite in a proportion of 1:1. In each pot, six seedlings were planted. Pots were maintained in a growth chamber with a light flux of 440 μmoles m^{-2} s^{-1}, photoperiod of 16 h light/8 h dark, and constant temperature of 18 °C. The substrate was maintained near field capacity during the experiment. The plants were fertilized with Steiner nutrient solution (1x) every 14 days (50 mL per pot).

To determine the minimum dose of herbicide that a plant can detect in the environment, we used the recommended rate of clodinafop-propargyl (60 g a. i. ha-1, 1x) and four 10-fold consecutive dilutions (0.1x, 0.01x, 0.001x and 0.0001x). Each dilution represents a treatment, and for the preparation of each dilution, the adjuvant concentration was the same. The control conditions consisted of the application of distilled water and adjuvant (0x). This spectrum of doses ranges from the recommended dose to an extremely diluted dose 10, 000 fold weaker than the recommended dose, which could represent the spray particle drift that occurs during herbicide application.

The herbicide treatments were applied when plants were at least 10 cm tall, using a pressurized CO_2 plot sprayer calibrated to a spray volume of 200 L per hectare. Four replicates were carried out for each dose. After the application of herbicides, pots were kept separate for 6 h to ensure the penetration of herbicide and avoid contamination among treatments.

Plant tissue was sampled for metabolomic analyses 24 h after the herbicide application; at that time, plants did not show damage symptoms in any of the treatments. Two plants per pot were taken, leaving four individuals in each pot for estimations of biomass and survival. The shoots were cut, washed in distilled water (1 min), placed inside aluminum foil bags, and flash-frozen in liquid nitrogen. Samples were lyophilized in a vacuum chamber at −50 °C for 72 h. The samples were stored in airtight bags in the dark to avoid the accumulation of moisture until their use.

4.2.1. Survival and Biomass Reduction

Twenty-one days after herbicide application, the number of live plants per pot was counted, and the plant shoots collected to determine biomass. Plant shoots were cut, placed in paper bags, and dried for 72 h at 80 °C until reaching constant weight. The percentage of biomass reduction (dry matter) at each dose was obtained by subtracting from the dry matter to untreated plants (0x) and multiplying by 100. Survival was calculated multiplying the number of live plants by 25; survival data were arcsine transformed prior to statistical analysis. Data were analyzed with ANOVA ($p \leq 0.05$), and when significant differences were found, a Tukey test ($p \leq 0.05$) was performed to compare among treatments.

4.2.2. Metabolic Fingerprinting Using GC-MS

Ten milligrams of lyophilized tissue was placed in a 2 mL capacity Eppendorf tube, and the tubes were submerged in liquid nitrogen for 2 min. Then, the plant tissue was ground inside the tube with a plastic pestle until obtaining a fine powder. 500 μL of methanol (Mass grade, Fisher Scientific, New Bedford, MA, USA) was added to each tube. To improve metabolite extraction, samples were homogenized for 1 h in a sonifier (Branson 1800, St. Louis, MO, USA). Tubes were centrifuged for 10 min at 13 rpm and the aqueous phase was filtered using a 0.2 μm nylon filter (Millipore, Burlington, MA, USA).

Samples were injected into a gas chromatograph (Clarus 680, Perkin-Elmer Inc., Waltham, MA, USA), equipped with a phase capillary column: 5% diphenyl 95% dimethylpolysiloxane 30 m long, 0.32 mm i.d., 0.25 μm film thickness, temperature limits between −60 a 320/350 °C (Elite-5 MS,

Perkin-Elmer Inc., Waltham, MA, USA). The injection was by autosampler. Helium gas was used at a flow rate of 1 mL/min, the flow remained constant, and there was an initial wait time of 0.5 min. The column temperature was initially maintained at 50 °C for 1 min and then ramped to 250 °C at 30 °C/min, remaining at this temperature for a further 10 min. The temperature of the injector was 230 °C. A mass spectrometer (Clarus SQ8T, Perkin-Elmer Inc., Waltham, MA, USA), with an electron impact ionization source (70 eV) in full scan mode was used. The analysis range was 40–500 m/z. The temperatures of the transfer line and ionization source were 230 and 250 °C, respectively.

4.2.3. Data Analyses

Original files of GC-MS were analyzed in the platform XCMS Online [35]. This platform provides the feature detection, retention time correction, peak alignment, and statistical analysis. Since the objective of this study was to obtain a global fingerprint, a non-target metabolomics scope was used. To avoid a wrong interpretation in the name of the metabolites, and for differentiating among metabolites, we only used metabolites with q-values ≤ 0.05, and the annotation of each one was according to the m/z and the retention time (RT) of each metabolite detected.

The results were represented in an heatmap-bicluster. An ion matrix was constructed using the metabolites with p-values ≤ 0.05. The construction of the heatmap was made using the platform Metaboanalyst (www.metaboanalyst.ca) [36]. In this platform, the data were normalized and auto-scaled. The dendrograms used Pearson correlation as a distance function, and the Ward clustering algorithm; the significance of the branches were $p \leq 0.05$. A supervised learning algorithm (Random forest) was used to measure the importance of each metabolite in the grouping and sample classification. When a metabolite represent an important difference among treatments, the identification of such molecule was made using the NIST library.

4.3. Experiment 2. Responses of Weeds Treated with Non-Lethal Dose to Posterior Herbicide Application

In order to answer the second question, a second experiment was carried out. The plant material and growth conditions were the same as for experiment 1. In this experiment, the treatments consisted of a first application of herbicides in doses of 0x, 0.0001x, 0.001x and 0.01x. Due to the high mortality observed in treatments with 0.1 and 1x we did not include these treatments in the second experiment. Each treatment had four pots, with six plants per plot.

Twenty-one days after the first herbicide application, the pots were sprayed for a second time using the same dose as the first exposure (0.0001x-T, 0.001x-T and 0.01x-T). At the same time, a set of plants of the same age that had never have been exposed to herbicides were sprayed with herbicides at the same doses (0.0001x-U, 0.001x-U and 0.01x-U). This second application included a control group that was sprayed only with distilled water and adjuvant. Twenty-four hours after the herbicide application, two plants per plot were sampled for metabolomic fingerprinting. The methods for sample collection, processing, injection into GC-MS and data analysis, were the same as described for experiment 1.

Author Contributions: Collecting plant material and growth—chambers experiments, J.A.T.-R.; metabolomic experiments and analysed the results, E.O.-Z.; Conceptualization and design of experiments, S.V.-M. and J.R.T.-G.; writing paper, J.R.T.-G.

Funding: This research received no external funding.

Acknowledgments: Jesus is grateful to CONACyT for affiliation as a member to Catedras-CONACyT program.

Conflicts of Interest: The authors declare no conflict of interest.

References

1. Powles, S.B.; Yu, Q. Evolution in Action: Plants Resistant to Herbicides. *Annu. Rev. Plant Biol.* **2010**, *61*, 317–347. [CrossRef] [PubMed]
2. Neve, P. Challenges for herbicide resistance evolution and management: 50 Years after Harper. *Weed Res.* **2007**, *47*, 365–369. [CrossRef]

3. Jang, S.; Marjanovic, J.; Gornicki, P. Resistance to herbicides caused by single amino acid mutations in acetyl-CoA carboxylase in resistant populations of grassy weeds. *New Phytol.* **2013**, *197*, 1110–1116. [CrossRef]
4. Maxwell, B.D.; Roush, M.L.; Radosevich, S.R. Predicting the Evolution and Dynamics of Herbicide Resistance in Weed Populations Resistance in Weed Populations. *Weed Technol.* **1990**, *4*, 2–13. [CrossRef]
5. Manalil, S. Evolution of herbicide resistance in *Lolium rigidum* under low herbicide rates: An Australian experience. *Crop Sci.* **2014**, *54*, 461–474. [CrossRef]
6. Yu, Q.; Powles, S. Metabolism-Based Herbicide Resistance and Cross-Resistance in Crop Weeds: A Threat to Herbicide Sustainability and Global Crop Production. *Plant Physiol.* **2014**, *166*, 1106–1118. [CrossRef]
7. Manalil, S.; Busi, R.; Renton, M.; Powles, S.B. Rapid Evolution of Herbicide Resistance by Low Herbicide Dosages. *Weed Sci.* **2011**, *59*, 210–217. [CrossRef]
8. Neve, P.; Powles, S. Recurrent selection with reduced herbicide rates results in the rapid evolution of herbicide resistance in *Lolium rigidum*. *Theor. Appl. Genet.* **2005**, *110*, 1154–1166. [CrossRef] [PubMed]
9. Neve, P.; Powles, S. High survival frequencies at low herbicide use rates in populations of *Lolium rigidum* result in rapid evolution of herbicide resistance. *Heredity* **2005**, *95*, 485–492. [CrossRef]
10. Conrath, U. Molecular aspects of defence priming. *Trends Plant Sci.* **2011**, *16*, 524–531. [CrossRef]
11. Martinez-Medina, A.; Flores, V.; Heil, M.; Mauch-Mani, B.; Pieterse, C.M.J.; Pozo, M.J.; Ton, J.; van Dam, N.M.; Conrath, U. Recognizing Plant Defense Priming. *Trends Plant Sci.* **2016**, *21*, 818–822. [CrossRef]
12. Belz, R.G.; Duke, S.O. Herbicides and plant hormesis. *Pest Manag. Sci.* **2014**, *70*, 698–707. [CrossRef]
13. Calabrese, E.J.; Bachmann, K.A.; Bailer, A.J.; Bolger, P.M.; Borak, J.; Cai, L.; Cedergreen, N.; Cherian, M.G.; Chiueh, C.C.; Clarkson, T.W.; et al. Biological stress response terminology: Integrating the concepts of adaptive response and preconditioning stress within a hormetic dose-response framework. *Toxicol. Appl. Pharmacol.* **2007**, *222*, 122–128. [CrossRef] [PubMed]
14. Gressel, J. Low pesticide rates may hasten the evolution of resistance by increasing mutation frequencies. *Pest Manag. Sci.* **2011**, *67*, 253–257. [CrossRef] [PubMed]
15. Brito, I.P.F.S.; Tropaldi, L.; Carbonari, C.A.; Velini, E.D. Hormetic effects of glyphosate on plants. *Pest Manag. Sci.* **2018**, *74*, 1064–1070. [CrossRef]
16. Cedergreen, N.; Streibig, J.C.; Kudsk, P.; Mathiassen, S.K.; Duke, S.O. The Occurrence of Hormesis in Plants and Algae. *Dose-Response* **2007**, *5*, 150–162. [CrossRef] [PubMed]
17. Cedergreen, N. Herbicides can stimulate plant growth. *Weed Res.* **2008**, *48*, 429–438. [CrossRef]
18. Gressel, J.; Dodds, J. Commentary: Hormesis can be used in enhancing plant productivity and health; but not as previously envisaged. *Plant Sci.* **2013**, *213*, 123–127. [CrossRef]
19. Belz, R.G.; Cedergreen, N.; Duke, S.O. Herbicide hormesis—can it be useful in crop production? *Weed Res.* **2011**, *51*, 321–332. [CrossRef]
20. Vieira, B.C.; Luck, J.D.; Amundsen, K.L.; Gaines, T.A.; Werle, R.; Kruger, G.R. Response of *Amaranthus spp.* following exposure to sublethal herbicide rates via spray particle drift. *PLoS ONE* **2019**, *14*, e0220014. [CrossRef]
21. Saito, K.; Matsuda, F. Metabolomics for Functional Genomics, Systems Biology, and Biotechnology. *Annu. Rev. Plant Biol.* **2010**, *61*, 463–489. [CrossRef]
22. Viant, M.R. Applications of metabolomics to the environmental sciences. *Metabolomics.* **2009**, *5*, 1–2. [CrossRef]
23. Brunetti, C.; George, R.M.; Tattini, M.; Field, K.; Davey, M.P. Metabolomics in plant environmental physiology. *J. Exp. Bot.* **2013**, *64*, 4011–4020. [CrossRef]
24. Trenkamp, S.; Eckes, P.; Busch, M.; Fernie, A.R. Temporally resolved GC-MS-based metabolic profiling of herbicide treated plants treated reveals that changes in polar primary metabolites alone can distinguish herbicides of differing mode of action. *Metabolomics* **2009**, *5*, 277–291. [CrossRef]
25. García-Flores, M.; Juárez-Colunga, S.; García-Casarrubias, A.; Trachsel, S.; Winkler, R.; Tiessen, A. Metabolic profiling of plant extracts using direct-injection electrospray ionization mass spectrometry allows for high-throughput phenotypic characterization according to genetic and environmental effects. *J. Agric. Food Chem.* **2015**, *63*, 1042–1052. [CrossRef]
26. Torres-García, J.R.; Tafoya-Razo, J.A.; Velázquez-Márquez, S.; Tiessen, A. Double herbicide-resistant biotypes of wild oat (*Avena fatua*) display characteristic metabolic fingerprints before and after applying ACCase- and ALS-inhibitors. *Acta Physiol. Plant.* **2018**, *40*, 119. [CrossRef]
27. Delye, C. Weed Resistance to Acetyl Coenzyme A Carboxylase Inhibitors: An Update. *Weed Sci.* **2005**, *53*, 728–746. [CrossRef]

28. Somody, C.N.; Nalewaja, J.D.; Miller, S. Wild Oat (*Avena fatua*) and *Avena sterilis* Morphological Characteristics and Response to Herbicides. *Weed Sci.* **1984**, *32*, 353–359. [CrossRef]
29. Fukusaki, E.; Kobayashi, A. Plant Metabolomics: Potential for Practical Operation. *J. Biosci. Bioeng.* **2005**, *100*, 347–354. [CrossRef]
30. Aliferis, K.A.; Chrysayi-Tokousbalides, M. Metabolomics in pesticide research and development: Review and future perspectives. *Metabolomics* **2011**, *7*, 35–53. [CrossRef]
31. Yu, Q.; Han, H.; Cawthray, G.R.; Wang, S.F.; Powles, S.B. Enhanced rates of herbicide metabolism in low herbicide-dose selected resistant *Lolium rigidum*. *Plant Cell Environ.* **2013**, *36*, 818–827. [CrossRef]
32. Heap, I. Global perspective of herbicide-resistant weeds. *Pest Manag. Sci.* **2014**, *70*, 1306–1315. [CrossRef]
33. International Survey of Herbicide Resistant Weeds. Available online: www.weedscience.org (accessed on 23 July 2019).
34. Tafoya-Razo, J.A.; Núñez-Farfán, J.; Torres-García, J.R. Migration by seed dispersal of ACCase-inhibitor-resistant *Avena fatua* in north-western Mexico. *Pest Manag. Sci.* **2017**, *73*, 167–173. [CrossRef]
35. Tautenhahn, R.; Patti, G.J.; Rinehart, D.; Siuzdak, G. XCMS online: A web-based platform to process untargeted metabolomic data. *Anal. Chem.* **2012**, *84*, 5035–5039. [CrossRef]
36. Chong, J.; Soufan, O.; Li, C.; Caraus, I.; Li, S.; Bourque, G.; Wishart, D.S.; Xia, J. MetaboAnalyst 4.0: Towards more transparent and integrative metabolomics analysis. *Nucleic Acids Res.* **2018**, *46*, W486–W494. [CrossRef]

plants

MDPI

Review

Target-Site Mutations Conferring Herbicide Resistance

Brent P. Murphy and Patrick J. Tranel *

Department of Crop Sciences, University of Illinois, Urbana, IL 61801, USA; brentpm2@illinois.edu

* Correspondence: tranel@illinois.edu; Tel.: +1-217-333-1531

Received: 4 September 2019; Accepted: 26 September 2019; Published: 28 September 2019

Abstract: Mutations conferring evolved herbicide resistance in weeds are known in nine different herbicide sites of action. This review summarizes recently reported resistance-conferring mutations for each of these nine target sites. One emerging trend is an increase in reports of multiple mutations, including multiple amino acid changes at the glyphosate target site, as well as mutations involving two nucleotide changes at a single amino acid codon. Standard reference sequences are suggested for target sites for which standards do not already exist. We also discuss experimental approaches for investigating cross-resistance patterns and for investigating fitness costs of specific target-site mutations.

Keywords: D1 protein; acetolactate synthase; tubulin; ACCase; EPSPS; phytoene desaturase; PPO; glutamine synthetase; auxin

1. Introduction

Herbicide-resistance mechanisms broadly fall under two categories: target-site mechanisms and non-target-site mechanisms [1,2]. The former involves a change to the molecular target of the herbicide (usually an enzyme) that decreases its affinity for the herbicide. Although much less common, target-site resistance can also occur via increased expression of the target, which results in more herbicide required to achieve a lethal effect [3,4]. Non-target-site resistance encompasses any mechanism that reduces the amount of herbicide that reaches the target site, or that ameliorates the effect of the herbicide despite its inhibition of the target site.

Our understanding of specific DNA changes that confer non-target-site resistance is still in its infancy [5]. In contrast, the first DNA change conferring evolved target-site resistance (to triazines) was identified over three and a half decades ago [6]. Since then, numerous resistance-conferring mutations have been identified from dozens of weed species and now span nine herbicide target sites (Table 1). The purpose of this review is to provide an update of new mutations that have been recently identified for each of these nine target sites. In this review, we consider "new mutations" to be those that confer an amino acid change that has not been reported previously from any weed species. Notably, gene duplication is not within the scope of this review. While gene duplication events have been observed to confer resistance to herbicide Groups 1 and 9 [3,4], the underlying genetic mechanisms for resistance evolution lie outside of the applicable target-site coding region. We discuss each target site sequentially, beginning with a brief description of the target. We then highlight the most recent reviews of mutations for each target before reviewing new mutations associated with each target. Resistance cases are discussed in order of WSSA group number, with the HRAC classification listed in parenthesis [7].

Table 1. Identification of mutations conferring target-site resistance to herbicides.

Target Site	Representative Herbicide	Year [1]
D1 protein	atrazine	1983 [6]
acetolactate synthase	chlorimuron	1992 [8]
tubulin	trifluralin	1998 [9]
acetyl CoA carboxylase	clethodim	2001 [10]
5-enolypyruvylshikimate-3-phosphate synthase	glyphosate	2002 [11]
phytoene desaturase	fluridone	2004 [12]
protoporphyrinogen oxidase	lactofen	2006 [13]
glutamine synthetase	glufosinate	2012 [14]
auxin receptor	2,4-D	2018 [15]

[1] Indicates first year of publication in peer-reviewed literature of a resistance-conferring mutation in the target-site from a field-evolved weed population.

2. Summary by Herbicide Group

2.1. *Acetyl-CoA Carboxylase Inhibitors: Group 1 (A)*

The basis of function for Group 1 chemistries was reviewed in detail by [16], and target-site resistance to Group 1 chemistries was last reviewed in 2014 [17]. Briefly, acetyl-CoA carboxylase (ACCase) catalyzes the carboxylation of acetyl-CoA to malonyl-CoA through a two-step, reversible reaction. First, biotin in complex with the enzyme is carboxylated, and second, the carboxyl group of biotin is transferred to acetyl-CoA, producing malonyl-CoA [16]. ACCase is composed of three domains: the biotin carboxylase domain, biotin carboxyl carrier protein domain, and the carboxyltransferase domain. Inhibitors of ACCase are classified within three chemical families: the aryloxyphenoxypropionates (FOPs), cyclohexanediones (DIMs), and phenylpyrazolin (DEN) [17]. Although ACCase is present within the cytoplasm and chloroplast, ACCase inhibitors affect only the homomeric, plastidic ACCase isoform specific to the Poaceae family, through nearly competitive, reversible inhibition [17]. By convention, amino acid numbering follows the *Alopecurus myosuroides* sequence, CAC84161 [17]. The crystal structures of the ACCase carboxyltransferase domain, as derived from yeast, have been produced both in a 'native' state [18] and in complex with representative chemicals from all three herbicide families within the group [19–21].

Target-site ACCase resistance mutations commonly evolve in grass weed species in response to selection and show great diversity in terms of mutation sites. Codon changes at positions 1781, 1999, 2027, 2041, 2078, 2088, and 2096 have been previously reviewed [17]. Subsequently, two new substitutions have been reported in weedy species, including one at a new site.

The Ile-2041 position has been well characterized for resistance to FOP chemistries. In China, a new substitution, Ile-2041-Thr, was observed in *Alopecurus aequalis* [22]. Dose–responses conducted on segregants of a single, heterozygous plant revealed resistance to a number of FOP chemistries and sensitivity to several DIMs. In addition, reduced sensitivity towards pinoxaden, a DEN chemistry, was reported. Reduced sensitivity, and even resistance, caused by substitutions at the Ile-2041 position have been observed to pinoxaden, although this was inconsistent between species. For instance, the Ile-2041-Asn substitution has been reported to cause a high level of resistance to pinoxaden in *Beckmannia syzigachne* [23], while providing reduced sensitivity in *Lolium multiflorum* [24]. Functional validation is required to reach a consensus on the effects of Ile-2041 position on DEN chemistries.

First identified in *Eleusine indica* from Malaysia, the Asn-2097-Asp substitution is suggested to provide resistance to fluazifop, a FOP [25]. While the functional effects of substitutions at the Asn-2097 position are uncharacterized, substitutions in the proximal Gly-2096 position have been characterized to confer FOP-specific resistance [26]. However, the contribution of this new substitution to the resistance was not explored further. As heterozygous individuals within the *E. indica* population of interest were identified, cosegregation studies would provide greater evidence of the importance of Asn-2097 towards Group 1 inhibitor resistance.

2.2. *Acetolactate Synthase Inhibitors: Group 2 (B)*

Resistance to acetolactate synthase (ALS) inhibitors was most recently reviewed in 2014 [27]. Acetolactate synthase, also referred to as acetohydroxyacid synthase (AHAS), is a dual-functioning enzyme. The enzyme catalyzes both the synthesis of acetolactate from two pyruvate molecules, and the synthesis of acetohydroxybutyrate from ketobutyrate and pyruvate, where both reactions require thiamin diphosphate, FAD, and Mg^{2+} [28]. ALS is the first enzyme in the synthesis pathway for the branched-chain amino acids valine, leucine, and isoleucine, and the depletion of these amino acids is the mode of action for ALS-inhibiting herbicides [27]. Encoded in the nucleus, the enzyme is localized to plastids. Group 2 is composed of five herbicide families: sulfonylurea (SU), imidazolinone (IMI), triazolopyrimidine (TP), pyrimidinyl-thiobenzoate (PTB), and sulfonyl-aminocarbonyl-triazolinone (SCT). Crystal structures of *Arabidopsis thaliana* in complex with SU, IMI, TP, PTB, and SCT have been generated [29–31]. The naming convention for amino acid substitutions is based on the *A. thaliana* sequence AY124092, and is in agreement with the available crystal structures.

The relative ease at which weeds evolve resistance to ALS inhibitors is the Achilles heel of these herbicides, and resistance is often due to target-site substitutions [32]. Evolved resistance in weeds has been attributed to substitutions at each of the following eight different amino acids: Ala-122, Pro-197, Ala-205, Asp-376, Arg-377, Trp-574, Ser-653, and Gly-654. Often, several different substitutions at each of these eight sites are able to confer resistance. Since the last review, six new substitutions (at previously reported sites) have been reported in weedy species and are summarized herein.

The Ala-122-Asn substitution was first reported in *Echinochloa crus-galli* from Italy [33]. This substitution has been reported to confer resistance to SU chemistries in yeast [34], and substitutions at this position can result in resistance to both SU and IMI chemistries [27]. Resistant and sensitive plants from the same field location were identified and bulked, and resistance was characterized to be representative of SU, IMI, TP, and PTB chemistries. Sequence analysis was used to identify the amino acid substitution, and in vitro ALS enzyme activity bioassays supported observed resistance, indicative of a target-site resistance mechanism. A fitness cost under ideal growth conditions, but not under competition, was identified through comparative growth analysis.

The Ala-122-Ser substitution was first reported in *Amaranthus palmeri* from Argentina [35]. Previously, the Ala-122-Ser mutation has been reported to confer resistance to SU chemistries in yeast [34]. Greenhouse screening identified consistent resistance to representative SU, IMI, and TP chemistries when compared to an unrelated sensitive population. In vitro ALS enzyme activity bioassays support resistance to the representative SU and IMI chemistry, but not to the representative TP chemistry. Sequence analysis on a set of eight resistant plants revealed that the population was a composite between the Ala-122-Ser, Ser-653-Asn, and several uncharacterized substitutions (Pro-84-His and Ala-282-Asp). The Ala-122-Ser mutation was observed only in combination with the uncharacterized Ala-282-Asp substitution. Functional characterization in the absence of other substitutions is necessary to confirm this potential resistance mechanism.

As first published in 2015 by Liu et al. [36] in *Myosoton aquaticum* from China, the Pro-197-Glu substitution confers resistance to representative SU, IMI, TP and PTB chemistries. Substitutions at the Pro-197 site are generally considered to be SU-specific [27]. A homozygous-resistant population was developed through genotypic selection, and dose–response compared to an unrelated sensitive population identified uniform resistance. In vitro ALS enzyme activity bioassays confirmed reduced sensitivity of the ALS enzyme when compared to the sensitive population for the SU, IMI, TP, and PTB chemical families. The response of the Pro-197-Glu substitution to the SCT chemistries is unknown.

The Pro-197-Phe substitution was first reported in *Sisymbrium orientale* in Southern Australia [37]. A population survey of the species was conducted for resistance to SU and IMI chemistries, where sequence analysis of survivors was conducted. Of the 65 populations under investigation, one population was reported to consistently possess the Pro-197-Phe substitution, the result of a double nucleotide substitution when compared to the wild-type sequence. This population was resistant to a representative SU chemistry, while sensitive to an IMI chemistry. To date, functional validation of the

Pro-197-Phe has not been conducted. The effect of the Pro-197-Phe substitution on the TP, PTB, and SCT chemical families is unknown.

The Ala-205-Phe substitution was first reported in *Poa annua* in Tennessee, US [38]. Substitutions at the Ala-205 position are considered to confer SU-specific resistance [27]. Dose–response with representative SU chemistry [39], and delimiting rate of a representative IMI chemistry, identified in planta resistance when compared to an unrelated sensitive population. Sequence analysis revealed the Ala-205-Phe substitution, as derived from a double nucleotide substitution. These substitutions were artificially introduced to the *A. thaliana* gene sequence, expressed, and purified from *Escherichia coli* for use in in vitro ALS enzyme activity assays. These in vitro assays revealed resistance responses when compared to wild-type to a wide range of chemistries from each of the SU, IMI, TP, PTB, and SCT chemical families. Interestingly, no resistance was observed to florasulam, though this observation was not confirmed in planta, perhaps due to the lack of efficacy of florasulam on grasses. The in planta response of the Ala-205-Phe substitution to the TP, PTB, and SCT chemical families is unknown.

The Trp-574-Arg substitution was first reported in *Digitaria sanguinalis* in China [40]. Substitutions at the Trp-574 site have been documented to confer broad cross-resistance to the ALS-inhibiting chemistries. Dose–response revealed resistance to representative SU, IMI, and TP chemistries when compared to an unrelated sensitive population. Reduced sensitivity of the ALS enzyme to the SU, IMI, and TP chemistries was characterized with in vitro ALS enzyme activity assays. The response of the Trp-574-Arg substitution to the PTB and SCT chemical families is unknown.

2.3. *Microtubule Inhibitors: Group 3 (K_1)*

Resistance to microtubule inhibitors was last reviewed in 2010, though the dinitroaniline trifluralin was reviewed in 2013 [1,41]. Tubulin heterodimers, composed of α- and β-tubulin, polymerize to form microtubules. Microtubules are key structural polymers, which mediate multiple cellular processes through the dynamic reorganization of microtubules [42]. Microtubule inhibitors bind to the α-tubulin subunit in a reversible, competitive manner, deregulating the organization of microtubules [41–43]. Currently, there are five classes of microtubule inhibitors: dinitroaniline, phosphoroamidate, pyridine, benzamide, and benzoic acid. The crystal structure of bovine tubulin heterodimer is available, which has been used to computationally determine dinitroaniline binding domains and validated through mutagenesis studies [42,44,45]. Because α-tubulin is highly conserved, the need for a reference sequence is minimal. If necessary, the authors recommend the *Setaria viridis* sequence CAE52514.

Resistance to Group 3 herbicides is relatively rare (reported in just 12 weed species). As of 2013, α-tubulin substitutions implicated in resistance included Leu-125-Met, Leu-136-Phe, Val-202-Phe, Thr-239-Ile, and Met-268-Thr [1,41]. Target-site resistance to microtubule inhibitors is unique among herbicides in that it is a recessive trait. This recessive nature likely accounts in part for the relative rarity of target-site resistance to these herbicides (especially, in out-crossed species) [1], despite the fact that there appear to be numerous substitutions that can confer resistance.

The substitution of Arg-243 to Met or Lys was first reported in *Lolium rigidum* from Western Australia [46]. Dose–responses of multiple dinitroaniline compounds revealed resistance when compared to an unrelated sensitive population [47]. Sequence analysis revealed a subset of the population possessed Arg-243 substitutions, though in combination with the known substitutions Thr-239-Ile or Val-202-Phe. Transgenic α-tubulin genes carrying wild-type (Arg-243), Arg-243-Met, and Arg-243-Lys were introduced into *Oryza sativa* and resistance to multiple dinitroaniline chemistries in calli lines with similar recombinant protein abundance levels were observed for the mutant but not the wild-type genes. Structural modelling revealed the substitutions are expected to reduce binding efficiency between trifluralin and the tubulin subunit. The interaction of these Arg-243 substitutions with the other substitutions were not documented.

2.4. Synthetic Auxins: Group 4 (O)

Herbicides within Group 4, the synthetic auxins, are synthetic analogues of the endogenous plant hormone indole-3-acetic acid (IAA) [48]. The deregulation of auxin-dependent plant signalling pathways results in the efficacy of these herbicides for weed control. Synthetic auxin chemistries are largely preferential towards dicots, with the exception of quinclorac, which has grass activity. A more comprehensive review of the mode of action of synthetic auxin chemistries is provided by Grossmann [49]. Resistance to synthetic auxin chemistries was last reviewed in 2018, though only non-target-site resistance mechanisms had been published in weedy species at that time [48]. Currently, synthetic auxins are separated into seven classes: phenoxy-carboxylates, quinolone-carboxylates, pyrimidine-carboxylates, benzoates, pyridine-carboxylates, pyridyloxy-carboxylates, and arylpicolinates [48]. Unlike other herbicide groups, which target a specific protein, synthetic auxins interact with numerous proteins, including from the following families: TIR1 and Auxin F-Box [50], AUX/IAA protein, AUX1/LAX influx carrier, PIN efflux carrier, and ABCB [51]. To date, however, evolved resistance to synthetic auxins has only been reported in the AFB and AUX/IAA families, and as such, these are discussed here.

In *A. thaliana*, there are six members of the AFB family, TIR1 and AFB1-5 [52]. All members of the AFB family are nuclear-encoded and localize to the nucleus [52]. Selective, or perhaps preferencial, binding of synthetic auxin chemistries to subsets of these protein targets has been reported [53], which suggests the loss of sensitivity of a few, or even one, of these receptors may result in herbicide resistance. In *A. thaliana*, there are 29 AUX/IAA genes [54]. AUX/IAA proteins are co-receptors, which pair with AFB proteins in the presence of auxin to form a larger co-receptor complex, termed the SCF^TIR1 [55]. The crystal structure of the TIR1-ASK1 has been developed in the native state and in complex with IAA, 1-naphthalene acetic acid (1-NAA), and 2,4-dichlorophenoxyacetic acid (2,4-D) [56]. As, prior to 2018, target-site resistance to synthetic auxins had not been reported, no convention has been established for amino acid numbering.

In 2018, LeClere et al. [15] were the first to identify a target-site resistance mechanism in *Kochia scoparia* to representatives of three classes of synthetic auxins: benzoates, phenoxy-carboxylates, and pyridine-carboxylates. Root length assays and dose–response analysis identified resistance to the aforementioned chemical classes. Transcriptome sequencing and sequence analysis identified a two-nucleotide substitution that results in a single amino acid substitution in the highly conserved GWPPV/I region (GWPPV/I→NWPPV/I) of KsIAA16. Yeast Two-Hybrid assays demonstrated a loss of interaction between the KsIAA16 and KsTIR1 in the presence of representative benzoates, phenoxy-carboxylates, and pyridine-carboxylates as a result of the Gly-73-Asn substitution (as numbered based on the *A. thaliana* IAA16 sequence). Co-segregation analysis in the F2 generation suggests benzoate resistance is linked to the observed nucleotide substitutions. Resistance mediated through the Gly-73-Asn substitution was confirmed through complementation in *A. thaliana*. A fitness cost associated with the Gly-73-Asn substitution was observed within F2 segregants.

As LeClere et al. [15] were the first to document target-site resistance to synthetic auxins, the amino acid numbering convention should follow their example. Within their research, LeClere et al. [15] utilized the *A. thaliana* sequence AT3G04730 to number amino acids. If resistance-conferring mutations are found on other auxin receptors, we propose amino acid numbering should occur, following the most appropriate gene within *A. thaliana*.

2.5. Photosystem II Inhibitors: Groups 5 (C_1), 6 (C_3) and 7 (C_2)

Herbicide resistance to photosystem II inhibitors was last reviewed in 2010 [1]. Photosystem II inhibitors, while spread over three herbicide groups, are reversible, competitive inhibitors of the Qb-binding niche of the D1 protein in the photosystem II complex. Photosystem II inhibitors are further subclassified into numerous chemical classes [57]. Group 5 contains the phenyl-carbamates, pyridazinones, triazines, triazinones, triazolinones, and uracils. Group 6 contains the benzothiadiazinones, nitriles, and phenyl-pyridazines. Group 7 contains the amides and ureas. Inhibition results in a disruption of the chloroplastic electron transport chain, resulting in the build-up of reactive oxygen species. The D1 protein is encoded by the chloroplastic psbA gene, and is expressed within the chloroplast. The crystal structure of the D1 protein was derived from the L-protein of *Chlamydomonas reinhardtii* [58], and more recently, from other purple bacteria [59]. Amino acid numbering is based on *A. thaliana*.

As mentioned in the Introduction, the first identified resistance-conferring target-site mutation was to the photosystem II inhibitors. This mutation caused a Ser-264-Gly change and, since, has emerged as the dominant mutation for resistance to these herbicides. Other mutations, known as of 2010, included Val-219-Ile, Ala-251-Val, Phe-255-Ile, Ser-264-Thr, and Asn-266-Thr. Target-site resistance to photosystem II inhibitors is unique in that, because the target site is encoded by a plastidic gene, resistance is expected to be maternally inherited in most, if not all, weed species.

A Leu-218-Val substitution was first reported in *Chenopodium album* from Germany [60]. Dose–response identified resistance to triazinone, but not triazine chemistries. Sequence analysis identified the Leu-218-Val substitution unique to the resistant biotype. While the Leu-218 amino acid is part of the Qb-binding niche, modelling was not conducted to determine the impact of the observed substitution. Although Leu-218-Val has not been functionally validated, it is adjacent to the Val-219 site, where substitutions have been documented to confer resistance to Groups 5 and 7 [61]. Inheritance of the observed resistance was not conducted, but it is expected to be maternally inherited.

A Phe-274-Val substitution was reported in *Raphanus raphanistrum* from Western Australia [62]. This is the first report of a resistance-conferring substitution at or near the Phe-274 site. Delimiting dose and dose–response analysis identified a resistant response to representative Group 5 and 7 chemistries, but increased sensitivity to a representative Group 6 chemistry. Structural modelling suggests that the Phe-274-Val substitution results in a weakening in binding efficiency of representative Group 5 and 7 chemistries. Inheritance of the observed resistance was not conducted and the substitution has not been functionally validated.

2.6. EPSP Synthase Inhibitors: Group 9 (G)

5-enolpyruvylshikimate-3-phosphate (EPSP) synthase catalyzes the reaction between phosphoenol pyruvate and shikimate-3-phosphate to produce EPSP and inorganic phosphate [63]. Resistance to EPSP synthase inhibitors has been well reviewed in 2014 [64] and 2018 [65]. EPSP synthase is a nuclear-encoded, chloroplast-localized enzyme required for aromatic amino acid production [66]. Glyphosate, the only chemistry within Group 9, is a slowly reversible to irreversible competitive inhibitor at the phosphoenol pyruvate binding pocket [63]. The crystal structure of EPSP synthase, as derived from *E. coli* in complex with glyphosate, is known [63]. The numbering of amino acids is referenced against the start of the mature EPSP synthase enzyme of plants, such as the *A. thaliana* sequence, AT2G45300 [64].

Target-site resistance to Group 9 is unique due to the apparent necessity for multiple amino acid substitutions to confer a strong phenotypic response. Substitutions at Pro-106 have been observed in isolation; however, the Thr-102-Ile substitution has only been observed in combination with Pro-106-Ser. Here, two new combinations of amino acid substitutions, including a triple-substitution event, are reviewed.

The double-substitution Thr-102-Ile + Pro-106-Thr, termed TIPT, was first observed in *Bidens subalternanas* in Paraguay [67]. The Pro-106-Thr substitution has been reported to confer low-level resistance, and the Thr-102-Ile has been reported to confer resistance only when in combination with Pro-106-Ser [65]. Dose–response conducted between the putative resistant accession and a geographically proximal sensitive accession identified the resistance response. Shikimate accumulation assay results suggest that the resistant accession has an insensitive target-site. No significant variation for non-target-site mechanisms, such as reduced absorption and translocation, metabolism, gene amplification, and vacuolar sequestration, was observed. Gene sequencing revealed the presence of the TIPT substitutions in the resistant accession. No segregation or inheritance studies were conducted.

Patent literature describing the TIPT double-substitution was previously reviewed by Sammons and Gaines [64].

The triple-substitution Thr-102-Ile + Ala-103-Val + Pro-106-Ser, termed TAP-IVS, was first identified in *A. hybridus* in Argentina [68]. The effect of substitutions at the Ala-103 position are uncharacterized. Dose–response against a geographically proximal sensitive population identified the resistance response. Shikimate accumulation assay suggests that the resistant population has an insensitive target site. Gene sequencing identified the TAP-IVS triple substitution. While gene amplification was observed within the resistant accession, mRNA levels of EPSP synthase did not correlate with resistance. Structural modelling suggests that TAP-IVS results in a rearrangement of the glyphosate binding domain, and a reduction in interaction sites is observed when compared to wild-type and other resistance-endowing substitutions. Co-segregation and inheritance of the TAP-IVS was not conducted. The impact of Ala-103-Val alone, or in other combinations of substitutions, was not characterized. Functional validation of the TAP-IVS substitutions remains necessary.

2.7. Glutamine Synthetase Inhibitors: Group 10 (H)

Resistance to glutamine synthetase inhibitors was last reviewed in 2002 [69]. In short, glutamine synthetase catalyzes the formation of L-glutamine from L-glutamate and ammonia [70]. Two primary isoforms of glutamine synthetase exist in plants: the nuclear-encoded and cytosol-targeted GS1, and the nuclear-encoded and chloroplast-targeted GS2. While inhibition of glutamine synthetase results in an accumulation of ammonium, the primary cause of plant death was thought to occur via inhibition of photorespiration [71]. The transamination of glyoxylate to glycine is a necessary step of photorespiration, where the amino group originates from the fixation of ammonia into glutamine [71,72]. Recently, rapid accumulation of reactive oxygen species has been proposed as the primary reason for glufosinate toxicity [73]. Phosphinotricin (glufosinate) is the only chemistry within Group 10, and appears to inhibit both GS1 [74] and GS2 [75]. The crystal structure of GS1 was developed from *Zea mays* [76] and both GS1 and GS2 from *Medicago truncatula* [77].

An Asn-171-Asp substitution was identified in *Lolium perenne* L. spp. *multiflorum* from the US [14]. This is the first report of herbicide resistance due to an altered target site within Group 10. Dose–response revealed resistance when compared to two unrelated sensitive populations. Glutamine synthetase activity assays identified a reduced-sensitivity target site within the resistant accession. Sequence analysis of GS2 identified the Asn-171-Asp substitution. Inheritance and co-segregation analysis were not conducted. Fitness costs associated with Asn-171-Asp were not characterized.

2.8. Phytoene Desaturase Inhibitors: Group 12 (F_1)

Resistance to inhibitors of phytoene desaturase, termed Group 12, was last reviewed in 2014 [78]. Phytoene desaturase mediates the second step of the carotenoid biosynthesis pathway. The enzyme catalyzes the desaturation of 15-cis-phytoene, creating two of the four double bonds required for lycopene synthesis [79]. Phytoene desaturase is a nuclear-encoded, chloroplast-localized enzyme [80]. Herbicides in Group 12 are classified into two classes: pyridazinones and pyridinecarboxamides [57]. In addition, multiple chemistries are unclassified. Recently, the crystal structure of phytoene desaturase alone and in complex with a representative pyridazinone was reported from *Oryza sativa* [79], though no standardized amino acid numbering system appears to be established. The *O. sativa* sequence AAD02489 would make an excellent reference for amino acid numbering.

As of 2014, evolved target-site resistance was attributed only to substitutions of Arg-304, to either Ser, Cys, or His [78]. Resistance to Group 12 herbicide was only reported within five weed species, including the aquatic weed, *Hydrilla verticillata*.

A Leu-498-Val, (originally numbered as Leu-526-Val) substitution was first observed in *S. orientale* from Australia [81]. Dose–response revealed a resistant phenotype to a representative pyridinecarboxamide when compared to unrelated sensitive populations. Segregation analysis suggested that the resistant phenotype was mediated by a single, co-dominant to dominant locus.

Sequence analysis identified the Leu-498-Val substitution, which is equivalent to the Leu-538 position in *O. sativa*. The Leu-538 position in *O. sativa* corresponds to the causative mechanism of resistance to pyridazinones in numerous species [79]. No fitness costs associated with the Leu-498-Val substitution were observed in the F2 generation when grown in monoculture or in competition with wheat [82].

A double-substitution Glu-425-Asp + Leu-498-Val was first observed in *S. orientale* from Australia [83]. Dose–response revealed a resistant phenotype to diflufenican and picolinafen, which are both pyridinecarboxamides. The single-substitution Leu-498-Val was documented as highly resistant to diflufenican, but not picolinafen. Sequence analysis identified the presence of the Glu-425-Asp + Leu-498-Val double substitution. Segregation analysis suggests the observed resistance is mediated by a single, codominant to dominant locus. No fitness costs associated with double substitution were observed in the F2 generation when grown in monoculture or in competition with wheat [82]. Functional validation of the Glu-425-Asp remains necessary.

2.9. *Protoporphyrinogen Oxidase Inhibitors: Group 14 (E)*

Resistance to Group 14 was last reviewed in 2014 [78] and 2018 [84]. Protoporphyrinogen oxidase (PPO) catalyzes the oxidation of protoporphyrinogen IX to protoporphyrin IX [78]. The enzyme is nuclear encoded with two main isoforms: PPO1, which is largely localized to the chloroplast, and PPO2, which is largely localized to the mitochondria [84]. Inhibition of PPO results in an accumulation of protoporphyrin IX within the cytoplasm as a result of protoporphyrinogen IX 'leaking' out of organelles. Inhibitors of PPO, termed Group 14, are organized into numerous classes: diphenyl ethers, N-phenylphthalimides, oxadiazoles, oxazolidinediones, phenylpyrazoles, pyridinediones, thiadiazoles, triazinones, and triazolinones [57]. Several chemistries remain unclassified. The crystal structure of PPO2 is characterized from *Nicotiana tabacum* in complex with a phenylpyrazole chemistry [85].

To date, all the evolved substitutions reported in peer-reviewed literature for resistance to PPO inhibitors are in PPO2, perhaps because this isoform is dual targeted to both organelles in at least some species [84]. Because PPO inhibitors possess target sites in both mitochondria and chloroplasts, resistance gained at both localizations (e.g., by a single mutation in the gene encoding PPO2) may be required for a resistant phenotype. Resistance-conferring mutations, reported as of 2018, include a deletion of a Gly codon at postion 210, and substitutions of Arg-128 (or 98, depending on the numbering system) to Leu, Gly, or Met. The amino acid numbering system has not been consistent. For instance, Arg-128 describes the position reflective of the *N. tabacum* crystal structure, while Arg-98 describes the position reflective of the *Ambrosia artemisiifolia* enzyme, in which the substitution was first identified. We recommend that the naming convention follow the *N. tabacum* sequence because of the available crystal structure.

A Gly-399-Ala substitution was first reported in *A. palmeri* from Arkansas, US [86]. Dose–response revealed a resistant phenotype against a representative diphenyl ether chemistry when compared to an unrelated sensitive population. F1 inheritance suggests that the resistant phenotype is dominant. Sequence analysis revealed the Gly-399-Ala substitution. In silico modeling with the available crystal structure suggests that the Gly-399-Ala substitution decreases the binding-pocket size. In vitro PPO enzyme activity assays suggest resistance to representative diphenyl ether, pyrimidinedione, triazolinone, N-phenylphthalimide, phenylpyrazole, thiadiazole, and oxadiazole chemistries and pyraclonil. However, the enzyme containing the Gly-399-Ala had notably reduced enzyme activity in the absence of inhibitors and, therefore, the concentration of the Gly-399-Ala was increased relative to wild-type. As enzyme concentrations were not constant between the Gly-399-Ala and wild-type, further validation in vivo is necessary.

An Arg-128-Ile substitution was first reported in *Amaranthus tuberculatus* in the US [87]. Substitutions at the Arg-128 position have been reported to confer resistance to diphenyl ether, pyrimidinedione, and triazolinones [88]. Resistant plants from a population survey were identified through delimiting rate screening. DNA of each population was bulked and the gene encoding PPO2, termed *PPX2*, was amplified, barcoded, and subjected to next-generation sequencing. Single-nucleotide polymorphism (SNP) calling revealed the Arg-128-Ile substitution, which was supported by single plant sequence analysis. Pseudo-in vivo enzyme assays using the BT3 hemG system suggest that the Arg-128-Ile substitution confers resistance to a representative diphenyl ether chemistry.

3. Discussion

Prior to this review, causal variants for herbicide resistance have largely been considered to be the result of a single nucleotide change from wild-type. With the exception of the TIPS double-substitution in EPSP synthase of *E. indica* [89], which was reported in 2015, nearly all characterized herbicide resistance mechanisms have been explained through a single modification. However, of the 19 target-site mechanisms under review, six cases spanning four herbicide groups require more than one modification from wild-type. The question of which substitutions at a given amino acid position can mediate a resistant phenotype is highly pertinent for resistance management. Previously, this question has been focused to the subset of amino acids which may result from single-nucleotide substitution, as in the case of ALS- and PPO-inhibitors [27,87]. Perhaps in response to the increase in these 'multiple-modification' mechanisms, the idea of screening all amino acid substitutions may gain traction.

A primary goal driving the need to characterize herbicide resistance mechanisms is the management of herbicide-resistant weeds. Successful management of herbicide resistance is largely dependent on (a) to which chemistries resistance is conferred, (b) the distribution of the resistance mechanism, and, if present, (c) the associated fitness cost.

3.1. *Cross-Resistance Patterns*

The exhaustive characterization of cross-resistance patterns associated with a given resistance mechanism is a daunting challenge. For instance, there are 32 published, unique amino acid substitutions reported to confer resistance to at least one Group 2 chemistry. Fifty-seven chemistries within Group 2 have been reported [57], resulting in over 1800 potential unique interactions. While not universal, the use of in vitro enzyme activity assays provides rapid and excellent support for the characterization of herbicide resistance. Furthermore, the high-throughput nature of in vitro assays allows for rapid screening of both multiple herbicide chemical families and multiple members of each chemical family. Together with in silico crystal structure predictions, in vitro assays could be of great use for the development of new herbicidal compounds within existing families. For instance, the observation that the Ala-205-Phe mediates cross-resistance to all tested chemistries, with the exception of florasulam, in vitro could inform the production of new active compounds.

Translational issues, from in vitro to in planta, have been reported [17]. While having a much lower throughput, functional validation in planta is the gold standard for demonstrating resistance and susceptibility. A uniform and replicable system for in planta functional validation is necessary to facilitate high-throughput screening initiatives. The easily transformable model organism *A. thaliana* is compatible with numerous herbicide groups [15,90]. For herbicide groups which do not provide control of *A. thaliana*, such as ACCase-inhibitors, *S. viridis*, a model system used for the study of millets [91], may be an excellent target. *S. viridis* is a grassy weed readily controlled by numerous Group 1 herbicides [92]. In addition, transformation systems within the species are well characterized and utilize the simple floral-dip technique [93]. While transformation of *S. viridis*, as mediated through floral dip, results in a notably low yield of transformants, floral dip does not require specialized tissue culture capabilities. Should such capabilities be available, *O. sativa* may be another excellent choice [94].

A key problem with the use of transformation-based strategies for the functional validation of herbicide resistance is dosage effects. In this case, dosage effects can be thought of as the increase in protein abundance relative to wild-type. As exemplified by the gene amplification resistance mechanism towards glyphosate, increased expression of a sensitive target protein can result in a resistant phenotype. Even the use of the native promoter can greatly affect herbicide efficacy, as observed by LeClere et al. [15]. Mutagenesis of the target gene would eliminate these concerns. As a particular variant of interest may not be present within mutagenesis collections, a targeted approach is required.

Fortunately, targeted mutagenesis techniques, such as those mediated through CRISPR/Cas technologies, have become increasingly popular. Through the use of flanking CRIPSR sites, homology-directed repair could be exploited to simply introduce the desired substitutions into the native copy of the target enzyme [95], eliminating dosage-effect concerns. Alternatively, directed base editing through cytosine base editors, which mediate C-to-T nucleotide substitutions, or adenine base editors, which mediate A-to-G substitutions, could avoid the relatively low frequency of homology-directed repair [96]. Through these advances, we believe that the risks involved with in planta functional validation have been greatly diminished and could form the basis for screening resistance-by-chemistry interactions.

3.2. *Distribution of Resistance Mechanisms*

The distribution of resistance mechanisms is often quantified through routine surveillance across broad geographies. Surveys that monitor herbicide resistance within the same geography over time can provide insights into the effectiveness of management practices in controlling, or encouraging, herbicide resistance. An excellent example is the weed resistance monitoring program in the Northern Great Plains of Canada, which has been routinely conducted since the mid-1990s [97]. While numerous examples of these resistance surveys exist throughout the literature, they are not the primary subject of this review. However, these surveys can be utilized to identify new resistance mechanisms, as illustrated through Nie et al.'s work for the characterization of the Arg-128-Ile, PPO2 substitution of *A. tuberculatus* [87].

3.3. *Fitness Cost Analysis: A Major Knowledge Gap*

Fitness cost analysis has been the subject of past reviews [98,99]. A primary challenge with fitness cost studies is how to control for the genetic background between the R and S plant. From the perspective of this review, of the 19 resistance mechanisms characterized, 13 were compared to unrelated sensitive populations and three to geographically proximal sensitive populations. Dose–responses were not conducted on the remaining three resistance mechanisms. While the generation of nearly-isogenic lines (NILs) or transgenic lines provides the greatest control of genetic background, multiple factors prevent application. NILs are time consuming and impractical to produce in species where backcrosses are challenging. Transgenic methods often require the use of a model system, such as A. thaliana, which abstracts the experiment from a direct field application. Furthermore, issues related to the positional insertion of a given transgene and dosage effects may provide confounding factors, complicating the fitness cost analysis. Finally, the concept that a genetic background itself can compensate for the fitness cost of resistance has support [100].

We believe that the use of transgenic methods for fitness cost analysis is undervalued. Targeted mutagenesis techniques, as previously discussed, mitigate many of the systematic errors. Here, we provide some thoughts on the necessity of direct, in-field measurements for fitness cost analysis.

The main goal of a fitness cost analysis is to determine fitness costs associated with a given allele. As fitness cost analysis is resource-intensive, the fitness cost observed is often extrapolated to multiple different species [98] and environments that are subjected to different evolutionary pressures. Therefore, fitness cost analysis within a model system may be as appropriate as such analysis within the native system. As previously discussed, targeted mutagenesis strategies can augment or eliminate issues related to positional and dosage effects. Compensatory effects of the genetic background are poorly understood at a functional level. From a theoretical perspective, Liebig's law of the minimum [101] may be relevant. Following Liebig's law, fecundity is determined by the most limiting factor. Within the context of a fitness cost analysis, a cost will only be observed when the given allele results in a greater limitation than observed within the wild-type. Therefore, a genetic background that 'compensates' for a resistance trait may be less fit in the greater environment, in the absence of selection, than a genetic background where the fitness cost is observable. An alternative form of compensation would exist within allopolyploids, which can mitigate a present fitness cost of a mutant allele by carrying the wild-type allele within a separate genome. Finally, technical issues, such as positional and dosage effects, can be largely eliminated through targeted mutagenesis, as described previously.

Author Contributions: Conceptualization, B.P.M. and P.J.T.; writing—original draft preparation, B.P.M.; writing—review and editing, B.P.M. and P.J.T.

Funding: This research received no external funding.

Conflicts of Interest: The authors declare no conflict of interest.

References

1. Powles, S.B.; Yu, Q. Evolution in action: Plants resistant to herbicides. *Ann. Rev. Plant Biol.* **2010**, *61*, 317–347. [CrossRef] [PubMed]
2. Délye, C.; Jasieniuk, M.; Le Corre, V. Deciphering the evolution of herbicide resistance in weeds. *Trends Genet.* **2013**, *29*, 649–658. [CrossRef] [PubMed]
3. Laforest, M.; Soufiane, B.; Simard, M.J.; Obeid, K.; Page, E.; Nurse, R.E. Acetyl-CoA carboxylase overexpression in herbicide-resistant large crabgrass (*Digitaria sanguinalis*). *Pest Manag. Sci.* **2017**, *73*, 2227–2235. [CrossRef] [PubMed]
4. Gaines, T.A.; Zhang, W.; Wang, D.; Bukun, B.; Chisholm, S.T.; Shaner, D.L.; Nissen, S.J.; Patzoldt, W.L.; Tranel, P.J.; Culpepper, A.S.; et al. Gene amplification confers glyphosate resistance in *Amaranthus palmeri*. *Proc. Natl. Acad. Sci. USA* **2010**, *107*, 1029–1034. [CrossRef]
5. Délye, C. Unravelling the genetic bases of non-target-site-based resistance (NTSR) to herbicides: A major challenge for weed science in the forthcoming decade. *Pest Manag. Sci.* **2013**, *69*, 176–187. [CrossRef] [PubMed]
6. Hirschberg, J.; Mcintosh, L. Molecular basis of herbicide resistance in *Amaranthus hybridus*. *Science* **1983**, *222*, 1346–1349. [CrossRef]
7. Shaner, D.L. *Herbicide Handbook*, 10th ed.; Weed Science Society of America: Lawrence, KS, USA, 2014.
8. Guttieri, M.J.; Eberlein, C.V.; Mallory-Smith, C.A.; Thill, D.C.; Hoffman, D.L. DNA sequence variation in domain A of the acetolactate synthase genes of herbicide-resistant and -susceptible weed biotypes. *Weed Sci.* **1992**, *40*, 670–677. [CrossRef]
9. Anthony, R.G.; Waldin, T.R.; Ray, J.A.; Bright, S.W.J.; Hussey, P.J. Herbicide resistance caused by spontaneous mutation of the cytoskeletal protein tubulin. *Nature* **1998**, *393*, 260. [CrossRef]
10. Zagnitko, O.; Jelenska, J.; Tevzadze, G.; Haselkorn, R.; Gornicki, P. An isoleucine/leucine residue in the carboxyltransferase domain of acetyl-CoA carboxylase is critical for interaction with aryloxyphenoxypropionate and cyclohexanedione inhibitors. *Proc. Natl. Acad. Sci. USA* **2001**, *98*, 6617–6622. [CrossRef]
11. Baerson, S.R. Glyphosate-resistant goosegrass. Identification of a mutation in the target enzyme 5-enolpyruvylshikimate-3-phosphate synthase. *Plant Physiol.* **2002**, *129*, 1265–1275. [CrossRef]
12. Michel, A.; Arias, R.S.; Scheffler, B.E.; Duke, S.O.; Netherland, M.; Dayan, F.E. Somatic mutation-mediated evolution of herbicide resistance in the nonindigenous invasive plant hydrilla (*Hydrilla verticillata*). *Mol. Ecol.* **2004**, *13*, 3229–3237. [CrossRef] [PubMed]
13. Patzoldt, W.L.; Hager, A.G.; McCormick, J.S.; Tranel, P.J. A codon deletion confers resistance to herbicides inhibiting protoporphyrinogen oxidase. *Proc. Natl. Acad. Sci. USA* **2006**, *103*, 12329–12334. [CrossRef] [PubMed]
14. Avila-Garcia, W.V.; Sanchez-Olguin, E.; Hulting, A.G.; Mallory-Smith, C. Target-site mutation associated with glufosinate resistance in Italian ryegrass (*Lolium perenne* L. ssp. multiflorum). *Pest Manag. Sci.* **2012**, *68*, 1248–1254. [CrossRef] [PubMed]
15. LeClere, S.; Wu, C.; Westra, P.; Sammons, R.D. Cross-resistance to dicamba, 2,4-D, and fluroxypyr in *Kochia scoparia* is endowed by a mutation in an *AUX/IAA* gene. *Proc. Natl. Acad. Sci. USA* **2018**, *115*, E2911–E2920. [CrossRef] [PubMed]
16. Délye, C.; Zhang, X.-Q.; Michel, S.; Matéjicek, A.; Powles, S.B. Molecular bases for sensitivity to acetyl-coenzyme A carboxylase inhibitors in black-grass. *Plant Physiol.* **2005**, *137*, 794–806. [CrossRef] [PubMed]

17. Kaundun, S.S. Resistance to acetyl-CoA carboxylase-inhibiting herbicides. *Pest Manag. Sci.* **2014**, *70*, 1405–1417. [CrossRef] [PubMed]
18. Zhang, H.; Yang, Z.; Shen, Y.; Tong, L. Crystal structure of the carboxyltransferase domain of acetyl-coenzyme A carboxylase. *Science* **2003**, *299*, 2064–2067. [CrossRef]
19. Zhang, H.; Tweel, B.; Tong, L. Molecular basis for the inhibition of the carboxyltransferase domain of acetyl-coenzyme-A carboxylase by haloxyfop and diclofop. *Proc. Natl. Acad. Sci. USA* **2004**, *101*, 5910–5915. [CrossRef]
20. Xiang, S.; Callaghan, M.M.; Watson, K.G.; Tong, L. A different mechanism for the inhibition of the carboxyltransferase domain of acetyl-coenzyme A carboxylase by tepraloxydim. *Proc. Natl. Acad. Sci. USA* **2009**, *106*, 20723–20727. [CrossRef]
21. Yu, L.P.C.; Kim, Y.S.; Tong, L. Mechanism for the inhibition of the carboxyltransferase domain of acetyl-coenzyme A carboxylase by pinoxaden. *Proc. Natl. Acad. Sci. USA* **2010**, *107*, 22072–22077. [CrossRef]
22. Guo, W.; Zhang, L.; Wang, H.; Li, Q.; Liu, W.; Wang, J. A Rare Ile-2041-Thr Mutation in the ACCase gene confers resistance to ACCase-inhibiting herbicides in shortawn foxtail (*Alopecurus aequalis*). *Weed Sci.* **2017**, *65*, 239–246. [CrossRef]
23. Pan, L.; Li, J.; Zhang, T.; Zhang, D.; Dong, L.Y. Cross-resistance patterns to acetyl coenzyme A carboxylase (ACCase) inhibitors associated with different ACCase mutations in *Beckmannia syzigachne*. *Weed Res.* **2015**, *55*, 609–620. [CrossRef]
24. Ghanizadeh, H.; Harrington, K.C.; Mesarich, C.H. The target site mutation Ile-2041-Asn is associated with resistance to ACCase-inhibiting herbicides in *Lolium multiflorum*. *N. Z. J. Agric. Res.* **2019**, 1–14. [CrossRef]
25. Cha, T.S.; Najihah, M.G.; Sahid, I.B.; Chuah, T.S. Molecular basis for resistance to ACCase-inhibiting fluazifop in *Eleusine indica* from Malaysia. *Pestic. Biochem. Physiol.* **2014**, *111*, 7–13. [CrossRef] [PubMed]
26. Délye, C. Weed resistance to acetyl coenzyme A carboxylase inhibitors: An update. *Weed Sci.* **2005**, *53*, 728–746. [CrossRef]
27. Yu, Q.; Powles, S.B. Resistance to AHAS inhibitor herbicides: Current understanding. *Pest Manag. Sci.* **2014**, *70*, 1340–1350. [CrossRef] [PubMed]
28. Chipman, D.M.; Duggleby, R.G.; Tittmann, K. Mechanisms of acetohydroxyacid synthases. *Curr. Opin. Chem. Biol.* **2005**, *9*, 475–481. [CrossRef]
29. McCourt, J.A.; Pang, S.S.; King-Scott, J.; Guddat, L.W.; Duggleby, R.G. Herbicide-binding sites revealed in the structure of plant acetohydroxyacid synthase. *Proc. Natl. Acad. Sci. USA* **2006**, *103*, 569–573. [CrossRef]
30. Garcia, M.D.; Nouwens, A.; Lonhienne, T.G.; Guddat, L.W. Comprehensive understanding of acetohydroxyacid synthase inhibition by different herbicide families. *Proc. Natl. Acad. Sci. USA* **2017**, *114*, E1091–E1100. [CrossRef]
31. Lonhienne, T.; Garcia, M.D.; Pierens, G.; Mobli, M.; Nouwens, A.; Guddat, L.W. Structural insights into the mechanism of inhibition of AHAS by herbicides. *Proc. Natl. Acad. Sci. USA* **2018**, *115*, E1945–E1954. [CrossRef]
32. Tranel, P.J.; Wright, T.R. Resistance of weeds to ALS-inhibiting herbicides: What have we learned? *Weed Sci.* **2002**, *50*, 700–712. [CrossRef]
33. Panozzo, S.; Scarabel, L.; Rosan, V.; Sattin, M. A new Ala-122-Asn amino acid change confers decreased fitness to ALS-resistant *Echinochloa crus-galli*. *Front. Plant Sci.* **2017**, *8*, 2042. [CrossRef] [PubMed]
34. Duggleby, R.G.; McCourt, J.A.; Guddat, L.W. Structure and mechanism of inhibition of plant acetohydroxyacid synthase. *Plant Physiol. Biochem.* **2008**, *46*, 309–324. [CrossRef] [PubMed]
35. Larran, A.S.; Palmieri, V.E.; Perotti, V.E.; Lieber, L.; Tuesca, D.; Permingeat, H.R. Target-site resistance to acetolactate synthase (ALS)-inhibiting herbicides in *Amaranthus palmeri* from Argentina. *Pest Manag. Sci.* **2017**, *73*, 2578–2584. [CrossRef] [PubMed]
36. Liu, W.; Yuan, G.; Du, L.; Guo, W.; Li, L.; Bi, Y.; Wang, J. A novel Pro197Glu substitution in acetolactate synthase (ALS) confers broad-spectrum resistance across ALS inhibitors. *Pestic. Biochem. Physiol.* **2015**, *117*, 31–38. [CrossRef]
37. Long, W.; Malone, J.; Boutsalis, P.; Preston, C. Diversity and extent of mutations endowing resistance to the acetolactate synthase (AHAS)-inhibiting herbicides in Indian hedge mustard (*Sisymbrium orientale*) populations in Australia. *Pestic. Biochem. Physiol.* **2019**, *157*, 53–59. [CrossRef]

38. Brosnan, J.T.; Vargas, J.J.; Breeden, G.K.; Grier, L.; Aponte, R.A.; Tresch, S.; Laforest, M. A new amino acid substitution (Ala-205-Phe) in acetolactate synthase (ALS) confers broad spectrum resistance to ALS-inhibiting herbicides. *Planta* **2016**, *243*, 149–159. [CrossRef] [PubMed]
39. Brosnan, J.T.; Breeden, G.K.; Vargas, J.J.; Grier, L. A biotype of annual bluegrass (*Poa annua*) in Tennessee is resistant to inhibitors of ALS and Photosystem II. *Weed Sci.* **2015**, *63*, 321–328. [CrossRef]
40. Li, J.; Li, M.; Gao, X.; Fang, F. A novel amino acid substitution Trp574Arg in acetolactate synthase (ALS) confers broad resistance to ALS-inhibiting herbicides in crabgrass (*Digitaria sanguinalis*). *Pest Manag. Sci.* **2017**, *73*, 2538–2543. [CrossRef] [PubMed]
41. Fernandes, T.C.C.; Pizano, M.A.; Marin-Morales, M.A. Characterization, modes of action and effects of trifluralin: A review. In *Herbicides-Current Research and Case Studies in Use*; IntechOpen: London, UK, 2013.
42. Mitra, A.; Sept, D. Binding and interaction of dinitroanilines with apicomplexan and kinetoplastid α-tubulin. *J. Med. Chem.* **2006**, *49*, 5226–5231. [CrossRef]
43. Hugdahl, J.D.; Morejohn, L.C. Rapid and reversible high-affinity binding of the dinitroaniline herbicide oryzalin to tubulin from *Zea mays* L. *Plant Physiol.* **1993**, *102*, 725–740. [CrossRef] [PubMed]
44. Löwe, J.; Li, H.; Downing, K.H.; Nogales, E. Refined structure of alpha beta-tubulin at 3.5 A resolution. *J. Mol. Biol.* **2001**, *313*, 1045–1057. [CrossRef] [PubMed]
45. Morrissette, N.S.; Mitra, A.; Sept, D.; Sibley, L.D. Dinitroanilines bind α-tubulin to disrupt microtubules. *Mol. Biol. Cell* **2004**, *15*, 1960–1968. [CrossRef] [PubMed]
46. Chu, Z.; Chen, J.; Nyporko, A.; Han, H.; Yu, Q.; Powles, S. Novel α-tubulin mutations conferring resistance to dinitroaniline herbicides in *Lolium rigidum*. *Front. Plant Sci.* **2018**, *9*, 97. [CrossRef] [PubMed]
47. Chen, J.; Yu, Q.; Owen, M.; Han, H.; Powles, S. Dinitroaniline herbicide resistance in a multiple-resistant *Lolium rigidum* population. *Pest Manag. Sci.* **2018**, *74*, 925–932. [CrossRef] [PubMed]
48. Busi, R.; Goggin, D.E.; Heap, I.M.; Horak, M.J.; Jugulam, M.; Masters, R.A.; Napier, R.M.; Riar, D.S.; Satchivi, N.M.; Torra, J.; et al. Weed resistance to synthetic auxin herbicides. *Pest Manag. Sci.* **2018**, *74*, 2265–2276. [CrossRef] [PubMed]
49. Grossmann, K. Auxin herbicides: Current status of mechanism and mode of action. *Pest Manag. Sci.* **2010**, *66*, 113–120. [CrossRef]
50. Dharmasiri, N.; Dharmasiri, S.; Estelle, M. The F-box protein TIR1 is an auxin receptor. *Nature* **2005**, *435*, 441. [CrossRef]
51. Vieten, A.; Sauer, M.; Brewer, P.B.; Friml, J. Molecular and cellular aspects of auxin-transport-mediated development. *Trends Plant Sci.* **2007**, *12*, 160–168. [CrossRef]
52. Dharmasiri, N.; Dharmasiri, S.; Weijers, D.; Lechner, E.; Yamada, M.; Hobbie, L.; Ehrismann, J.S.; Jürgens, G.; Estelle, M. Plant development is regulated by a family of auxin receptor F box proteins. *Dev. Cell* **2005**, *9*, 109–119. [CrossRef]
53. Walsh, T.A.; Neal, R.; Merlo, A.O.; Honma, M.; Hicks, G.R.; Wolff, K.; Matsumura, W.; Davies, J.P. Mutations in an auxin receptor homolog AFB5 and in SGT1b confer resistance to synthetic picolinate auxins and not to 2,4-dichlorophenoxyacetic acid or indole-3-acetic acid in Arabidopsis. *Plant Physiol.* **2006**, *142*, 542–552. [CrossRef] [PubMed]
54. Tiwari, S.B.; Wang, X.-J.; Hagen, G.; Guilfoyle, T.J. AUX/IAA proteins are active repressors, and their stability and activity are modulated by auxin. *Plant Cell* **2001**, *13*, 2809–2822. [CrossRef] [PubMed]
55. Lee, S.; Sundaram, S.; Armitage, L.; Evans, J.P.; Hawkes, T.; Kepinski, S.; Ferro, N.; Napier, R.M. Defining binding efficiency and specificity of auxins for SCFTIR1/AFB-Aux/IAA co-receptor complex formation. *ACS Chem. Biol.* **2014**, *9*, 673–682. [CrossRef] [PubMed]
56. Tan, X.; Calderon-Villalobos, L.I.A.; Sharon, M.; Zheng, C.; Robinson, C.V.; Estelle, M.; Zheng, N. Mechanism of auxin perception by the TIR1 ubiquitin ligase. *Nature* **2007**, *446*, 640–645. [CrossRef] [PubMed]
57. Heap, I. The International Survey of Herbicide Resistant Weeds. Available online: http://weedscience.org/ (accessed on 16 July 2019).
58. Tietjen, K.G.; Kluth, J.F.; Andree, R.; Haug, M.; Lindig, M.; Müller, K.H.; Wroblowsky, H.J.; Trebst, A. The herbicide binding niche of photosystem II-a model. *Pestic. Sci.* **1991**, *31*, 65–72. [CrossRef]
59. Lambreva, M.D.; Russo, D.; Polticelli, F.; Scognamiglio, V.; Antonacci, A.; Zobnina, V.; Campi, G.; Rea, G. Structure/Function/Dynamics of Photosystem II Plastoquinone Binding Sites. *Curr. Protein Pept. Sci.* **2014**, *15*, 285–295. [CrossRef] [PubMed]

60. Thiel, H.; Varrelmann, M. Identification of a new PSII target site psbA mutation leading to D1 amino acid Leu218 Val exchange in the *Chenopodium album* D1 protein and comparison to cross-resistance profiles of known modifications at positions 251 and 264. *Pest Manag. Sci.* **2014**, *70*, 278–285. [CrossRef] [PubMed]
61. Mengistu, L.W.; Christoffers, M.J.; Lym, R.G. A psbA mutation in *Kochia scoparia* (L) Schrad from railroad rights-of-way with resistance to diuron, tebuthiuron and metribuzin. *Pest Manag. Sci.* **2005**, *61*, 1035–1042. [CrossRef] [PubMed]
62. Lu, H.; Yu, Q.; Han, H.; Owen, M.J.; Powles, S.B. A novel psbA mutation (Phe274-Val) confers resistance to PSII herbicides in wild radish (*Raphanus raphanistrum*). *Pest Manag. Sci.* **2019**, *75*, 144–151. [CrossRef]
63. Schönbrunn, E.; Eschenburg, S.; Shuttleworth, W.A.; Schloss, J.V.; Amrhein, N.; Evans, J.N.; Kabsch, W. Interaction of the herbicide glyphosate with its target enzyme 5-enolpyruvylshikimate 3-phosphate synthase in atomic detail. *Proc. Natl. Acad. Sci. USA.* **2001**, *98*, 1376–1380. [CrossRef] [PubMed]
64. Sammons, R.D.; Gaines, T.A. Glyphosate resistance: State of knowledge. *Pest Manag. Sci.* **2014**, *70*, 1367–1377. [CrossRef] [PubMed]
65. Heap, I.; Duke, S.O. Overview of glyphosate-resistant weeds worldwide. *Pest Manag. Sci.* **2018**, *74*, 1040–1049. [CrossRef] [PubMed]
66. Della-Cioppa, G.; Bauer, S.C.; Klein, B.K.; Shah, D.M.; Fraley, R.T.; Kishore, G.M. Translocation of the precursor of 5-enolpyruvylshikimate-3-phosphate synthase into chloroplasts of higher plants in vitro. *Proc. Natl. Acad. Sci. USA* **1986**, *83*, 6873–6877. [CrossRef] [PubMed]
67. Takano, H.K.; Fernandes, V.N.A.; Adegas, F.S.; Oliveira, R.S.; Westra, P.; Gaines, T.A.; Dayan, F.E. A novel TIPT double mutation in EPSPS conferring glyphosate resistance in tetraploid *Bidens subalternans*. *Pest Manag. Sci.* **2019**. [CrossRef] [PubMed]
68. Perotti, V.E.; Larran, A.S.; Palmieri, V.E.; Martinatto, A.K.; Alvarez, C.E.; Tuesca, D.; Permingeat, H.R. A novel triple amino acid substitution in the EPSPS found in a high-level glyphosate-resistant *Amaranthus hybridus* population from Argentina. *Pest Manag. Sci.* **2019**, *75*, 1242–1251. [CrossRef]
69. Donn, G.; Köcher, H. Inhibitors of glutamine synthetase. In *Herbicide Classes in Development: Mode of Action, Targets, Genetic Engineering, Chemistry*; Böger, P., Wakabayashi, K., Hirai, K., Eds.; Springer: Berlin/Heidelberg, Germany, 2002; pp. 87–101. ISBN 978-3-642-59416-8.
70. Edwards, J.W.; Walker, E.L.; Coruzzi, G.M. Cell-specific expression in transgenic plants reveals nonoverlapping roles for chloroplast and cytosolic glutamine synthetase. *Proc. Natl. Acad. Sci. USA* **1990**, *87*, 3459–3463. [CrossRef]
71. Wild, A.; Wendler, C. Inhibitory action of glufosinate on photosynthesis. *Z. Naturforsch. C* **1993**, *48*, 369–373. [CrossRef]
72. Wendler, C.; Barniske, M.; Wild, A. Effect of phosphinothricin (glufosinate) on photosynthesis and photorespiration of C3 and C 4 plants. *Photosynth. Res.* **1990**, *24*, 55–61. [CrossRef]
73. Takano, H.K.; Beffa, R.; Preston, C.; Westra, P.; Dayan, F.E. Reactive oxygen species trigger the fast action of glufosinate. *Planta* **2019**, *249*, 1837–1849. [CrossRef]
74. Avila, C.; García-Gutiérrez, A.; Crespillo, R.; Cánovas, F.M. Effects of phosphinotricin treatment on glutamine synthetase isoforms in Scots pine seedlings. *Plant Physiol. Biochem.* **1998**, *36*, 857–863. [CrossRef]
75. Logusch, E.W.; Walker, D.M.; McDonald, J.F.; Franz, J.E. Inhibition of plant glutamine synthetases by substituted phosphinothricins. *Plant Physiol.* **1991**, *95*, 1057–1062. [CrossRef] [PubMed]
76. Unno, H.; Uchida, T.; Sugawara, H.; Kurisu, G.; Sugiyama, T.; Yamaya, T.; Sakakibara, H.; Hase, T.; Kusunoki, M. Atomic structure of plant glutamine synthetase: A key enzyme for plant productivity. *J. Biol. Chem.* **2006**, *281*, 29287–29296. [CrossRef] [PubMed]
77. Torreira, E.; Seabra, A.R.; Marriott, H.; Zhou, M.; Llorca, Ó.; Robinson, C.V.; Carvalho, H.G.; Fernández-Tornero, C.; Pereira, P.J.B. The structures of cytosolic and plastid-located glutamine synthetases from *Medicago truncatula* reveal a common and dynamic architecture. *Acta Crystallogr. D Biol. Crystallogr.* **2014**, *70*, 981–993. [CrossRef] [PubMed]
78. Dayan, F.E.; Owens, D.K.; Tranel, P.J.; Preston, C.; Duke, S.O. Evolution of resistance to phytoene desaturase and protoporphyrinogen oxidase inhibitors—State of knowledge. *Pest Manag. Sci.* **2014**, *70*, 1358–1366. [CrossRef] [PubMed]
79. Brausemann, A.; Gemmecker, S.; Koschmieder, J.; Ghisla, S.; Beyer, P.; Einsle, O. Structure of phytoene desaturase provides insights into herbicide binding and reaction mechanisms involved in carotene desaturation. *Structure* **2017**, *25*, 1222–1232. [CrossRef] [PubMed]

80. Al-Babili, S.; Lintig, J.V.; Haubruck, H.; Beyer, P. A novel, soluble form of phytoene desaturase from *Narcissus pseudonarcissus* chromoplasts is Hsp70-complexed and competent for flavinylation, membrane association and enzymatic activation. *Plant J.* **1996**, *9*, 601–612. [CrossRef] [PubMed]
81. Dang, H.T.; Malone, J.M.; Boutsalis, P.; Gill, G.; Preston, C. The mechanism of diflufenican resistance and its inheritance in oriental mustard (*Sisymbrium orientale* L.) from Australia. *Pest Manag. Sci.* **2018**, *74*, 1279–1285. [CrossRef] [PubMed]
82. Dang, H.T.; Long, W.; Malone, J.M.; Preston, C.; Gill, G. No apparent fitness costs associated with phytoene desaturase mutations conferred resistance to diflufenican and picolinafen in oriental mustard (*Sisymbrium orientale* L.). *Pestic. Biochem. Physiol.* **2019**, *155*, 51–57. [CrossRef] [PubMed]
83. Dang, H.T.; Malone, J.M.; Gill, G.; Preston, C. Cross-resistance to diflufenican and picolinafen and its inheritance in oriental mustard (*Sisymbrium orientale* L.). *Pest Manag. Sci.* **2019**, *75*, 195–203. [CrossRef]
84. Dayan, F.E.; Barker, A.; Tranel, P.J. Origins and structure of chloroplastic and mitochondrial plant protoporphyrinogen oxidases: Implications for the evolution of herbicide resistance. *Pest Manag. Sci.* **2018**, *74*, 2226–2234. [CrossRef]
85. Koch, M.; Breithaupt, C.; Kiefersauer, R.; Freigang, J.; Huber, R.; Messerschmidt, A. Crystal structure of protoporphyrinogen IX oxidase: A key enzyme in haem and chlorophyll biosynthesis. *EMBO J.* **2004**, *23*, 1720–1728. [CrossRef] [PubMed]
86. Rangani, G.; Salas-Perez, R.A.; Aponte, R.A.; Knapp, M.; Craig, I.R.; Mietzner, T.; Langaro, A.C.; Noguera, M.M.; Porri, A.; Roma-Burgos, N. A novel single-site mutation in the catalytic domain of protoporphyrinogen oxidase IX (PPO) confers resistance to PPO-inhibiting herbicides. *Front. Plant Sci.* **2019**, *10*, 568. [CrossRef] [PubMed]
87. Nie, H.; Mansfield, B.C.; Harre, N.T.; Young, J.M.; Steppig, N.R.; Young, B.G. Investigating target-site resistance mechanism to the PPO-inhibiting herbicide fomesafen in waterhemp and interspecific hybridization of *Amaranthus* species using next generation sequencing. *Pest Manag. Sci.* **2019**. [CrossRef] [PubMed]
88. Rousonelos, S.L.; Lee, R.M.; Moreira, M.S.; Vangessel, M.J.; Tranel, P.J. Characterization of a common ragweed (*Ambrosia artemisiifolia*) population resistant to ALS- and PPO-inhibiting herbicides. *Weed Sci.* **2012**, *60*, 335–344. [CrossRef]
89. Yu, Q.; Jalaludin, A.; Han, H.; Chen, M.; Sammons, R.D.; Powles, S.B. Evolution of a double amino acid substitution in the 5-enolpyruvylshikimate-3-phosphate synthase in *Eleusine indica* conferring high-level glyphosate resistance. *Plant Physiol.* **2015**, *167*, 1440–1447. [CrossRef] [PubMed]
90. Sathasivan, K.; Haughn, G.W.; Murai, N. Molecular basis of imidazolinone herbicide resistance in *Arabidopsis thaliana* var Columbia. *Plant Physiol.* **1991**, *97*, 1044–1050. [CrossRef] [PubMed]
91. Huang, P.; Shyu, C.; Coelho, C.P.; Cao, Y.; Brutnell, T.P. *Setaria viridis* as a model system to advance millet genetics and genomics. *Front. Plant Sci.* **2016**, *7*, 1781. [CrossRef]
92. Douglas, B.J.; Morrison, I.N.; Thomas, A.G.; Maw, M.G. The biology of Canadian weeds: 70. *Setaria viridis* (L.) Beauv. *Can. J. Plant Sci.* **1985**, *65*, 669–690. [CrossRef]
93. Martins, P.K.; Nakayama, T.J.; Ribeiro, A.P.; Cunha, B.A.D.B.; Nepomuceno, A.L.; Harmon, F.G.; Kobayashi, A.K.; Molinari, H.B.C. *Setaria viridis* floral-dip: A simple and rapid *Agrobacterium*-mediated transformation method. *Biotechnol. Rep.* **2015**, *6*, 61–63. [CrossRef]
94. Hiei, Y.; Ohta, S.; Komari, T.; Kumashiro, T. Efficient transformation of rice (*Oryza sativa* L.) mediated by *Agrobacterium* and sequence analysis of the boundaries of the T-DNA. *Plant J.* **1994**, *6*, 271–282. [CrossRef]
95. Cong, L.; Ran, F.A.; Cox, D.; Lin, S.; Barretto, R.; Habib, N.; Hsu, P.D.; Wu, X.; Jiang, W.; Marraffini, L.A.; et al. Multiplex genome engineering using CRISPR/Cas systems. *Science* **2013**, *339*, 819–823. [CrossRef] [PubMed]
96. Molla, K.A.; Yang, Y. CRISPR/Cas-mediated base editing: Technical considerations and practical applications. *Trends Biotechnol.* **2019**. [CrossRef] [PubMed]
97. Beckie, H.J.; Leeson, J.Y.; Thomas, A.G.; Brenzil, C.A.; Hall, L.M.; Holzgang, G.; Lozinski, C.; Shirriff, S. Weed resistance monitoring in the Canadian prairies. *Weed Technol.* **2008**, *22*, 530–543. [CrossRef]
98. Vila-Aiub, M.M.; Neve, P.; Powles, S.B. Fitness costs associated with evolved herbicide resistance alleles in plants. *New Phytol.* **2009**, *184*, 751–767. [CrossRef] [PubMed]
99. Keshtkar, E.; Abdolshahi, R.; Sasanfar, H.; Zand, E.; Beffa, R.; Dayan, F.E.; Kudsk, P. Assessing fitness costs from a herbicide-resistance management perspective: A review and insight. *Weed Sci.* **2019**, *67*, 137–148. [CrossRef]

100. Paris, M.; Roux, F.; Bérard, A.; Reboud, X. The effects of the genetic background on herbicide resistance fitness cost and its associated dominance in *Arabidopsis thaliana*. *Heredity* **2008**, *101*, 499–506. [CrossRef] [PubMed]
101. Paris, Q. The return of von Liebig's "law of the minimum". *Agron. J.* **1992**, *84*, 1040–1046. [CrossRef]

plants

MDPI

Review

Non-Target-Site Resistance to Herbicides: Recent Developments

Mithila Jugulam * and Chandrima Shyam

Department of Agronomy, Kansas State University, Manhattan, KS 66506, USA; chandrima@ksu.edu
* Correspondence: mithila@ksu.edu; Tel.: +1-785-532-2755

Received: 26 August 2019; Accepted: 12 October 2019; Published: 15 October 2019

Abstract: Non-target-site resistance (NTSR) to herbicides in weeds can be conferred as a result of the alteration of one or more physiological processes, including herbicide absorption, translocation, sequestration, and metabolism. The mechanisms of NTSR are generally more complex to decipher than target-site resistance (TSR) and can impart cross-resistance to herbicides with different modes of action. Metabolism-based NTSR has been reported in many agriculturally important weeds, although reduced translocation and sequestration of herbicides has also been found in some weeds. This review focuses on summarizing the recent advances in our understanding of the physiological, biochemical, and molecular basis of NTSR mechanisms found in weed species. Further, the importance of examining the co-existence of TSR and NTSR for the same herbicide in the same weed species and influence of environmental conditions in the altering and selection of NTSR is also discussed. Knowledge of the prevalence of NTSR mechanisms and co-existing TSR and NTSR in weeds is crucial for designing sustainable weed management strategies to discourage the further evolution and selection of herbicide resistance in weeds.

Keywords: non-target-site resistance; herbicide mode of action; co-existence; environmental conditions

1. Introduction

Herbicide use is indispensable in modern agriculture as it offers exceptional tool for weed management and also facilitates no-till crop production to conserve soil and moisture. However, repeated field applications of herbicides with the same mechanism of action resulted in the selection of herbicide-resistant weeds. The Weed Science Society of America (http//www.wssa.net) defines herbicide resistance as the inherited ability of a plant to survive and reproduce following exposure to a dose of herbicide normally lethal to the wild type. Under continuous selection pressure, i.e., the repeated use of herbicides with the same mode of action, the resistant plants increase in frequency over time, resulting in the domination by individuals resistant to that herbicide. In addition to the selection pressure of herbicides, biological and genetic factors of weed species, properties of herbicides, and agronomic practices also play an important role in the evolution and spread of herbicide resistance [1]. Biological characteristics of highly troublesome weeds, including prolific seed production, high germination percentage, a wide window of emergence, seed dispersal, and longevity, help to maintain a high frequency of resistant individuals in the population. Genetic factors, such as natural mutations conferring herbicide resistance, inheritance of herbicide-resistant genes in the weed population, and fitness cost of resistance genes in the presence or absence of the herbicide, also play an important role in the evolution and spread of herbicide resistance [2,3].

2. Mechanisms of Herbicide Resistance

A key aspect in predicting the evolutionary trajectory of herbicide-resistance traits is understanding the mechanism(s) of herbicide resistance. Mechanisms of herbicide resistance in weeds can be broadly classified into target-site resistance (TSR) and/or non-target-site resistance (NTSR). The TSR mechanisms largely involve mutation(s) in the target site of action of an herbicide, resulting in an insensitive or less sensitive target protein of the herbicide [1]. In such cases, the TSR is primarily determined by monogenic traits [3]. Additionally, TSR can also evolve as a result of the over-expression or amplification of the target gene [4]. NTSR mechanisms include reduced herbicide uptake/translocation, increased herbicide metabolism, decreased rate of herbicide activation, and/or sequestration [5]. Metabolism-based NTSR involves the increased activity of enzyme complexes such as esterases, cytochrome P450s (CYP450s), glutathione S-transferases (GSTs), and/or Uridine 5′-diphospho (UDP)-glucosyl transferases [1]. NTSR, especially if it involves herbicide detoxification by these enzymes, is usually governed by many genes (polygenic) and may confer resistance to herbicides with completely different modes of action [3,6]. However, monogenic inheritance of NTSR has also been reported in several herbicide-resistant weeds [7–9]. The evolution of NTSR via herbicide detoxification is a serious threat to weed management as it can bestow resistance to multiple herbicides, leaving limited herbicide options for weed control, as well as potential resistances to herbicides not yet commercially available [10]. Comprehensive information on the evolution of TSR-based resistance in weeds are discussed elsewhere in this special issue. In this review recent advances in understanding the mechanisms of NTSR to herbicides with different modes of action are discussed.

3. Known NTSR Mechanisms in Weed Species for Different Herbicide Modes of Actions

3.1. *Acetyl CoA Carboxylase (ACCase)-Inhibitors*

ACCase is a crucial enzyme that catalyzes the formation of malonyl CoA via the carboxylation of acetyl CoA while using bicarbonate as the source of carbon [11]. Malonyl CoA is needed for de novo fatty acid biosynthesis, and thus, is essential for plant survival. ACCase-inhibitors impede malonyl CoA formation in sensitive grass species, ultimately leading to plant death [11,12]. These herbicides are used as important post-emergence options for managing grass weeds in dicotyledonous crops. To date, 48 weeds have been reported to have evolved resistance to these herbicides [13] via both TSR and NTSR mechanisms. Predominantly, TSR has been reported as the leading mechanism, caused by amino acid substitutions in the carboxyl transferase domain of the ACCase enzyme [14,15].

Metabolic resistance to ACCase-inhibiting herbicides has been documented in Asia minor bluegrass [16], barnyard grass [17], blackgrass [18], Italian ryegrass [19,20], Japanese foxtail [21], rigid ryegrass [22–24], and wild oat [25]. In the majority of these cases, enhanced metabolism mediated by CYP450s was reported. For instance, rapid degradation of diclofop-methyl was observed in rigid ryegrass populations from Australia [22,24]. Interestingly, exposure to low doses of diclofop-methyl acid application rapidly selected for metabolic resistance in rigid ryegrass [26]. Moreover, the metabolites produced in these resistant plants were found to be similar to those in wheat formed via ring hydroxylation and sugar conjugation [26]. This result suggests that in resistant grasses, the metabolism of ACCase-inhibitors occurs through a wheat-like detoxification pathway mediated by CYP450s [25,26]. Studies involving CYP450 or GST inhibitors, such as malathion and piperonyl butoxide (PBO), have been used to indicate the involvement of these detoxification systems. The organophosphate insecticide, malathion can decrease the rate of metabolism and increase the metabolic half-life of herbicides by inhibiting CYP450-based hydroxylation in corn [27,28]. Pre-treatment with CYP450 inhibitors like PBO or malathion has been shown to reduce resistance to ACCase-inhibitors in Asia minor bluegrass [16] and Japanese foxtail [21], indicating the role of CYP450s in enhancing metabolism in these resistant weeds. Conversely, pre-treatment with 2,4-D, a CYP450 inducer, increased the rate of metabolism of diclofop-methyl in susceptible rigid ryegrass populations [29]. Apart from CYP450s, involvement of GSTs and glucosyltransferases (GTs) have also been documented to govern the metabolic resistance to ACCase-inhibitors. Transcriptome analysis of diclofop-resistant rigid ryegrass, led to the identification of four contigs, including two *CYP450s*, one *GT*, and one nitronate monooxygenase (*NMO*) as potential candidate genes for metabolic resistance to diclofop [24]. Similarly, researchers have reported greater GST activity in resistant plants following ACCase-inhibitor application [16,17,30].

3.2. Acetolactate Synthase (ALS)-Inhibitors

ALS inhibitors were first commercialized in 1982, and by 1998, the number of weed species with resistance to this group of herbicides had surpassed other herbicides [31]. These herbicides, also referred as acetohydroxy acid synthase (AHAS) inhibitors, inhibit ALS or AHAS enzyme, which is vital for the biosynthesis of branched-chain amino acids isoleucine, leucine, and valine [32]. In general, these are broad-spectrum, post-emergence herbicides used for controlling weeds in a variety of crops like wheat and soybean. However, few ALS inhibitors, such as trifloxysulfuron, are also used as pre-emergence options to control weeds. Currently, resistance to ALS inhibitors is reported in 161 weeds globally [13]. TSR caused by single amino acid substitutions has been reported in most of these ALS-inhibitor-resistant weeds. Until recently, detection of these mutations have highlighted more importance on identifying TSR mechanisms compared to NTSR, even though they can co-exist in the same population [33,34]. However, the identification of plants lacking mutations in the ALS domain and surviving herbicide application has led researchers to focus on elucidating the NTSR mechanisms.

Enhanced metabolism conferring resistance to ALS inhibitors has been documented in several grass and broadleaf weeds, such as barnyard grass [35], common waterhemp [36], Palmer amaranth [37], rice barnyard grass [38,39], rigid brome [40], short awn foxtail (*Alopecurus aequalis*) [41,42], and water chickweed (*Myosoton aquaticum*) [33]. Numerous studies have also elucidated the molecular basis of metabolic resistance to ALS inhibitors. Though genes involved in metabolic resistance can be different depending on the weed species and history of herbicide application [43], most of these studies have predominantly identified multiple *CYP450* genes that are either constitutively expressed or upregulated following ALS inhibitor application [38,41,44]. For example, the mechanism of mesosulfuron-methyl resistance in short awn foxtail was studied and two *CYP450* genes, i.e., CYP94A1 and CYP71A4, were identified to be constitutively overexpressed in the resistant plants [41]. In a similar study, two *CYP450* genes, i.e., *CYP81A12* and *CYP81A21*, were identified as candidate genes conferring resistance to bensulfuron-methyl and penoxsulam in rice barnyard grass [38]. Several *CYP450* genes mediating NTSR to ALS inhibitors have been identified in water chickweed [44], ryegrass [45], flixweed [46], and blackgrass [18,47]. In addition to CYP450s, involvement of GSTs, GTs, and ATP-binding cassette (ABC) transporters have also been reported [42,44–46]. For instance, in ALS-inhibitor-resistant water chickweed, four genes—including three *CYP450s* (having homology to *CYP734A1*, *CYP76C1*, and *CYP86B1*) and an *ABC* transporter (having homology to *ABCC10*)—were identified as being highly expressed in all resistant plants [44]. Another commonly used procedure to test the CYP450 mediated metabolic resistance to ALS inhibitors has been the increase in sensitivity upon pre-treatment with CYP450 inhibitors, such as PBO, phorate, and malathion. Such increased sensitivity was observed in rigid ryegrass [48], short awn foxtail [41], Palmer amaranth [37], common waterhemp [36], barnyard grass [35], and rigid brome [40]. Malathion application also reversed 2,4-D-induced protection against chlorsulfuron in susceptible rigid ryegrass from Australia, suggesting the involvement of CYP450s in metabolizing chlorsulfuron [29].

3.3. Synthetic Auxinic Herbicides

Synthetic auxinic herbicides (SAH) are known to mimic the natural plant hormone, indole 3-acetic acid (IAA) [49]. These auxin analogs are mostly used for controlling broadleaf weeds in monocot crops, except quinclorac and quinmerac, which are known to have some grass activity [50]. Despite being introduced as early as 1945, the evolution of resistance to SAH has been slow, and so far, 39 weeds are reported to have developed resistance [13]. In the majority of weeds, NTSR mechanism(s) via (i) reduced uptake, (ii) decreased translocation, and (iii) increased metabolism has been documented. The reduced uptake of SAH is often affected by the properties of the leaf cuticle or other structural barriers that prevent absorption of the herbicide into mesophyll after herbicide application [51]. However, reduced uptake is a minor mechanism and has been shown to impart resistance in fewer weeds, such as ground ivy [51] and prickly lettuce [52]. However, reduced translocation resulting in the decreased movement of SAH to the site of action is common. Such a reduction in translocation was reported in several weed species, such as wild radish [53], oriental mustard [54], corn poppy [55], and prickly lettuce [52]. For instance, reduced translocation was observed in oriental mustard where approximately 77% of 2,4-D (2,4-dichlorophenoxyacetic acid) was retained in the treated leaves of resistant plants compared to 32% in susceptible plants at 72 h after treatment (HAT) [54]. In another study, the application of auxin efflux inhibitors 1-naphthylphthalamic (NPA) and 2,3,5-triiodobenzoic acid (TIBA) via roots of 2,4-D-susceptible wild radish plants significantly inhibited the translocation of 2,4-D out of the treated leaves, mimicking the 2,4-D-resistant wild radish [53]. Application of the same inhibitors did not affect the translocation of 2,4-D in the resistant biotype, suggesting alteration of the activity of ABC-transporters present in the plasma-membrane that usually facilitate the long-distance transport of 2,4-D [53]. MCPA (2-methyl-4-chlorophenoxyacetic acid) resistance in wild radish from Australia has been attributed to the rapid translocation to the roots [56]. At 48 HAT, a significantly lower amount of MCPA was recovered from resistant plants compared to susceptible plants, suggesting a possible root exudation of MCPA out of the plants [56]. Rapid metabolism of SAH is another major NTSR mechanism reported in several dicot weed species, where similar to the naturally tolerant monocot species, detoxification of herbicides occurs via ring-hydroxylation followed by conjugation, mediated by CYP450s. Such rapid detoxification of SAH has been reported in common waterhemp [36,57] and corn poppy [58]. In 2,4-D-resistant common waterhemp, 2,4-D was found to metabolize at a much faster rate compared to the susceptible plants, resulting in a lower metabolic half-life of 2,4-D [57]. In two 2,4-D-resistant corn poppy populations from Spain, enhanced metabolism was reported [58]. Two hydroxy metabolites were detected in the roots and shoots of the resistant plants, but not in the susceptible plants, suggesting a possible enhanced metabolism of herbicide due to CYP450-based hydroxylation in resistant plants [58]. Increased sensitivity of SAH-resistant biotypes was observed when pre-treated with CYP450-inhibitor malathion followed by herbicide application [36,57,58].

3.4. Photosystem II (PS-II)-Inhibitors

PS-II inhibitors act by competitively binding to the plastoquinone binding site (Q_B) on the D1 protein in the PS-II complex of the chloroplast [59]. This blockage disrupts photosynthesis since plastoquinone is vital for the electron transfer from PS-II to PS-I, and for generating nicotinamide adenine dinucleotide phosphate (NADPH) and ATP. The D1 protein is encoded by chloroplastic *psbA* gene, and hence, TSR to PS-II inhibitors can only be inherited maternally [60]. So far, 74 weed species have been reported to evolve resistance to PS-II inhibitors globally, via both TSR and NTSR mechanisms [13]. TSR to PS-II inhibition as a result of point mutations in the Q_B binding site has been reported in several weeds, such as kochia [61] and wild radish [62].

NTSR to PS-II inhibitors have been documented in annual bluegrass [63], common ragweed [64], common waterhemp [65,66], Palmer amaranth [67,68], and wild radish [62]. In the majority of these cases, the metabolism of PS-II inhibitors was catalyzed by the enhanced activity of GST enzymes [69] and/or CYP450 enzymes [70]. For example, atrazine-resistant Palmer amaranth from Kansas was found to conjugate atrazine 24 times faster than the susceptible plants via enhanced GST-activity [67]. Similarly, the enhanced metabolism of atrazine was found in two common waterhemp populations from Illinois [10] and Nebraska [65]. In the atrazine-resistant common waterhemp from Nebraska, at 6 HAT, approximately 92% of the atrazine was found to be conjugated by GSTs, whereas 92% of atrazine was still retained as a parent compound in susceptible plants [65]. Involvement of a phi-class GST, i.e., AtuGSTF2, was identified in mediating atrazine resistance in common waterhemp from Illinois [66]. Application of GST-inhibitors like 4-chloro-7-nitro-1,2,3-benzoxadiazole (NBD-cl) prior to atrazine application in resistant common waterhemp has resulted in greater sensitivity to atrazine [10]. Similarly, pre-treatment with a CYP450 inhibitor, 1-aminobenzotriazole, has shown increased sensitivity to resistant rigid ryegrass to simazine [70]. Apart from increased metabolism, reduced absorption and translocation can also impart PS-II-inhibitor resistance. Reduced absorption, translocation and increased metabolism of atrazine were observed in a PS-II-inhibitor (atrazine, diuron, semicarbazone)-resistant annual bluegrass population where known mutations in the *psbA* gene were lacking [63].

3.5. Enolpyruvyl Shikimate-3-Phosphate Synthase (EPSPS)-Inhibitors

Glyphosate, a non-selective, broad-spectrum herbicide, inhibits EPSPS in the shikimate pathway by acting as a transition state analog of phosphoenolpyruvate (PEP), which is a substrate for EPSPS [71]. The shikimate pathway produces the aromatic amino acids tryptophan, tyrosine, and phenylalanine, which are vital for plant growth and development. Additionally, glyphosate can increase carbon flow to the shikimate pathway, resulting in a shortage of carbon for other essential pathways [72]. Currently, there are 44 weeds reported to have evolved resistance to glyphosate [13]. Many of these resistance cases are either by alteration in the target (*EPSPS*) gene [73,74] or amplification and over-expression of the target gene [75–77].

Reduced translocation of glyphosate to meristematic sinks has been reported as the most common NTSR mechanism [78–80]. This mechanism has been reported in Palmer amaranth [81–83], horseweed [84], hairy fleabane [84], Italian ryegrass [85], rigid ryegrass [74,86], common waterhemp [73], Johnsongrass [87,88], sourgrass [89], and giant ragweed [90]. Reduction in translocation has been attributed to the evolution of a transporter that sequesters glyphosate inside the plant vacuole, thus preventing it from reaching the chloroplast [78]. In glyphosate-resistant horseweed, more (>85%) glyphosate was present in the vacuole of the resistant compared to only 15% in the susceptible plants [80]. Such sequestration was irreversible at least up to several days following the glyphosate application [80,91,92]. Similar modified sub-cellular distribution of glyphosate was found in glyphosate-resistant *Conyza bonariensis* [79]. ABC transporter proteins have been proposed to sequester glyphosate via active glyphosate transport [78,93]. Through GS-FLX 454 pyrosequencing, an increased expression of several *ABC* transporter genes was found in glyphosate-resistant horseweed following glyphosate application [94]. However, the role of a specific gene or gene family mediating glyphosate sequestration resulting in NTSR is still unknown. The reduced uptake of glyphosate has also been shown to impart low-level resistance to glyphosate in several weeds, such as Palmer amaranth [81,83], sourgrass [89], and Johnsongrass [87]. An enhanced metabolism of glyphosate is another mechanism responsible for high tolerance to glyphosate and was observed in some biotypes of sourgrass [95], horseweed [96], and *Echinochloa colona* [97]. In sourgrass biotypes with a greater tolerance to glyphosate, more than 56% of glyphosate was metabolized into aminomethylphosphonic acid (AMPA), glyoxylate, and sarcosine at 168 HAT compared to 10% in susceptible biotypes [95]. Similar, rapid metabolism of glyphosate was observed in resistant horseweed populations where almost 100% of glyphosate metabolized into glyoxylate, sarcosine, and AMPA within 96 HAT [96]. Through RNA-seq analysis, an aldo-keto reductase (AKR) contig with greater expression and activity, exhibiting metabolic resistance to glyphosate was identified in an *Echinochloa colona* population from Australia [97]. Further, glyphosate metabolites, such as AMPA and glyoxalate, were also found in *Escherichia coli* expressing the *AKR* gene (*EcAKR4-1*), which was similar to the resistant *Echinochloa colona* plants [97].

3.6. 4-Hydroxyphenylpyruvate Dioxygenase (HPPD) and Carotenoid-Inhibitors

HPPD enzyme is required for catalyzing the conversion of 4-hydroxyphenylpyruvate (HPP) to 2,5-dihydroxyphenylacetate (homogentisate) to produce plastoquinone and tocopherols in the carotenoid biosynthesis pathway [98]. Plastoquinone is essential for the electron transfer from PS-II to PS-I and also as a co-factor of phytoene desaturase (PDS), required for carotenoid formation [99]. Hence, most of these herbicides inhibit carotenoid formation, ultimately resulting in photo-oxidation of chlorophyll molecules and lipid peroxidation of the cell membranes by forming singlet oxygen [100]. So far, NTSR to HPPD-inhibitors has been reported most often. However, TSR caused by higher *HPPD* gene and protein expression has been reported in mesotrione-resistant Palmer amaranth [99].

Enhanced metabolism was the primary NTSR mechanism reported in HPPD-inhibitors-resistant Palmer amaranth [99,101,102], common waterhemp [10,36,103–105], and rice barnyardgrass [106]. In mesotrione-resistant Palmer amaranth, more than 90% of mesotrione was metabolized at 24 HAT [99]. Rapid 4-hydroxylation, followed by glycosylation and a higher expression of certain CYP450 enzymes, were identified in tembotrione-resistant Palmer amaranth compared to the susceptible biotype [102]. Similarly, increased mesotrione metabolism via 4-hydroxylation of the dione ring was reported in mesotrione-resistant common waterhemp from Nebraska [105]. Pre-treatment with CYP450-inhibitors has been shown to increase the sensitivity of resistant common waterhemp populations to mesotrione [10,36,103]. CYP450s belonging to sub-family CYP81A were found to impart metabolic-resistance to clomazone in rice barnyard grass [106]. *Arabidopsis* lines transformed with CYP81A12, CYP81A21, CYP81A15, and CYP81A24 showed increased resistance to clomazone, indicating their involvement in metabolizing clomazone [106].

3.7. Protoporphyrinogen Oxidase (PPO)-Inhibitors

PPO inhibitors are important broad-spectrum herbicides that growers can use to control weeds resistant to ALS inhibitors and glyphosate [107]. PPO inhibitors impede the PPO enzyme, which is required for catalyzing the conversion of protoporphyrinogen IX to protoporphyrin IX in the last step of plant heme and chlorophyll biosynthesis [108,109]. The inhibition of the PPO enzyme leads to the accumulation of intermediates in the cytosol, which are photoactively oxidized, ultimately leading to the production of highly reactive oxygen species (ROS). These ROS attack lipids and proteins in cell membranes and cause lipid peroxidation, leading to plant death [110]. So far, 13 weeds have evolved resistance to PPO inhibitors [13]. The most common mechanism of resistance reported was a single amino acid deletion (Gly210) or substitution in the *PPX2* (e.g., Arg98Leu) gene [107].

NTSR-based PPO inhibitor resistance has been reported in two pigweed species: Palmer amaranth [111] and common waterhemp [112]. A common waterhemp population resistant to carfentrazone-ethyl lacked the presence of known mutations previously reported to confer TSR to PPO inhibitors but exhibited increased sensitivity to carfentrazone-methyl when pre-treated with malathion [112]. This suggests the possible involvement of CYP450 in conferring resistance to carfentrazone-ethyl in common waterhemp [112]. Similarly, the absence of known mutations was reported in fomesafen-resistant Palmer amaranth [111]. The same population was further found to be cross-resistant to flumioxazin, acifluorfen, and saflufenacil [113]. Involvement of both CYP450s and GSTs was reported to mediate fomesafen resistance in Palmer amaranth due to increased sensitivity when pre-treated with malathion or NBD-cl [111,113].

3.8. Photosystem I (PS-I)-Inhibitors

Paraquat is a non-selective, fast-acting herbicide that diverts electrons from PS-I, leading to the inhibition of photosynthesis. Paraquat accepts a single electron in order to generate a reduced cation radical, that on further reaction with oxygen, generates a superoxide ion [114]. In the presence of light, paraquat catalyzes the production of superoxide ions, which eventually form hydroxy radicals and result in lipid peroxidation [114,115]. Only a single NTSR mechanism, i.e., reduced translocation via vacuolar sequestration, has been reported in horseweed [84], hairy fleabane [84], rigid ryegrass [116], and Italian ryegrass [117]. The amount of paraquat present in the leaf protoplast of resistant and susceptible rigid ryegrass was estimated and a 2–3-fold higher retention of herbicide was found in leaves of resistant plants, indicating possible sequestration of the herbicide in vacuoles [116]. A similar mechanism of paraquat resistance was reported in Italian ryegrass [117]. However, the molecular basis of such sequestration is still unknown.

3.9. *Very Long Chain Fatty Acid (VLCFA) Synthesis-Inhibitors*

VLCFA inhibiting herbicides are known to affect several steps in the elongation of the carbon chain of very long chain fatty acids [118]. VLCFAs are required for the formation of triacylglycerols, waxes, phospholipids, and complex sphingolipids, which are essential for various plant functions [119]. For instance, phospholipids and sphingolipids are required during cell division, as well as for maintaining membrane trafficking pathways [119–121]. The ever-increasing occurrence of ACCase and ALS-inhibitor resistance has led growers to rely more on VLCFA inhibitors, which are important pre-emergence herbicides for controlling grasses.

NTSR mechanisms to very long chain fatty acid (VLCFA) inhibitors have been studied in ryegrass populations from Australia [122], U.K., and France [123], as well as Palmer amaranth from Arkansas [124]. Metabolic resistance to pyroxasulfone was reported in a rigid ryegrass population from Australia, where approximately 88% of parental applied herbicides was metabolized within 24 HAT [122]. Pyroxasulfone metabolites were formed via glutathione conjugation and two putative *GST* genes were 2–6-fold constitutively overexpressed in resistant ryegrass populations [122]. Interestingly, continuous sub-optimal application of pyroxasulfone can rapidly select for resistant biotypes of rigid ryegrass [125]. Application of herbicides at sub-optimal doses can favor the selection of several minor resistance alleles and facilitate their accumulation in cross-pollinating weeds like rigid ryegrass, leading to rapid evolution of polygenic NTSR [126,127]. Moreover, such selection can promote the evolution of cross-resistance to other VLCFA inhibitors like prosulfocarb and triallate [128]. In flufenacet-resistant ryegrass populations from the U.K. and France, enhanced metabolism due to conjugation by GSTs was reported [123]. Similarly, enhanced metabolism of flufenacet was observed in resistant blackgrass [129]. Recently, resistance to *s*-metolachlor was documented in Palmer amaranth from Arkansas. A reduction in root growth was observed when the resistant accessions were kept for germination in agar solution containing a GST-inhibitor, NBD-cl, indicating the role of GSTs in mediating the resistance [124].

4. Influence of Environmental Factors on NTSR Mechanisms

NTSR mechanisms can be affected by changes in environmental conditions [130,131]. Factors like the mode of action of herbicides and the physiology of weed species can contribute significantly in the alteration of NTSR under different environmental conditions. Both herbicide-resistant and susceptible biotypes have shown increased and decreased tolerance to herbicides under different environmental conditions. NTSR mechanisms are suspected to develop gradually in response to biotic and abiotic stresses, which enable them to adapt to the growing conditions [132]. Changing environmental conditions can seriously affect herbicide efficacy and favor the selection of more tolerant biotypes. Hence, information on how NTSR mechanisms behave in different weed species in varying environmental conditions can be very crucial to mitigate such selection. The effect of environmental conditions, such as temperature, CO_2 concentration, and relative humidity, on NTSR mechanisms has been studied in several weed species. Altered temperature regimes were shown to impact herbicide absorption [133–136], translocation and sequestration [133,134,136–141], and metabolism [30,139,142]. For example, exposure to high temperature was found to reduce pinoxaden sensitivity of grass species, such as *Brachypodium hybridum* [142]. A higher level of inactive glucose-conjugated pinoxaden metabolites in these grasses was observed under high- versus low-temperature conditions [142]. Such increased detoxification of pinoxaden in *Brachypodium hybridum* was associated with a possible increased enzymatic activity of reactive oxygen species scavengers [30]. In another study, the suppression of vacuolar sequestration of glyphosate at low temperature was found to result in the increased sensitivity of glyphosate-resistant horseweed [137]. Recently, poor control of kochia due to the reduced absorption of glyphosate and translocation of dicamba at high temperature was reported [133]. Decreased efficacy of mesotrione in controlling Palmer amaranth due to rapid metabolism at high temperature was documented [139]. Conversely, improved 2,4-D efficacy at high temperature due to increased translocation was found in both glyphosate-resistant and -susceptible common and giant ragweed [134]. Additionally, increased absorption of glyphosate in common ragweed, and increased absorption and translocation of glyphosate in giant ragweed, improved glyphosate efficacy at high temperature [134]. Apart from temperature, changes in CO_2 concentrations can affect herbicide translocation and sequestration in weeds [140,143]. In horseweed and lambsquarter, an increase in glyphosate translocation was found at elevated CO_2 levels and increased temperature, leading to a reduced glyphosate sensitivity [140]. Similarly, in *Echinochloa colona*, high CO_2 and increased temperature reduced the efficacy of cyhalofop-butyl by decreasing translocation [143]. Altering relative humidity (RH) can also affect herbicide translocation in pigweeds [144]. For instance, in Palmer amaranth, redroot pigweed, and common waterhemp, a higher translocation of glufosinate was found at high, compared to low, RH [144]. These findings indicate the need to further elucidate and evaluate the impact of environmental conditions on the sensitivity of weeds to herbicides to slow evolution of herbicide resistance.

5. Coexistence of TSR and NTSR Mechanisms

Numerous cases of TSR have been reported as a result of single nucleotide polymorphisms resulting in amino acid substitutions in the target sites of several herbicides, such as PS-I, ALS, and ACCase inhibitors, and glyphosate. Novel mechanisms of TSR in weeds, such as the deletion of codons, leading to PPO-inhibitor resistance in common waterhemp and Palmer amaranth [107,145], as well as gene amplification-based resistance to glyphosate [75,77] and ACCase inhibitors [146], were reported. These findings will help in identifying the precise genetic elements involved in the evolution of TSR in resistant weeds. Similarly, recent advances have also helped to understand the physiological and molecular basis of NTSR in weed species. Interestingly, several cases of coexistence of these mechanisms have been reported. For instance, ALS-inhibitor-resistant water chickweed [33] and barnyard grass [147], ACCase-inhibitor-resistant *Vulpia bromoides* [148], Italian ryegrass [19], short awn foxtail [41], PS-II inhibitor-resistant wild radish [62], EPSPS-inhibitor-resistant rigid ryegrass [74], common waterhemp [73], HPPD-inhibitor-resistant Palmer amaranth [99], and microtubule inhibitor-resistant rigid ryegrass [34] have been identified with TSR and NTSR mechanisms in the same populations. Therefore, if TSR mechanisms are found to contribute to herbicide resistance, it is also necessary to investigate the NTSR mechanisms and vice versa. Although deciphering both types of resistances in the same weed species can be challenging, understanding the coexistence of TSR and NTSR mechanisms for the same herbicide is valuable for predicting possible cross-resistance to other herbicides, thereby assisting in management of resistance.

6. Conclusions and Prospects

Weed management practices can impact the mechanisms by which weeds evolve resistance to herbicides. Additionally, a key aspect in predicting the evolutionary trajectory of herbicide-resistant traits is understanding the mechanism(s) of herbicide resistance. More importantly, understanding the relationship between the weed management tactics and their influence on evolutionary mechanisms (TSR or NTSR) that determine herbicide resistance in weed species will help to formulate effective future strategies to manage these increasingly problematic weeds. It has been proposed that a lower rate of herbicides result in the evolution of polygenic traits, whereas high herbicide doses may favor monogenic target-site-based resistances [127]. Likewise, Gressel [149] proposed that suboptimal herbicide use rates can result in the evolution of polygenic herbicide resistance. Understanding the type of selection pressure leading to the evolution of NTSR mechanisms, especially metabolic resistance, is extremely valuable and needed to sustain the limited herbicide portfolio and develop integrated weed management strategies.

Funding: This research received no external funding.

Conflicts of Interest: The authors declare no conflict of interest.

References

1. Powles, S.B.; Yu, Q. Evolution in action: Plants resistant to herbicides. *Annu. Rev. Plant Biol.* **2010**, *61*, 317–347. [CrossRef] [PubMed]
2. Jasieniuk, M.; Brûlé-Babel, A.L.; Morrison, I.N. The evolution and genetics of herbicide resistance in weeds. *Weed Sci.* **1996**, *44*, 176–193. [CrossRef]
3. Délye, C.; Jaseniuk, M.; Le Corre, V. Deciphering the evolution of herbicide resistance in weeds. *Trends Genet.* **2013**, *29*, 649–658. [CrossRef] [PubMed]
4. Sammons, R.D.; Gaines, T.A. Glyphosate resistance: State of knowledge. *Pest Manag. Sci.* **2014**, *70*, 1367–1377. [CrossRef] [PubMed]
5. Devine, M.D.; Eberlein, C.V. Physiological, biochemical and molecular aspects of herbicide resistance based on altered target sites. In *Herbicide Activity: Toxicology, Biochemistry and Molecular Biology*; Roe, R.M., Burton, J.D., Kuhr, R.J., Eds.; IOS: Amsterdam, The Netherlands, 1997; pp. 159–185.
6. Preston, C. Inheritance and linkage of metabolism-based herbicide cross-resistance in rigid ryegrass (*Lolium rigidum* Gaud.). *Weed Sci.* **2003**, *51*, 4–12. [CrossRef]
7. Yamada, T.; Kambara, Y.; Imaishi, H.; Ohkawa, H. Molecular cloning of novel Cytochrome P450 species induced by chemical treatments in cultured tobacco cells. *Pestic. Biochem. Physiol.* **2000**, *68*, 11–25. [CrossRef]
8. Gion, K.; Inui, H.; Takakuma, K.; Yamada, T.; Kambara, Y.; Nakai, S.; Fijuwara, H.; Miyamura, T.; Imaishi, H.; Ohkawa, H. Molecular mechanisms of herbicide-inducible gene expression of tobacco CYP71AH11 metabolizing the herbicide chlorotoluron. *Pestic. Biochem. Physiol.* **2014**, *108*, 49–57. [CrossRef]
9. Huffman, J.; Hausman, N.; Hager, A.; Riechers, D.; Tranel, P. Genetics and inheritance of nontarget-site resistances to atrazine and mesotrione in a waterhemp (*Amaranthus tuberculatus*) population from Illinois. *Weed Sci.* **2015**, *63*, 799–809. [CrossRef]
10. Ma, R.; Kaundun, S.S.; Tranel, P.J.; Riggins, C.W.; McGinness, D.L.; Hager, A.G.; Hawkes, T.; McIndie, E.; Riechers, D.E. Distinct detoxification mechanisms confer resistance to mesotrione and atrazine in a population of waterhemp. *Plant Physiol.* **2013**, *163*, 363–377. [CrossRef]
11. Kaundun, S.S. Resistance to acetyl-CoA carboxylase-inhibiting herbicides. *Pest. Manag. Sci.* **2014**, *70*, 1405–1417. [CrossRef]
12. Devine, M.D. Acetyl-CoA carboxy lase inhibitors. In *Herbicide Classes in Development*; Boger, P., Wakabayashi, K., Hirai, K., Eds.; Springer: Berlin, Germany, 2002; pp. 103–137.
13. Heap, I. The International Survey of Herbicide Resistant Weeds. Available online: www.weedscience.org (accessed on 7 August 2019).
14. Yu, L.P.; Kim, Y.S.; Tong, T. Mechanism for the inhibition of the carboxyltransferase domain of acetyl-coenzyme A carboxylase by pinoxaden. *Proc. Natl. Acad. Sci. USA* **2010**, *107*, 22072–22077. [CrossRef] [PubMed]
15. Zhang, Z.Y.H.; Shen, Y.; Tong, L. Crystal structure of the carboxyltransferase domain of acetyl-coenzyme A carboxylase. *Science* **2003**, *299*, 2064–2067. [CrossRef] [PubMed]
16. Zhao, N.; Ge, L.; Yan, Y.; Bai, S.; Wang, D.; Liu, W.; Wang, J. Trp-1999-Ser mutation of acetyl CoA carboxylase and cytochrome P450s-involved metabolism confer resistance to fenoxaprop-*P*-ethyl in *Polypogon fugax*. *Pest Manag. Sci.* **2019**. [CrossRef] [PubMed]
17. Huan, Z.; Zhang, H.; Hou, Z.; Zhang, S.; Zhang, Y.; Liu, W.T.; Bi, Y.; WANG, J. Resistance level and metabolism of barnyard-grass (*Echinochloa crusgalli* (L.) Beauv.) populations to quizalofop-p-ethyl in Heilongjiang province. *China Agric. Sci. China* **2011**, *10*, 1914–1922. [CrossRef]
18. Délye, C.; Gardin, J.A.C.; Bucansaud, K.; Chauvel, B.; Pett, C. Non-target-site-based resistance should be the centre of attention for herbicide resistance research: *Alopecurus myosuroides* as an illustration. *Weed Res.* **2011**, *51*, 433–437. [CrossRef]
19. Kaundan, S.S. An aspartate to glycine change in the carboxyl transferase domain of acetyl CoA carboxylase and non-target-site mechanism(s) confer resistance to ACCase inhibitor herbicides in a *Lolium multiflorum* population. *Pest Manag. Sci.* **2010**, *66*, 1249–1256. [CrossRef] [PubMed]
20. Kaundan, S.S.; Bailly, G.C.; Dale, R.P.; Hutchings, S.J.; McIndoe, E. A Novel W1999S Mutation and Non-Target Site Resistance Impact on Acetyl-CoA Carboxylase Inhibiting Herbicides to Varying Degrees in a UK *Lolium multiflorum* Population. *PLoS ONE* **2013**, *8*, e58012. [CrossRef]
21. Feng, Y.; Gao, Y.; Zhang, Y.; Dong, L.; Li, J. Mechanisms of resistance to pyroxsulam and ACCase inhibitors in Japanese foxtail (*Alopecurus japonicus*). *Weed Sci.* **2016**, *64*, 695–704. [CrossRef]

22. Busi, R.; Vila-Aiub, M.M.; Powles, S.B. Genetic control of a cytochrome P450 metabolism-based herbicide resistance mechanism in *Lolium rigidum*. *Heredity* **2011**, *106*, 817–824. [CrossRef]
23. Han, H.; Yu, Q.; Owen, M.J.; Cawthray, G.R.; Powles, S.B. Widespread occurrence of both metabolic and target-site herbicide resistance mechanisms in *Lolium rigidum* populations. *Pest Manag. Sci.* **2016**, *72*, 255–263. [CrossRef]
24. Gaines, T.A.; Lorentz, L.; Figge, A.; Herrmann, J.; Maiwald, F.; Ott, M.C.; Han, H.; Busi, R.; Yu, Q.; Powles, S.B. RNA-Seq transcriptome analysis to identify genes involved in metabolism-based diclofop resistance in *Lolium rigidum*. *Plant J.* **2014**, *78*, 865–876. [CrossRef] [PubMed]
25. Ahmad-Hamdani, M.S.; Yu, Q.; Han, H.; Cawthray, G.R.; Wang, S.F.; Powles, S.B. Herbicide resistance endowed by enhanced rates of herbicide metabolism in wild oat (*Avena* spp.). *Weed Sci.* **2013**, *61*, 55–62. [CrossRef]
26. Yu, Q.; Han, H.; Cawthray, G.R.; Wang, S.F.; Powles, S.B. Enhanced rates of herbicide metabolism in low herbicide-dose selected resistant *Lolium rigidum*. *Plant Cell Environ.* **2013**, *36*, 818–827. [CrossRef] [PubMed]
27. Kreuz, K.; Fonné-Pfister, R. Herbicide-insecticide interaction in maize: Malathion inhibits cytochrome P450-dependent primisulfuron metabolism. *Pestic. Biochem. Phys.* **1992**, *43*, 232–240. [CrossRef]
28. Baerg, R.J.; Barrett, M.; Polge, N.D. Insecticide and Insecticide Metabolite Interactions with Cytochrome P450 Mediated Activities in Maize. *Pestic. Biol. Phys.* **1996**, *55*, 10–20. [CrossRef]
29. Han, H.; Yu, Q.; Cawthray, G.R.; Powles, S.B. Enhanced herbicide metabolism induced by 2,4-D in herbicide susceptible *Lolium rigidum* provides protection against diclofop-methyl. *Pest Manag. Sci.* **2013**, *69*, 996–1000. [CrossRef]
30. Matzrafi, M.; Shaar-Moshe, L.; Rubin, B.; Peleg, Z. unraveling the transcriptional basis of temperature-dependent pinoxaden resistance in *Brachypodium hybridum*. *Front. Plant Sci.* **2017**, *8*, 1064. [CrossRef]
31. Tranel, P.; Wright, T.R. Resistance of Weeds to ALS-Inhibiting Herbicides: What Have We Learned? *Weed Sci.* **2002**, *50*, 700–712. [CrossRef]
32. Umbarger, H.E. Amino acid biosynthesis and its regulation. *Annu. Rev. Biochem.* **1978**, *47*, 533–606. [CrossRef]
33. Bai, S.; Zhang, F.; Li, Z.; Wang, H.; Wang, Q.; Wang, J.; Liu, W.; Bai, L. Target-site and non-target-site-based resistance to tribenuron-methyl in multiply-resistant *Myosoton aquaticum* L. *Pestic. Biochem. Phys.* **2019**, *155*, 8–14. [CrossRef]
34. Chen, J.; Chu, Z.; Han, H.; Goggin, D.E.; Yu, Q.; Sayer, C.; Powles, S.B. A Val-202-Phe α-tubulin mutation and enhanced metabolism confer dinitroaniline resistance in a single *Lolium rigidum* population. *Pest. Manag. Sci.* **2019**. [CrossRef]
35. Riar, D.S.; Norsworthy, J.K.; Bond, J.A.; Bararpour, M.T.; Wilson, M.J.; Scott, R.C. Resistance of *Echinochloa crus-galli* populations to acetolactate synthase-inhibiting herbicides. *Int. J. Agron.* **2012**, *2012*, 893953. [CrossRef]
36. Shergill, L.S.; Bish, M.D.; Jugulam, M.; Bradley, K.W. Molecular and physiological characterization of six-way resistance in an *Amaranthus tuberculatus* var. rudis biotype from Missouri. *Pest. Manag. Sci.* **2018**, *74*, 2688–2698. [CrossRef]
37. Nakka, S.; Thompson, C.R.; Peterson, D.E.; Jugulam, M. Target site–based and non-target site based resistance to ALS inhibitors in Palmer amaranth (*Amaranthus palmeri*). *Weed Sci.* **2017**, *65*, 681–689. [CrossRef]
38. Iwakami, S.; Endo, M.; Saika, H.; Okuno, J.; Nakamura, N.; Yokoyama, M.; Watanabe, H.; Toki, S.; Uchino, A.; Inamura, T. Cytochrome P450 CYP81A12 and CYP81A21 are associated with resistance to two acetolactate synthase inhibitors in *Echinochloa phyllopogon*. *Plant Physiol.* **2014**, *165*, 618–629. [CrossRef]
39. Iwakami, S.; Kamidate, Y.; Yamaguchi, T.; Ishizaka, M.; Endo, M.; Suda, H.; Nagai, K.; Sunohara, Y.; Toki, S.; Uchino, A.; et al. CYP81A P450s are involved in concomitant cross-resistance to acetolactate synthase and acetyl-CoA carboxylase herbicides in *Echinochloa phyllopogon*. *New Phytol.* **2019**, *221*, 2112–2122. [CrossRef]
40. Owen, M.J.; Goggin, D.E.; Powles, S.B. Non-target-site-based resistance to ALS-inhibiting herbicides in six *Bromus rigidus* populations from Western Australian cropping fields. *Pest Manag. Sci.* **2012**, *68*, 1077–1082. [CrossRef]
41. Zhao, N.; Yan, Y.; Ge, L.; Zhu, B.; Liu, W.; Wang, J. Target site mutations and cytochrome P450s confer resistance to fenoxaprop-*P*-ethyl and mesosulfuron-methyl in *Alopecurus aequalis*. *Pest Manag. Sci.* **2019**, *75*, 204–214. [CrossRef]

42. Zhao, N.; Li, W.; Bai, S.; Guo, W.; Yuan, G.; Wang, F.; Liu, W.; Wang, J. Transcriptome profiling to identify genes involved in mesosulfuron-methyl resistance in *Alopecurus aequalis*. *Front. Plant Sci.* **2017**, *8*, 1391. [CrossRef]
43. Yu, Q.; Powles, S. Metabolism-Based Herbicide Resistance and Cross-Resistance in Crop Weeds: A Threat to Herbicide Sustainability and Global Crop Production. *Plant Physiol.* **2014**, *166*, 1106–1118. [CrossRef]
44. Liu, W.; Bai, S.; Zhao, S.; Li, W.; Zhang, L.; Wang, J. Non-target site-based resistance to tribenuron-methyl and essential involved genes in *Myosoton aquaticum* (L.). *BMC Plant Biol.* **2018**, *18*, 225. [CrossRef]
45. Duhoux, A.; Carrere, S.; Gouzy, J.; Bonin, L.; Delye, C. RNA-Seq analysis of rye-grass transcriptomic response to an herbicide inhibiting acetolactate-synthase identifies transcripts linked to non-target-site-based resistance. *Plant Mol. Biol.* **2015**, *87*, 473–487. [CrossRef]
46. Yang, Q.; Deng, W.; Li, X.; Yu, Q.; Bai, L.; Zheng, M. Target-site and non-target-site based resistance to the herbicide tribenuron-methyl in flixweed (*Descurainia sophia* L.). *BMC Genom.* **2016**, *17*, 551. [CrossRef]
47. Gardin, J.A.C.; Gouzy, J.; Carrere, S.; Delye, C. ALOMY base, a resource to investigate non-target-site-based resistance to herbicides inhibiting acetolactate-synthase (ALS) in the major grass weed *Alopecurus myosuroides* (black-grass). *BMC Genom.* **2015**, *16*, 590. [CrossRef]
48. Busi, R.; Gaines, T.A.; Powles, S.B. Phorate can reverse P450 metabolism-based herbicide resistance in *Lolium rigidum*. *Pest Manag. Sci.* **2017**, *73*, 410–417. [CrossRef]
49. Busi, R.; Goggin, D.E.; Heap, I.M.; Horak, M.J.; Jugulam, M.; Masters, R.A.; Napier, R.M.; Riar, D.S.; Satchivi, N.M.; Torra, J.; et al. Weed resistance to synthetic auxin herbicides. *Pest Manag. Sci.* **2018**, *74*, 2265–2276. [CrossRef]
50. Grossmann, K. Auxin herbicides: Current status of mechanism and mode of action. *Pest. Manag. Sci.* **2010**, *66*, 113–120. [CrossRef]
51. Kohler, E.A.; Throssell, C.S.; Reicher, Z.J. 2,4-D rate response, absorption, and translocation of two ground ivy (*Glechoma hederacea*) populations. *Weed Technol.* **2004**, *18*, 917–923. [CrossRef]
52. Riar, D.S.; Burke, I.C.; Yenish, J.P.; Bell, J.; Gill, K. Inheritance and physiological basis for 2,4-D resistance in prickly lettuce (*Lactuca serriola* L.). *J Agric. Food Chem.* **2011**, *59*, 9417–9423. [CrossRef]
53. Goggin, D.E.; Cawthray, C.B. 2,4-D resistance in wild radish: Reduced herbicide translocation via inhibition of cellular transport. *J. Exp. Bot.* **2016**, *67*, 3223–3323. [CrossRef]
54. Dang, H.T.; Malone, J.M.; Boutsalis, P.; Krishnan, M.; Gill, G.; Preston, C. Reduced translocation in 2,4-D-resistant oriental mustard populations (*Sisymbrium orientale* L.) from Australia. *Pest Manag. Sci.* **2018**, *74*, 1524–1532. [CrossRef]
55. Rey-Caballero, J.; Menendez, J.; Gine-Bordonaba, J.; Salas, M.; Alcantara, R.; Torra, J. Unravelling the resistance mechanisms to 2,4-D (2,4-dichlorophenoxyacetic acid) in corn poppy (*Papaver rhoeas*). *Pestic Biochem. Phys.* **2016**, *133*, 67–72. [CrossRef]
56. Jugulam, M.; DiMeo, N.; Veldhuis, L.J.; Walsh, M.; Hall, J.C. Investigation of MCPA (4-Chloro-2-ethylphenoxyacetate) resistance in wild radish (*Raphanus raphanistrum* L.). *J. Agric. Food Chem.* **2013**, *61*, 12516–12521. [CrossRef]
57. Figueiredo, M.R.; Leibhart, L.J.; Reicher, Z.J.; Tranel, P.J.; Nissen, S.J.; Westra, P.; Bernards, M.L.; Kruger, G.R.; Gaines, T.A.; Jugulam, M. Metabolism of 2,4-dichlorophenoxyacetic acid contributes to resistance in a common waterhemp (*Amaranthus tuberculatus*) population. *Pest Manag. Sci.* **2018**, *74*, 2356–2362. [CrossRef]
58. Torra, J.; Rojano-Delgado, A.M.; Rey-Caballero, J.; Royo-Esnal, A.; Salas, M.L.; De Prado, R. Enhanced 2,4-D metabolism in two resistant *Papaver rhoeas* populations from Spain. *Front. Plant Sci.* **2017**, *8*, 1584. [CrossRef]
59. Gronwald, J. Resistance to photosystem II inhibiting herbicides. In *Herbicide Resistance in Plants: Biology and Biochemistry*; Lewis Publishers: Boca Raton, FL, USA, 1994; pp. 27–60.
60. Goloubinoff, P.; Edelman, M.; Hallick, R.B. Chloroplast-coded atrazine resistance in Solanum nigrum: *psbA* loci from susceptible and resistant biotypes are isogenic except for a single codon change. *Nucleic Acids Res.* **1984**, *12*, 9489–9496. [CrossRef]
61. Varanasi, V.K.; Godar, A.S.; Currie, R.S.; Dille, A.J.; Thompson, C.R.; Stahlman, P.W.; Jugulam, M. Field-evolved resistance to four modes of action of herbicides in a single kochia (*Kochia scoparia* L. Schrad.) population. *Pest. Manag. Sci.* **2015**, *71*, 1207–1212. [CrossRef]
62. Lu, H.; Yu, Q.; Han, H.; Owen, M.J.; Powles, S.B. Metribuzin resistance in a wild radish (Raphanus raphanistrum) population via both psbA Gene mutation and enhanced metabolism. *J. Agric. Food Chem.* **2019**, *67*, 1353–1359. [CrossRef]

63. Svyantek, A.; Aldahir, P.; Chen, S.; Flessner, M.; McCullough, P.; Sidhu, S.; McElroy, J. Target and nontarget resistance mechanisms induce annual bluegrass (*Poa annua*) resistance to atrazine, amicarbazone, and diuron. *Weed Tech.* **2016**, *30*, 773–782. [CrossRef]
64. Simard, M.J.; Laforest, M.; Soufiane, B.; Benoit, D.K.; Tardif, F.J. Linuron-resistant common ragweed (*Ambrosia artemisiifolia*) populations in Québec carrot fields: Presence and distribution of target site and non-target site resistant biotypes. *Can. J. Plant Sci.* **2018**, *98*, 345–352. [CrossRef]
65. Vennapusa, A.R.; Faleco, F.; Vieira, B.; Samuelson, S.; Kruger, G.R.; Werle, R.; Jugulam, M. Prevalence and mechanism of atrazine resistance in waterhemp (*Amaranthus tuberculatus*) from Nebraska. *Weed Sci.* **2018**, *66*, 595–602. [CrossRef]
66. Evans, A.F.; O'Brien, S.R.; Ma, R.; Hager, A.G.; Riggins, C.W.; Lambert, K.N.; Riechers, D.E. Biochemical characterization of metabolism-based atrazine resistance in *Amaranthus tuberculatus* and identification of an expressed GST associated with resistance. *Plant Biotechnol. J.* **2017**, *15*, 1238–1249. [CrossRef]
67. Nakka, S.; Godar, A.; Thompson, C.R.; Peteron, D.E.; Juglam, M. Rapid detoxification via Glutathione S-transferase (GST)-conjugation confers high level of atrazine resistance in Palmer amaranth (*Amaranthus palmeri*). *Pest Manag. Sci.* **2017**, *73*, 2236–2243. [CrossRef]
68. Chahal, P.S.; Jugulam, M.; Jhala, A.J. Mechanism of atrazine resistance in atrazine- and HPPD inhibitor-resistant Palmer amaranth (*Amaranthus palmeri* S. Watson) from Nebraska. *Can. J. Plant Sci.* **2019**. [CrossRef]
69. Anderson, M.P.; Gronwald, J.W. Atrazine resistance in a velvetleaf (*Abutilon theophrasti*) biotype due to enhanced glutathione S-transferase activity. *Plant Physiol.* **1991**, *96*, 104–109. [CrossRef]
70. Burnet, M.W.M.; Loveys, B.R.; Holtum, J.A.M.; Powles, S.B. Increased detoxification is a mechanism of simazine resistance in *Lolium rigidum*. *Pestic. Biochem. Physiol.* **1993**, *46*, 207–218. [CrossRef]
71. Duke, S.O.; Powles, S.B. Glyphosate: A once-in-a-century herbicide. *Pest. Manag. Sci.* **2008**, *64*, 319–325. [CrossRef]
72. Siehl, D.L. Inhibitors of EPSP synthase, glutamine synthase and histidine synthesis. In *Herbicide Activity: Toxicology, Biochemistry and Molecular Biology*; Roe, R.M., Burton, J.D., Kuhr, R.J., Eds.; IOS Press: Amsterdam, The Netherlands, 1997; pp. 37–67.
73. Nandula, V.K.; Ray, J.D.; Ribeiro, D.N.; Pan, Z.; Reddy, K.N. Glyphosate resistance in tall waterhemp (*Amaranthus tuberculatus*) from Mississippi is due to both altered target-site and nontarget-site mechanisms. *Weed Sci.* **2013**, *61*, 374–383. [CrossRef]
74. Bostamam, Y.; Malone, J.M.; Dolman, F.C.; Boutsalis, P.; Preston, C. Rigid ryegrass (*Lolium rigidum*) populations containing a target site mutation in EPSPS and reduced glyphosate translocation are more resistant to glyphosate. *Weed Sci.* **2012**, *60*, 474–479. [CrossRef]
75. Jugulam, M.; Niehues, K.; Godar, A.S.; Koo, D.H.; Danilova, T.; Friebe, B.; Sehgal, S.; Varanasi, V.K.; Wiersma, A.; Westra, P.; et al. Tandem amplification of a chromosomal segment harboring 5-Enolpyruvylshikimate-3-Phosphate synthase locus confers glyphosate resistance in *Kochia scoparia*. *Plant Physiol.* **2014**, *166*, 1200–1207. [CrossRef]
76. Koo, D.H.; Molin, W.T.; Saski, C.A.; Jiang, J.; Putta, K.; Jugulam, M.; Friebe, B.; Gill, B.S. Extra-chromosomal circular DNA (eccDNA) based amplification and transmission of herbicide resistance in crop weed *Amaranthus palmeri*. *Proc. Natl. Acad. Sci. USA* **2018**, *115*, 3332–3337. [CrossRef] [PubMed]
77. Gaines, T.A.; Zhang, W.; Wang, D.; Bukun, B.; Chisholm, S.T.; Shaner, D.L.; Nissen, S.J.; Patzoldt, W.L.; Tranel, P.J.; Culpepper, A.S.; et al. Gene amplification confers glyphosate resistance in *Amaranthus palmeri*. *Proc. Natl. Acad. Sci. USA* **2010**, *107*, 1029–1034. [CrossRef] [PubMed]
78. Shaner, D.L. The role of translocation as a mechanism of resistance to glyphosate. *Weed Sci.* **2009**, *57*, 118–123. [CrossRef]
79. Kleinman, Z.; Rubin, B. Non-target-site glyphosate resistance in *Conyza bonariensis* is based on modified subcellular distribution of the herbicide. *Pest Manag. Sci.* **2017**, *73*, 246–253. [CrossRef]
80. Ge, X.; d'Avignon, D.A.; Acerman, J.J.H.; Sammons, R.D. Rapid vacuolar sequestration: The horseweed glyphosate resistance mechanism. *Pest Manag. Sci.* **2010**, *66*, 345–348. [CrossRef]
81. Palma-Bautista, C.; Torra, J.; Garcia, M.J.; Bracamonte, E.; Rojano-Delgado, A.M.; la Cruz, R.A.; Prado, R.D. Reduced absorption and impaired translocation endows glyphosate resistance in *Amaranthus palmeri* harvested in glyphosate-resistant soybean from Argentina. *J. Agric. Food Chem.* **2019**, *67*, 1052–1060. [CrossRef]

82. Dominguez-Valenzuela, J.A.; Gherekhloo, J.; Fernandez-Moreno, P.T.; Cruz-Hipolito, H.E.; Cruz, R.A.L.; Sanchez-Gonalez, E.; Prado, R.D. First confirmation and characterization of target and non-target site re sistance to glyphosate in Palmer amaranth (*Amaranthus palmeri*) from Mexico. *Plant Physiol. Biochem.* **2017**, *115*, 212–218. [CrossRef]
83. Nandula, V.K.; Reddy, K.N.; Koger, C.H.; Poston, D.H.; Rimando, A.M.; Duke, S.O.; Bond, J.A.; Ribeiro, D.N. Multiple resistance to glyphosate and pyrithiobac in Palmer Amaranth (*Amaranthus palmeri*) from Mississippi and response to flumiclorac. *Weed Sci.* **2012**, *60*, 179–188. [CrossRef]
84. Moretti, M.L.; Hanson, B.D. Reduced translocation is involved in resistance to glyphosate and paraquat in *Conyza bonariensis* and *Conyza canadensis* from California. *Weed Res.* **2016**, *57*, 25–34. [CrossRef]
85. Gonzalez-Torralva, F.; Gil-Humanes, J.; Barro, F.; Brants, I.; Prado, R.D. Target site mutation and reduced translocation are present in a glyphosate-resistant *Lolium multiflorum* Lam. biotype from Spain. *Plant Physiol. Biochem.* **2012**, *58*, 16–22. [CrossRef]
86. Adu-Yeboah, P.; Malone, J.M.; Preston, C. Reduced glyphosate translocation in two glyphosate-resistant populations of rigid ryegrass (*Lolium rigidum*) from fence lines in South Australia. *Weed Sci.* **2014**, *62*, 4–10. [CrossRef]
87. Vila-Aiub, M.; Balbi, M.C.; Distefano, A.J.; Fernandez, L.; Hopp, E.; Yu, Q.; Powles, S.B. Glyphosate resistance in perennial *Sorghum halepense* (Johnsongrass), endowed by reduced glyphosate translocation and leaf uptake. *Pest Manag. Sci.* **2012**, *68*, 430–436. [CrossRef] [PubMed]
88. Riar, D.S.; Norsworthy, J.K.; Johnson, D.B.; Scott, R.C.; Bagavathiannan, M. Glyphosate resistance in a johnsongrass (*Sorghum halepense*) biotype from Arkansas. *Weed Sci.* **2011**, *59*, 299–304. [CrossRef]
89. Carvalho, L.B.D.; Alves, P.L.D.C.A.; Gonzalez-Torralva, F.; Cruz-Hipolito, H.E.; Rojano-Delgado, A.M.; Prado, R.D.; Gil-Humanes, J.; Barro, F.; Castro, M.D.L.D. Pool of resistance mechanisms to glyphosate in *Digitaria insularis*. *J. Agric. Food Chem.* **2012**, *602*, 615–622. [CrossRef]
90. Nandula, V.K.; Wright, A.A.; Horn, C.R.V.; Westra, P.; Reddy, K.N. Glyphosate resistance in giant ragweed (*Ambrosia trifida* L.) from Mississippi is partly due to reduced translocation. *Am. J. Plant Sci.* **2015**, *6*, 2104–2113. [CrossRef]
91. Ge, X.; d'Avignon, D.A.; Acerman, J.J.H.; Collavo, A.; Sattin, M.; Ostrander, E.L.; Hall, E.L.; Sammons, R.D.; Preston, C. Vacuolar glyphosate-sequestration correlates with glyphosate resistance in ryegrass (*Lolium spp.*) from Australia, South America, and Europe: A ^{31}P NMR investigation. *J. Agric. Food Chem.* **2012**, *605*, 1243–1250. [CrossRef]
92. Ge, X.; d'Avignon, D.A.; Acerman, J.J.H.; Sammons, R.D. In vivo ^{31}P-nuclear magnetic resonance studies of glyphosate uptake, vacuolar sequestration, and tonoplast pump activity in glyphosate-resistant horseweed. *Plant Physiol.* **2014**, *168*, 377. [CrossRef]
93. Yuan, J.S.; Abercrombie, L.G.; Cao, Y.; Halfhill, M.D.; Zhou, X.; Peng, Y. Functional genomics analysis of glyphosate resistance in *Conyza canadensis* (horseweed). *Weed Sci.* **2010**, *58*, 109–117. [CrossRef]
94. Peng, Y.; Abercrombie, L.L.G.; Yuan, J.S.; Riggins, C.W.; Sammons, R.D.; Tranel, P.J.; Stewart, C.N.S., Jr. Characterization of the horseweed (*Conyza canadensis*) transcriptome using GS-FLX 454 pyrosequencing and its application for expression analysis of candidate non-target herbicide resistance genes. *Pest Manag. Sci.* **2010**, *66*, 1053–1062. [CrossRef]
95. Carvalho, L.B.D.; Rojano-Delgado, A.M.; Alves, P.L.D.C.; Prado, R.D. Differential content of glyphosate and its metabolites in *Digitaria insularis* biotypes. *Commun. Plant Sci.* **2013**, *3*, 17–20. [CrossRef]
96. González-Torralva, F.; Rojano-Delgado, A.M.; de Castro, M.D.L.; Mülleder, N.; Prado, R.D. Two non-target mechanisms are involved in glyphosate-resistant horseweed (*Conyza canadensis* L. Cronq.) biotypes. *J. Plant Physiol.* **2012**, *169*, 1673–1679. [CrossRef] [PubMed]
97. Pan, L.; Yu, Q.; Han, H.; Mao, L.; Nyporko, A.; Fan, L.; Bai, L.; Stephen, B.; Powles, S.B. AKR metabolizes glyphosate and confers resistance in *Echinochloa colona*. *Plant Physiol.* **2019**. [CrossRef] [PubMed]
98. Beaudegnies, R.; Edmunds, A.J.F.; Fraser, T.E.M.; Hall, R.G.; Hawkes, T.R.; Mitchell, G.; Schaetzer, J.; Wendeborn, S.; Wibley, J. Herbicidal 4-hydroxyphenylpyruvate dioxygenase inhibitors—a review of the triketone chemistry story from a Syngenta perspective. *Bioorg. Med. Chem.* **2009**, *17*, 4134–4152. [CrossRef] [PubMed]
99. Nakka, S.; Godar, A.; Wani, P.S.; Thompson, C.R.; Peterson, D.E.; Roelofs, J.; Jugulam, M. Physiological and molecular characterization of hydroxyphenylpyruvate dioxygenase (HPPD)-inhibitor resistance in Palmer amaranth (*Amaranthus palmeri* S.Wats.). *Front. Plant Sci.* **2017**, *8*, 555. [CrossRef] [PubMed]

100. Siefermann, H.D. The light harvesting and protective functions of carotenoids in photosynthetic membranes. *Physiol. Plant.* **1987**, *69*, 561–568. [CrossRef]
101. Thompson, C.R.; Peterson, D.E.; Lally, N.G. Characterization of HPPD-resistant Palmer amaranth. In Proceedings of the Weed Science Society of America Annual Meetings, Waikoloa, Hawaii, 6–9 February 2012.
102. Küpper, A.; Peter, F.; Zöllner, P.; Lorentz, L.; Tranel, P.J.; Beffa, R.; Gaines, T.A. Tembotrione detoxification in 4-hydroxyphenylpyruvate dioxygenase (HPPD) inhibitor-resistant Palmer amaranth (*Amaranthus palmeri* S. Wats.). *Pest. Manag. Sci.* **2018**, *74*, 2325–2334. [CrossRef]
103. Oliveira, M.C.; Gaines, T.A.; Dayan, F.E.; Patterson, E.L.; Jhala, A.J.; Knezevic, S.Z. Reversing resistance to tembotrione in an *Amaranthus tuberculatus* (var. rudis) population from Nebraska, USA with cytochrome P450 inhibitors. *Pest Manag. Sci.* **2017**, *74*, 2296–2305. [CrossRef]
104. Hausman, N.E.; Singh, S.; Tranel, P.J.; Riechers, D.E.; Kaundun, S.S.; Polge, N.D.; Thomas, D.A.; Hager, A.G. Resistance to HPPD-inhibiting herbicides in a population of waterhemp (*Amaranthus tuberculatus*) from Illinois, United States. *Pest Manag. Sci.* **2011**, *67*, 258–261. [CrossRef]
105. Kaundun, S.S.; Hutchings, S.J.; Dale, R.P.; Howell, A.; Morris, J.A.; Kramer, V.C.; Shivrain, V.K.; Mcindoe, E. Mechanism of resistance to mesotrione in an *Amaranthus tuberculatus* population from Nebraska, USA. *PLoS ONE* **2017**, *12*, e0180095. [CrossRef]
106. Guo, F.; Iwakami, S.; Yamaguchi, T.; Uchino, A.; Sunohara, Y.; Matsumoto, H. Role of CYP81A cytochrome P450s in clomazone metabolism in *Echinochloa phyllopogon*. *Plant Sci.* **2019**, *283*, 31–328. [CrossRef]
107. Salas, R.A.; Burgos, N.R.; Tranel, P.J.; Singh, S.; Glasgow, L.; Scott, R.C.; Nichols, R.L. Resistance to PPO-inhibiting herbicide in Palmer amaranth from Arkansas. *Pest Manag. Sci.* **2016**, *72*, 864–869. [CrossRef] [PubMed]
108. Deybach, J.C.; Dasilva, V.; Grandchamp, B.; Nordmann, Y. The mitochondrial location of protoporphyrinogen oxidase. *Eur. J. Biochem.* **1985**, *149*, 431–435. [CrossRef] [PubMed]
109. Beale, S.I.; Weinstein, J.D. Tetrapyrrole metabolism in photosynthetic organisms. In *Biosynthesis of Heme and Chlorophylls*; Dailey, H.A., Ed.; McGraw-Hill: New York, NY, USA, 1990; pp. 287–391.
110. Sherman, T.D.; Becerril, J.M.; Matsumoto, H.; Duke, M.V.; Jacobs, J.M.; Jacobs, N.J.; Duke, S.O. Physiological basis for differential sensitivities of plant species to protoporphyrinogen oxidase-inhibiting herbicides. *Plant Physiol.* **1991**, *97*, 280–287. [CrossRef] [PubMed]
111. Varanasi, V.K.; Brabham, C.; Norsworthy, J.K. Confirmation and characterization of non–target site resistance to fomesafen in Palmer amaranth (*Amaranthus palmeri*). *Weed Sci.* **2018**, *66*, 702–709. [CrossRef]
112. Obenland, A.A.; Ma, R.; O'Brien, S.R.; Lygin, A.V.; Riechers, D.E. Carfentrazone-ethyl resistance in an *Amaranthus tuberculatus* population is not mediated by amino acid alterations in the PPO2 protein. *PLoS ONE* **2019**, *14*, e0215431. [CrossRef]
113. Varanasi, V.; Brabham, C.; Korres, N.; Norsworthy, J. Nontarget site resistance in Palmer amaranth [*Amaranthus palmeri* (S.) Wats.] confers cross-resistance to protoporphyrinogen oxidase-inhibiting herbicides. *Weed Technol.* **2019**, *33*, 349–354. [CrossRef]
114. Hawkes, T.R. Mechanisms of resistance to paraquat in plants. *Pest. Manag. Sci.* **2014**, *70*, 1316–1323. [CrossRef]
115. Gutteridge, J.M.C. Lipid peroxidation initiated by superoxide-dependent hydroxyl radicals using complexed iron and hydrogen peroxide. *FEBS Lett.* **1984**, *172*, 245–249. [CrossRef]
116. Yu, Q.; Huang, S.; Powles, S. Direct measurement of paraquat in leaf protoplasts indicates vacuolar paraquat sequestration as a resistance mechanism in *Lolium rigidum*. *Pestic. Biochem. Physiol.* **2010**, *98*, 104–109. [CrossRef]
117. Brunharo, C.A.C.G.; Hanson, B.D. Vacuolar sequestration of paraquat is involved in the resistance mechanism in *Lolium perenne* L. spp. *multiflorum*. *Front. Plant Sci.* **2017**, *8*, 1485. [CrossRef]
118. Tanetani, Y.; Kaku, K.; Kawai, K.; Fujioka, T.; Shimizu, T. Action mechanism of a novel herbicide, pyroxasulfone. *Pestic. Biochem. Phys.* **2009**, *95*, 47–55. [CrossRef]
119. Busi, R. Resistance to herbicides inhibiting the biosynthesis of very-long-chain fatty acids. *Pest. Manag. Sci.* **2014**, *70*, 1378–1384. [CrossRef] [PubMed]
120. Lechelt-Kunze, C.; Meissner, R.C.; Drewes, M.; Tietjen, K. Flufenacet herbicide treatment phenocopies the fiddlehead mutant in Arabidopsis thaliana. *Pest Manag. Sci.* **2003**, *59*, 847–856. [CrossRef] [PubMed]

121. Markham, J.E.; Molino, D.; Gissot, L.; Bellec, Y.; Hématy, K.; Marion, J.; Belcram, K.; Palauqui, J.C.; Satiat-JeuneMaitre, B.; Faure, J.D. Sphingolipids containing very-long-chain fatty acids define a secretory pathway for specific polar plasma membrane protein targeting in Arabidopsis. *Plant Cell* **2011**, *23*, 2362–2378. [CrossRef] [PubMed]
122. Busi, R.; Porri, A.; Gaines, T.A.; Powles, S.B. Pyroxasulfone resistance in *Lolium rigidum* is metabolism-based. *Pestic. Biochem. Physiol.* **2018**, *148*, 74–78. [CrossRef]
123. Dücker, R.; Zölner, P.; Lümmen, P.; Ries, S.; Collavo, A.; Beffa, R. Glutathione transferase plays a major role in flufenacet resistance of ryegrass (*Lolium* spp.) field populations. *Pest Manag. Sci.* **2019**. [CrossRef]
124. Brabham, C.; Norsworthy, J.; Houston, M.; Varanasi, V.; Barber, T. Confirmation of S-metolachlor resistance in Palmer amaranth (*Amaranthus palmeri*). *Weed Technol.* **2019**, 1–7. [CrossRef]
125. Busi, R.; Gaines, T.A.; Walsh, M.J.; Powles, S.B. Understanding the potential for resistance evolution to the new herbicide pyroxasulfone: Field selection at high doses versus recurrent selection at low doses. *Weed Res.* **2012**, *52*, 489–499. [CrossRef]
126. Manalil, S.; Busi, R.; Renton, M.; Powles, S. Rapid Evolution of herbicide resistance by low herbicide dosages. *Weed Sci.* **2011**, *59*, 210–217. [CrossRef]
127. Neve, P.; Powles, S. High survival frequencies at low herbicide use rates in populations of *Lolium rigidum* result in rapid evolution of herbicide. *Herdity* **2005**, *95*, 485–492. [CrossRef]
128. Busi, R.; Powles, S.B. Cross-resistance to prosulfocarb and triallate in pyroxasulfone-resistant *Lolium rigidum*. *Pest. Manag. Sci.* **2013**, *69*, 1379–1384. [CrossRef]
129. Dücker, R.; Zöllner, P.; Parcharidou, E.; Ries, S.; Lorentz, L.; Beffa, R. Enhanced metabolism causes reduced flufenacet sensitivity in black-grass (*Alopecurus myosuroides* Huds.) field populations. *Pest. Manag. Sci.* **2019**. [CrossRef]
130. Ramesh, K.; Matloob, A.; Aslam, F.; Florentine, S.K.; Chauhan, B.S. Weeds in a changing climate: Vulnerabilities, consequences, and implications for future weed management. *Front. Plant Sci.* **2017**, *8*, 95. [CrossRef] [PubMed]
131. Matzrafi, M. Climate change exacerbates pest damage through reduced pesticide efficacy. *Pest. Manag. Sci.* **2019**, *75*, 9–13. [CrossRef] [PubMed]
132. Cramer, G.R.; Urano, K.; Delrot, S.; Pezzotti, M.; Shinozaki, K. Effects of abiotic stress on plants: A systems biology perspective. *BMC Plant Biol.* **2011**, *11*, 163. [CrossRef]
133. Ou, J.; Stahlman, P.W.; Jugulam, M. Reduced absorption of glyphosate and decreased translocation of dicamba contribute to poor control of kochia (*Kochia scoparia*) at high temperature. *Pest Manag. Sci.* **2018**, *74*, 1134–1142. [CrossRef]
134. Ganie, Z.A.; Jugulam, M.; Jhala, A.J. Temperature influences efficacy, absorption, and translocation of 2,4-D or glyphosate in glyphosate-resistant and glyphosate-susceptible common ragweed (*Ambrosia artemisiifolia*) and giant ragweed (*Ambrosia trifida*). *Weed Sci.* **2017**, *65*, 588–602. [CrossRef]
135. Hatterman-Valenti, H.; Pitty, A.; Owen, M. Environmental effects on velvetleaf (*Abutilon theophrasti*) epicuticular wax deposition and herbicide absorption. *Weed Sci.* **2011**, *59*, 14–21. [CrossRef]
136. Palma-Bautista, C.; la Cruz, R.A.; Rojano-Delgado, A.M.; Dellaferrera, I.; Domínguez-Martínez, P.A.; Prado, R.D. Low temperatures enhance the absorption and translocation of ^{14}C-glyphosate in glyphosate-resistant *Conyza sumatrensis*. *J. Plant Physiol.* **2019**, *240*, 153009. [CrossRef]
137. Ge, X.; d'Avignon, D.A.; Acerman, J.J.H.; Duncan, B.; Spaur, M.B.; Sammons, R.D. Glyphosate-resistant horseweed made sensitive to glyphosate: Low-temperature suppression of glyphosate vacuolar sequestration revealed by ^{31}P NMR. *Pest Manag. Sci.* **2011**, *67*, 1215–1221. [CrossRef]
138. Purba, E.; Preston, C.; Powles, S.B. The mechanism of resistance to paraquat is strongly temperature dependent in resistant *Hordeum leporinum* Link and *H. glaucum* Steud. *Planta* **1995**, *196*, 464. [CrossRef]
139. Godar, A.S.; Varanasi, V.K.; Nakka, S.; Prasad, P.V.; Thompson, C.R.; Mithila, J. Physiological and molecular mechanisms of differential sensitivity of palmer amaranth (*Amaranthus palmeri*) to mesotrione at varying growth temperatures. *PLoS ONE* **2015**, *10*, e0126731. [CrossRef] [PubMed]
140. Matzrafi, M.; Brunharo, C.; Tehranchian, P.; Hanson, B.D.; Jasieniuk, M. Increased temperatures and elevated CO_2 levels reduce the sensitivity of *Conyza canadensis* and *Chenopodium album* to glyphosate. *Sci. Rep.* **2019**, *9*, 2228. [CrossRef]
141. Ghanizadeh, H.; Harrington, K.C.; James, T.K. Glyphosate-resistant population of *Lolium perenne* loses resistance at winter temperatures. *N. Z. J. Agric. Res.* **2015**, *58*, 423–431. [CrossRef]

142. Matzrafi, M.; Seiwert, B.; Reemtsma, T.; Rbin, B.; Peleg, Z. Climate change increases the risk of herbicide-resistant weeds due to enhanced detoxification. *Planta* **2016**, *244*, 1217–1227. [CrossRef] [PubMed]
143. Refatti, J.P.; de Avila, L.A.; Camargo, E.R.; Ziska, L.H.; Oliviera, C.; Salas-Perez, R.; Rouse, C.E.; Roma-Burgos, N. High [CO_2] and temperature increase resistance to cyhalofop-butyl in multiple-resistant *Echinochloa colona*. *Front. Plant Sci.* **2019**. [CrossRef]
144. Coetzer, E.; Al-Khatib, K.; Loughin, T. Glufosinate efficacy, absorption, and translocation in amaranth as affected by relative humidity and temperature. *Weed Sci.* **2001**, *49*, 8–13. [CrossRef]
145. Patzoldt, W.L.; Hager, A.G.; McCormick, J.S.; Tranel, P.J. A codon deletion confers resistance to herbicides inhibiting protoporphyrinogen oxidase. *Proc. Natl. Acad. Sci. USA* **2006**, *103*, 12329–12334. [CrossRef]
146. Laforest, M.; Soufiane, B.; Simard, M.; Obeid, K.; Page, E.; Nurse, R.E. Acetyl-CoA carboxylase overexpression in herbicide-resistant large crabgrass (*Digitaria sanguinalis*). *Pest. Manag. Sci.* **2017**, *73*, 2227–2235. [CrossRef]
147. Fang, J.; Zhang, Y.; Liu, T.; Yan, B.; Li, J.; Dong, L. Target-Site and metabolic resistance mechanisms to penoxsulam in barnyardgrass (*Echinochloa crus-galli* (L.) P. Beauv). *J. Agric. Food Chem.* **2019**, *67*, 8085–8095. [CrossRef]
148. Yu, Q.; Friesen, L.J.S.; Zhang, X.Q.; Powles, S.B. Tolerance to acetolactate synthase and acetyl-coenzyme A carboxylase inhibiting herbicides in *Vulpia bromoides* is conferred by two co-existing resistance mechanisms. *Pestic. Biochem. Phys.* **2004**, *78*, 21–30. [CrossRef]
149. Gressel, J. Low pesticide rates may hasten the evolution of resistance by increasing mutation frequencies. *Pest. Manag. Sci.* **2011**, *67*, 253–257. [CrossRef] [PubMed]

plants

Review

Population Genomic Approaches for Weed Science

Sara L. Martin [1,*], Jean-Sebastien Parent [1], Martin Laforest [2], Eric Page [3], Julia M. Kreiner [4] and Tracey James [1]

1 Ottawa Research and Development Centre, Agriculture and Agri-Food Canada, Ottawa, ON K1A 0C6, Canada; jean-sebastien.parent@canada.ca (J.-S.P.); tracey.james@canada.ca (T.J.)
2 Saint-Jean-sur-Richelieu Research and Development Centre, Agriculture and Agri-Food Canada, Saint-Jean-sur-Richelieu, QC J3B 3E6, Canada; Martin.Laforest@Canada.ca
3 Harrow Research and Development Centre, Agriculture and Agri-Food Canada, Harrow, ON N0R 1G0, Canada; Eric.Page@Canada.ca
4 Department of Ecology and Evolutionary Biology, University of Toronto, Toronto, ON M5S 3B2, Canada; julia.kreiner@mail.utoronto.ca
* Correspondence: Sara.Martin@Canada.ca; Tel.: +1-613-715-5406

Received: 16 August 2019; Accepted: 14 September 2019; Published: 19 September 2019

Abstract: Genomic approaches are opening avenues for understanding all aspects of biological life, especially as they begin to be applied to multiple individuals and populations. However, these approaches typically depend on the availability of a sequenced genome for the species of interest. While the number of genomes being sequenced is exploding, one group that has lagged behind are weeds. Although the power of genomic approaches for weed science has been recognized, what is needed to implement these approaches is unfamiliar to many weed scientists. In this review we attempt to address this problem by providing a primer on genome sequencing and provide examples of how genomics can help answer key questions in weed science such as: (1) Where do agricultural weeds come from; (2) what genes underlie herbicide resistance; and, more speculatively, (3) can we alter weed populations to make them easier to control? This review is intended as an introduction to orient weed scientists who are thinking about initiating genome sequencing projects to better understand weed populations, to highlight recent publications that illustrate the potential for these methods, and to provide direction to key tools and literature that will facilitate the development and execution of weed genomic projects.

Keywords: weeds; genomics; plant genome assembly; non-target site resistance; population genomics; genome scans; population genetics

1. Introduction

Biology is currently in the midst of a revolution caused by the advances in sequencing technology that allow us to examine genomes in detail [1]. Genomic information promises new insights for understanding the biology, evolutionary history, and adaptive potential in ways that were recently out of reach for laboratories studying organisms with genomes larger than model organisms (e.g., *Arabidopsis thaliana* (L.) Heyn. 135 Mb) [2–5]. Additionally, genomics at the population or species level are now possible in some species and will likely become practical for the majority of organisms in the near term. The huge potential of these advances has been exploited by some disciplines, such as those investigating bacteria [6,7], viruses [8,9] or humans [10–13], with greater alacrity than others. Notably, however, progress adopting genomic methods has been slow in weed science despite recognition of the power of these methods [14,15].

There are numerous impediments to a greater use of genomics in weed science. One of these elements is the lack of chromosome level reference genome sequences for weeds, as the majority of sequencing efforts have been focused on crops. Genome sequences are foundational for many approaches and the relatively early availability of the human genome sequence [16], model organisms such as *Arabidopsis* [17] and many crop species [18] have been essential to the rapid progress in applying genomic approaches to a wide range of disciplines. This issue has been noted by the weed science community and efforts such as the International Weed Science Consortium have been initiated [14]. However, an additional impediment to using these rapidly developing and expanding set of techniques is a lack of familiarity among weed scientists. As a result, our aim here is to provide a brief primer and introductory "how to guide" and "why would you guide" relevant to weed science. We briefly review *de novo* genome assembly and annotation as these methods are often fundamental for further work. Then we focus on how genomic approaches can be used to answer three key questions: 1) Where do agricultural weeds come from and why are they weedy; 2) what genes underlie herbicide resistance (HR); and, more speculatively, 3) can we alter weed populations to become easier to control? We highlight what resources would be needed for success and provide illustrative examples from both weed science and the broader scientific literature.

2. Developing Weed Genome Sequences as a Fundamental Tool

While some genomic approaches do not require a draft genome for the species of interest, the majority of techniques do, or benefit from the availability of at least a rough draft. Sequencing plant genomes is easier than ever before with the decreasing cost of sequencing and the increasing ease with which tools such as genome assembly programs can be installed and used. However, genome assembly remains a challenge that will require a significant investment of time and resources for the majority of weed species [19]. Here we provide a brief outline of how to approach a *de novo* genome sequencing project and provide an initial introduction to the steps required and some tools that could be used as a starting point. We do not attempt to provide a comprehensive list of resources or tools and in every case, there are often numerous alternatives that may be better suited to a particular weed species or easier to install in a specific computing environment. Further, new tools are continuously emerging (and older ones submerging) in this quickly evolving area. Various databases of these tools have been compiled such as omictools.com and bioinformaticssoftwareandtools.co.in. Valuably, a recent review by Jung et al. [20] is comprehensive with recommendations on the computational resources needed to complete these assemblies.

2.1. What Is a Draft Genome?

A draft genome of a plant species is a haploid representation of a portion of the total DNA and genes. As such, it is a simplified and limited representation of the total information contained in the genome of the individual sequenced. It will lack information on allelic variation and portions of the genome, especially repetitive elements and material near the centromeres [21]. A draft genome is comprised of a group, often a large group (Table 1), of contigs that vary in size and represent the portions of the genome assembled from overlapping and joining the smaller pieces provided by the sequencing reads and is often presented in a multi-fasta file. These contigs can be assembled into larger fragments, scaffolds. Finally, scaffolds can be assembled, ordered, and oriented into pseudomolecules. At the larger end, pseudomolecules may represent chromosomes, chromosome arms or smaller features such as the chloroplast's genome. In general, the fewer number of contigs an assembly has the better the assembly is considered. A metric used to compare continuity amongst genomes is the NG50 value. If one ordered all the contigs in an assembly from largest to smallest and added the length of each contig as you went down the list, the NG50 value would be the size of the contig when 50% of the species' expected genome size was reached [22]. The N50 value is similar and more frequently reported, but as the assembly size is used instead of the expected genome size, it can't be used to compare different assemblies even within species [22]. Drafts comprised of thousands of contigs can be sufficient for many purposes, including understanding evolutionary relationships among species, acting as a reference for studies of population biology, and for developing molecular identification tools. To understand fine-scale patterns of selection, however, a chromosomal level assembly is more desirable, allowing for the most detailed analysis and inferences that draw on correlated shifts in allele frequencies. In cases where a closely related species has been assembled to the chromosome level and chromosome number is conserved, this may, by assuming synteny (preserved order), be used to position scaffolds into pseudomolecules representing a first guess of what the genome may look like. However, this level of information has rarely been achieved for non-model, non-crop organisms.

2.2. Preparing and Assessing Plant Material

Important initial steps to help ensure the success of a project are assessment of the plant material to understand the species' genome size and composition and carefully considering the starting material including finding lower ploidy individuals or reducing heterozygosity through inbreeding or other genetic manipulations such as creating a doubled haploid.

It is preferable to know the size of the genome before the start of a sequencing project. Several databases have compiled information on the genome size and chromosome counts for plant species (see Rice et al. 2015 for a list of resources). A particularly useful resource for genome size information is the Plant DNA C-value Database (cvalues.science.kew.org) hosted by Kew Royal Botanic Gardens [23]. Similarly, chromosome counts are available from the Index to Plant Chromosome Numbers www.tropicos.org/project/ipcn hosted by the Missouri Botanical Garden and the Chromosome Counts Database ccdb.tau.ac.il [24].

In the absence of information from these sources, or in cases where multiple chromosome counts or DNA contents have been reported, analysis by flow cytometry can determine the DNA content of the material of interest [25–28]. This is relatively inexpensive and straight forward if you have access to a flow cytometer and can take as little as a week for an experienced laboratory. Fresh tissue is co-chopped in a buffer with the tissue of a species with known DNA content (internal standard), nuclei are stained with a fluorophore such as propidium iodide, and peaks in fluorescence are produced as a result of excitation by the flow cytometer's laser. Then the position of the sample's peak and the known standard are determined by analysis of the resulting histogram with appropriate software (e.g., [29]). The DNA content of the samples is then determined using these relative positions and the DNA content of the standard. Generally, at least three individuals should be tested and each analyzed with three technical replicates across three days. This provides the full 2C DNA content of the plant's nuclei in picograms. The 1C DNA content can then be calculated by dividing this value in half and convert to Mbp by multiplying by 978 Mbp/pg [30]. Difficulties with DNA content determination with flow cytometry typically centre around finding an extraction buffer that allows for the production of narrow peaks and low debris levels (coefficient of variation < 5), including enough nuclei in the sample peak (>1000), finding an appropriate standard, and understanding the data when it is complicated by extra peaks from contamination or endopolyploidy [31,32]. Methods using an external standard should not be used as they are less accurate. This information can be compared to DNA content and chromosome counts for species in the same genus to make educated guesses about the chromosome count for the material of interest, but a conclusive determination of chromosome number requires either the counting of chromosome spreads or the use of more advanced chromosome sorting techniques [33].

Table 1. Metrics of continuity and completion for weed genome assemblies available from GenBank. This list was compiled by search GenBank [34] in May 2019 for species included on one of the following five lists of weeds: 1) Species with herbicide resistance maintained at weedscience.org by Heap [35], 2) the United States Department of Agriculture's Federal Noxious Weed List [36], 3) Weeds of Nation Significance in Australia [37], 4) Weber and Gut's list of weeds spreading in Europe [38], or 5) the Canadian Weed Seed Order [39]. Year is the year the assembly was submitted to GenBank, and as with assembly level, coverage, sequencing technology used and assembly method were recorded from the assembly information page on GenBank. The number of contigs (greater than 500 bp long), assembled genome size, N50 and NG50 were determined by QUAST (v. 5.0.2). The number of BUSCOs that were (C)omplete, complete and (S)ingle-copy, complete and (D)uplicated, (F)ragmented, or (M)issing were determined using the eudicotyledons_odb10 set of 2121 conserved genes and BUSCO version 3.0.2. Where we could not locate a published for the genome, we have reported the lead author as listed as having submitted the genome to GenBank. Note that additional weed genomes may be available on CoGe, Phytozome, and the European nucleotide database.

Common Name	Latin Name	Year	Level of Assembly	No. of Contigs	Est. Genome Size (Mbp)	Assembled Size (Mbp)	N50	NG50	BUSCOS (Percentage of 2121 Genes)					Coverage	Sequencing Technology	Assembly Method	Reference or Lead Submitting Author
									C	S	D	F	M				
Milkweed	*Asclepias syriaca*	2017	Scaffold	221,885	411 [1]	237	2555	NA[2]	76	75	1	13	11	80.4	Illumina	Platanus SCUBAT	[40]
Winter Cress	*Barbarea vulgaris*	2016	Scaffold	7810	270	167	56,351	19,454	95	93	2	3	3	66.5	Illumina	Celera	[41]
Japanese Barberry	*Berberis thunbergii*	2018	Contig	11,815	1515 [1]	2241	397,058	654,137	88	30	57	3	9	104.8	PacBio	FALCON-Unzip	R. Bartaula
False Brome	*Brachypodium distachyon*	2018	Chromosome	11	355	271	59,130,575	59,130,575	80	76	4	6	15	9.4	ABI 3739	ARACHNE	[42]
Bird Rape	*Brassica rapa*	2017	Scaffold	70,673	485	386	3,737,062	2,395,810	98	80	18	1	1	212	Illumina PacBio	SOAPdenovo	[43]
Hemp	*Cannabis sativa*	2018	Chromosome	6653 [3]	820	892	60,968,100	62,039,859	88	72	16	4	7	79	PacBio	FALCON	[44]
Shepherd's Purse	*Capsella bursa-pastoris*	2017	Scaffold	8186	391 [1]	268	627,605	320,701	96	13	83	2	3	40	Illumina	Newbler Platanus	[45]
Horsetail Sheoak	*Casuarina equisetifolia* subsp. *incana*	2018	Scaffold	2936	340 [1]	301	1,020,118	894,734	97	93	4	1	2	546.9	Illumina PacBio	SOAPdenovo2 FALCON DISCOVAR	[46]
Swamp Oak	*Cauarina glauca*	2018	Scaffold	39,787	340	283	964,272	627,004	97	93	5	1	2	890	Illumina	SOAPdenovo	[47]
Mandarin Orange	*Citrus reticulata*	2018	Scaffold	67,725	460	344	1,376,405	577,147	98	96	2	1	1	200	Illumina	Platanus	[48]
Horseweed	*Conyza canadensis*	2014	Contig	20,075	335	326	20,748	20,226	66	44	22	10	24	350	Roche 454 Illumina PacBio	Newbler SOAPdenovo CLC NGS Cell	[49]
Jute Mallow	*Corchorus olitorius*	2017	Contig	52,373	450	377	16,573	13,050	93	90	3	3	4	47.7	Illumina	Newbler	[50]
Muskmelon	*Cucumis melo*	2012	Scaffold	10,823	450	375	4,428,067	3,741,400	94	92	2	2	4	13.5	Roche 454 Illumina	Newbler	[51]
Globe artichoke	*Cynara cardunculus*	2018	Chromosome	8283 [3]	1084	725	25,947,084	173,700	96	90	6	2	2	80	Illumina	AllPaths	[52]
Orchardgrass	*Dactylis glomerata*	2018	Scaffold	1,072,009	3327 [4]	840	3242	NA[2]	76	72	4	8	16	50	Illumina	SOAPdenovo	J. Li

Table 1. *Cont.*

Common Name	Latin Name	Year	Level of Assembly	No. of Contigs	Est. Genome Size (Mbp)	Assembled Size (Mbp)	N50	NG50	BUSCOS (Percentage of 2121 Genes) C	S	D	F	M	Coverage	Sequencing Technology	Assembly Method	Reference or Lead Submitting Author
Carrot	*Daucus carota* subsp. *sativus*	2016	Chromosome	4826 [3]	473	422	36,610,139	36,610,139	94	88	6	2	5	186	Roche 454 Illumina Sanger	SOAPdenovo GapCloser	[53]
Guinea yam	*Dioscorea rotundata*	2017	Chromosome	21	694 [1]	457	25,272,979	NA [2]	83	78	5	3	14	100	Illumina	Allpaths-LG SSPACE Premium	S. Natsume
Barnyardgrass	*Echinochloa crus-galli*	2017	Scaffold	4113	1400	486	705,200	NA [2]	89	26	63	2	10	170	Illumina PacBio	SOAPdenovo2 CANU	[54]
Paterson's curse	*Echium plantagineum*	2019	Chromosome	1091 [3]	333 [1]	349	1,429,328	1,517,519	96	46	50	1	3	115	Illumina PacBio	MECAT LACHESIS	C.-Y. Tang
Common sunflower	*Helianthus annuus*	2017	Chromosome	1528 [3]	3600	3028	178,899,001	174,509,413	89	80	9	3	8	100	PacBio	PBcR	[55]
Littlebell	*Ipomoea triloba*	2018	Chromosome	16	496 [1]	462	29,809,665	28,894,297	97	89	7	1	2	290	Illumina PacBio	SOAPdenovo2 SSPACE PBJelly, Pilon	[56]
Perennial Ryegrass	*Lolium perenne*	2016	Scaffold	666,180	2621 [1]	481	1361	NA [2]	31	29	2	22	47	5	Illumina	CLC Genomic Workbench	[57]
Horsemint	*Mentha longifolia*	2016	Scaffold	190,876	400	353	3915	3044	58	52	5	20	22	33	Illumina PacBio	MaSuRCA	[58]
Amur silver grass	*Miscanthus sacchariflorus*	2018	Chromosome	105,321 [3]	2513 [1]	2075	37,709	24,189	49	41	8	17	33	60	Illumina	ABySS SOAPdenovo2	J. De Vega
Longstamen Rice	*Oryza longistaminata*	2014	Scaffold	9688	782 [1]	362	30,401,905	NA [2]	86	80	6	4	10	52.5	Illumina	SOAPdenovo2	C. Brian
Red Rice	*Oryza punctata*	2014	Chromosome	12	586 [1]	394	31,244,610	28,494,620	81	74	7	6	13	130	Roche 454 Illumina	AllPaths	R. A. Wing
Brownbeard Rice	*Oryza rufipogon*	2015	Scaffold	3818	450 [1]	339	27,785,585	26,200,591	83	76	6	5	13	120			Q. Zhao
Rice	*Oryza sativa*	2019	Chromosome	367 [3]	489 [1]	415	28,085,715	26,003,091	88	81	6	3	9	148	PacBio	CANU	L. Wang
Broomcorn Millet	*Panicum miliaceum*	2018	Chromosome	466	923	848	48,259,421	45,112,342	83	25	58	4	13	160	Illumina PacBio	CANU	[59]
Opium Poppy	*Papaver somniferum*	2018	Chromosome	34,381 [3]	2870	2716	204,470,928	180,516,484	95	29	65	1	4	239	Illumina PacBio ONT	DeNovoMAGIC FALCON	[60]
White Poplar	*Populus alba*	2019	Contig	6087	508 [1]	707	248,703	390,844	95	52	43	1	3	130	Illumina PacBio	SMARTdenovo	[61]
Algarrobo blanco	*Prosopis alba*	2019	Contig	4454	391 [1]	500	237,044	357,710	70	49	21	3	27	30	PacBio	CANU	W. Kong
Wild Radish	*Raphanus raphistrum*	2014	Contig	64,732	515	254	10,333	NA [2]	95	82	12	3	2	47	Roche 454 Illumina	ABySS, Newbler, Celera Assembler, Minimus2	[62]
Radish	*Raphanus sativa*	2017	Chromosome	44,239 [3]	573	383	35,166,889	26,198,371	96	82	14	2	1	225	Illumina	SOAPdenovo2	[63]
Japanese Rose	*Rosa multiflora*	2017	Scaffold	83,189	711	740	90,830	95,085	91	66	25	4	5	327	Illumina	SOAPdenovo2 GapCloser	[64]
Wild Sugarcane	*Saccharum spontaneum*	2018	Chromosome	15,303 [3]	1565 [1]	3133	91,359,291	109,189,819	78	20	58	5	17	90	Illumina PacBio	CANU HiC	[65]

Table 1. *Cont.*

Common Name	Latin Name	Year	Level of Assembly	No. of Contigs	Est. Genome Size (Mbp)	Assembled Size (Mbp)	N50	NG50	BUSCOS (Percentage of 2121 Genes)					Coverage	Sequencing Technology	Assembly Method	Reference or Lead Submitting Author
									C	S	D	F	M				
Rye	*Secale cerale*	2017	Scaffold	1,581,707	7900	1685	2200	NA [2]	66	62	4	13	21	50	Illumina	CLC Assembly Cell, CarmA	[66]
Green Foxtail	*Setaria viridis*	2019	Chromosome	14	782 [1]	396	46,702,114	35,460,007	81	75	6	6	13	118	PacBio	MECAT	P. Huang
White Campion	*Silene latifolia*	2018	Scaffold	319,506	2640 [1]	1185	11,019	NA [2]	68	66	4	13	18	40	PacBio	SOAPdenovo2, CLC, PBJelly, SSPACE	[67]
Milk Thistle	*Silybum marianum*	2016	Contig	258,575	792 [1]	1478	6967	NA [2]	38	33	6	8	54	96	Illumina PacBio	Celera Assembler	Y. Lv
Sorghum	*Sorghum bicolor*	2017	Chromosome	869 [3]	730	709	68,658,214	68,658,214	86	80	5	4	10	8	Illumina Sanger	ARACHNE	[68]
Stinkweed	*Thlapsi arvense*	2015	Scaffold	6768	539	343	140,815	NA [2]	98	97	2	1	1	80	Illumina PacBio	CLC NGS Cell	[69]

[1] When not reported by the authors, we have estimated based the genome size based on the genome size available from Kew's C-DNA value database (see Section 2.2). In some cases, this has resulted in an estimate smaller than the assembled genome size. [2] In cases where the genome assemble size is not sufficiently higher than half of the expected genome size, an NG50 cannot be calculated (see Section 2.1). [3] In some cases chromosome-level genome assemblies have pieces left over and these increase the number of contigs included in the assembly files beyond the expected chromosome number. [4] Genome size estimate from DNA content analysis in Creber et al. [70].

Producing chromosome spreads is generally more accessible than chromosome sorting, but requires a significant amount of time, especially if the species' chromosomes are small or numerous. A pair of highly helpful videos on the technique, produced by the Beck Laboratory are available as an introduction (www.youtube.com/watch?v=iXqni6knH5A&t and www.youtube.com/watch?v=xVV4qBfSQLs&t) [71,72]. Several methods that inhibit spindle formation and increase the accumulation of metaphase cells can be used to facilitate chromosome counts. These include pre-treating material with pressurized nitrous oxide (NO_2), incubation in ice cold water, or exposing the cells to chemical inhibitors such as 8-hydroxyquinoline or colchicine [73]. For example, for a mitotic preparation, NO_2 pressurized to 8–10 atm (160 psi) can be applied for several hours to 1 cm long root tips in water using a specially constructed air sealed, iron pressure chamber with the regulator and hoses of the correct composition needed to attach and deliver NO_2 [74]. The water is then removed and replaced with fresh Carnoy's fixative for storage at 4 °C. Samples are then washed twice with distilled water and 1x citric buffer respectively. This buffer is replaced with enough 0.3% pectolytic enzyme solution [75] to ensure the material is fully submerged and incubated at 37 °C for 60 min. Digested root tips should form into a cell suspension when they are tapped with dissecting needles on a slide. If clumps form, or cells do not separate, incubation in the enzyme solution should be increased. Once cell suspension has been created, a drop of orcein stain [73] can be added [73], the drop carefully spread, and a coverslip placed on top. Then the slide is heated and squashed between filter paper using thumb pressure, ensuring no slippage. The slide can then be examined with a phase-contrast microscope for the quality of chromosome spread and count. If cytoplasm covers the nuclei then pepsin treatment may be effective [75]. Obtaining a good spread that will allow for certainty in chromosome number will generally take patience and practice.

An alternative to flow cytometry for determining genome size is to complete a k-mer plot of Illumina short read data (Illumina, San Diego, California, USA) [3] using a tool such as KmerGenie (kmergenie.bx.psu.edu) [76] or Jellyfish (www.cbcb.umd.edu/software/jellyfish), however it is preferable to have an estimate independent of the reference read data itself [77] (Figure 1). Following data generation with Jellyfish a script can be written in R to visualize the data or the data can be easily visualized using the website GenomeScope (qb.cshl.edu/genomescope). This data can also provide an indication of heterozygosity and can be used to determine the amount of the genome comprised of repetitive elements using tools such as RepeatExplorer (repeatexplorer.org) [78,79].

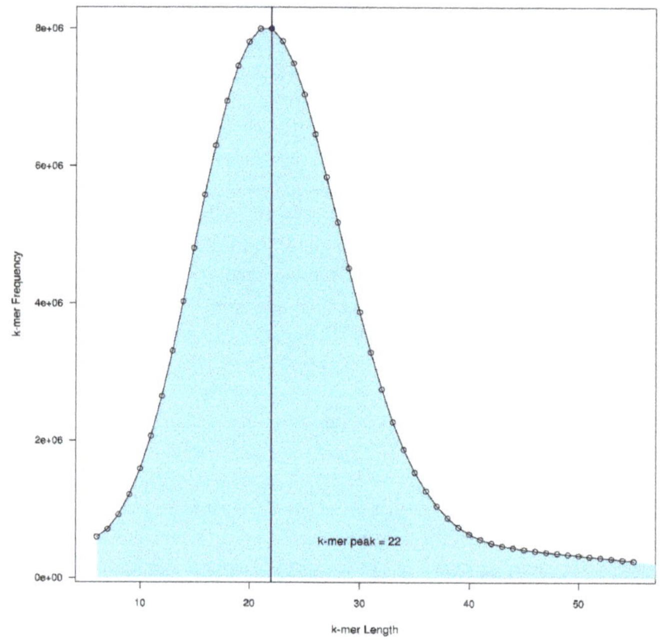

Figure 1. Plot of k-mer frequency by length produced for *Camelina neglecta* J.Brock, Mandáková, Lysak & Al-Shehbaz produced using Jellyfish and visualized using R. The position of the peak at a k-mer length of 22 is used to calculate genome size based on the area under the curve as represented by the light blue region. Here the genome size estimated is 248 Mb, while flow cytometry estimates indicate a genome size of 264 (±9) Mbp [80].

While the addition of long-read technology is making the assembly of highly heterozygous and repeat-rich genomes more feasible, genome assembly can be simplified by reducing heterozygosity and repetitive elements. In species that are self-compatible, repeated self-pollination can do both and result in a less redundant and more contiguous assembly [81]. In outcrossing or dioecious species or species with strong inbreeding depression reducing variation can be more difficult, requiring strategies such as repeated full sibling mating. Doubled haploids, generally produced via tissue culture of either male or female gametophytes, can solve this problem by completely eliminating heterozygosity, but are also a significant challenge and investment of time [82–84].

For genome sequencing, the 1C DNA content is perhaps the most important piece of information for designing the sequencing strategy, determining the quantity of sequencing required, and providing hints as to the species' degree of polyploidization or genome size inflation resulting from repetitive element proliferation.

Additional challenges await groups that wish to assemble genomes which have undergone recent or ancient polyploidization, which are notoriously more difficult to assemble, though long reads are making these genomes increasingly tractable. Many successfully sequenced crops fall in this category and specific strategies have been developed to assemble these genomes (reviewed by [85]). However, when a weed species is variable for ploidy the most feasible approach would be to select an individual with the lowest ploidy available for sequencing. However, the conclusions about the species' population genetics that are drawn from this genome would only be applicable to populations with this cytotype. In any case, a vouchered record of the material used for DNA extraction should be created and submitted to an herbarium to provide documentation of the species that has been sequenced [86].

2.3. DNA Extraction

Extraction of DNA of sufficient quality and quantity can be a surprisingly difficult hurdle. Technologies such as Pacific Biosciences' (PacBio) single molecule real-time (SMRT) sequencing (Pacific Biosciences, Menlo Park, California, USA) and Oxford Nanopore Technologies' sequencing systems (Oxford Nanopore Technologies, Oxford, UK) require high molecular weight (HMW) DNA at a high concentration (e.g., 10 µgs with an average size of 30–50 kbp for PacBio) [87]. This genomic HMW DNA needs to have little evidence of shearing, be free of contamination from protein, RNA, or polysaccharides and a 260/280 nm absorbance ratio of approximately 1.8–2.0. This is not always simple to achieve and time may need to be devoted to optimizing the DNA extraction protocol.

We have observed that the method of grinding the plant tissue appears to be the most critical step in obtaining HMW DNA with little shearing (Martin, unpublished). While many protocols suggest using bead mills with either ceramic, metal beads and/or sand, using the least time and speed reduces shearing. We have found that grinding tissue in 2 mL tubes with plastic pestles on dry ice, using wide bore tips, minimizing vortexing and pipetting will limit shearing and help ensure recovery of HMW DNA. Commercial kits are convenient and remove contaminants, but often an insufficient amount of DNA is obtained from a single extraction. However, multiple extractions can be pooled and concentrated to obtain the HMW at a sufficient concentration.

When sufficient tissue is available, many genome sequencing projects (e.g., [44,54,88,89]) have found success with variations on the traditional hexadecyltrimethylammonium bromide (CTAB) based method, described by Doyle and Doyle [90]. These methods often use a large quantity (g) of plant tissue ground in liquid nitrogen with a mortar and pestle. Many modifications of this original protocol are available, including Healy et al.'s [91] protocol for plants with large amounts of phenolics and polysaccharides. These compounds can inhibit downstream library preparations and are particularly important to eliminate. If required, further purification can be done with additional ethanol precipitations or magnetic beads (Agilent, Santa Clara, California, USA). For example, a strategy to prepare fragments for sequencing is to shear the DNA into large fragments of 20 kb in size using g-TUBES (Covaris, Woburn, MA, USA) and then selecting fragments of appropriate size with an apparatus such as the Blue Pippin (Sage Science, Beverly, MA, USA). In addition, some laboratories have found specially designed tips, such as Qiagen Genomic Tips (Qiagen, Hilden, Germany) to be helpful during preparation of the samples. Other technologies such as the Short Read Eliminator Kit (Circulomics, Baltimore, MD, USA) can be used to optimize sequencing by removing shorter fragments. Following extraction, DNA integrity and concentration need to be assessed. A variety of tools exist to complete these steps including the Tapestation or Bioanalyzer system (Agilent Genomics) [87]. However, it has been noted that DNA quantities should be measured on a Qubit Fluorometer (Thermo Fisher Scientific, Waltham, Massachusetts, USA) or similar as Nanodrop (Thermo Fisher Scientific) can overestimate quantity [87].

2.4. Sequencing Strategies

Assembling a genome using large pieces is much easier than using small pieces. Therefore, the majority of sequencing projects now combine long read (e.g., PacBio or ONT) and short read data. Long reads, which generally average 10 kb or more in length, make assembling plant genomes comparatively easier and general result in a more contiguous assembly. Genome assembly is sensitive to repeated sequences and these can only be resolved if the sequencing technology spans the regions. However, the error rate for long reads maybe as high as 15% and therefore require greater depth (30× per haploid genome, see below) to allow a consensus to be called from the data [20]. While short read Illumina data is unable to resolve long repeats, it has higher accuracy and can be used to correct long read data [4] either before or after assembly to improve the accuracy or completeness of the genome [20].

The recommended coverage for genome assembly varies from 40× to 60× at a minimum. For example, Li and Harkness [3] suggest 40–50× and Del Angel et al. [4] and Jung et al. [20] suggest a minimum of 60× for small, inbred, diploid genomes. Coverage is generally estimated based on the Lander-Waterman equation [92] as read length multiplied by read number divided by the haploid genome size for the species. Perhaps more simply for project planning, the amount of sequencing data needed for a project can be calculated by multiplying the estimated size of the plant's haploid genome by the coverage needed. However, it is important to note that coverage will be reduced by quality control and filtering steps compared to the raw coverage. Additionally, the coverage will not be uniform across the nuclear genome. For example, up to 20% of the raw data may be DNA from the chloroplast resulting in relatively deep coverage of the relatively small chloroplast genome, but less coverage of the nuclear genome [93]. After generating sequence data, there are generally five, often iterative, steps before the "final" genome is ready for downstream analysis: 1) Data assessment and filtering, 2) assembly (often by multiple assemblers), 3) error correction and polishing, 4) scaffolding and/or the placement of scaffolds on chromosome sized pseudomolecules, and 5) annotation.

2.5. Data Assessment, Correction and Filtering

Before starting with the assembly process, it is advisable to assess the quality of the sequencing data and filter the reads based on this quality. However, some assemblers integrate quality filtering and correction as early steps in their assembly process and additional steps with alternative software may or may not improve the final assembly. Read length can also be a consideration as, for example, some long read assemblers will refuse to work if reads shorter than 500 bp are included in the input data. The software FastQC (www.bioinformatics.babraham.ac.uk/projects/fastqc/) provides a summary of quality parameters that is very helpful to assess the quality of short or long read data: Average per base quality, per tile quality, per sequence quality, per base content, per sequence GC content, per base N content, sequence length distribution, sequence duplication level, overrepresented sequences, adapter content and k-mer content. Overall quality of long read data can be also assessed with tools such as Nanoplot (github.com/wdecoster/NanoPlot) [94]. Correction of long read data with short reads can be done prior to assembly with tools such as LoRDEC (www.atgc-montpellier.fr/lordec) [95]. Filtering can be done with a variety of tools available online such as Trimmomatic (www.usadellab.org/cms/?page=trimmomatic) [96]. This type of software will generally remove reads or regions in the reads that are below a certain quality threshold as well as sequencing adapters or the "bar codes" of specific sequences that allow for identification of particular reads following multiplexing. Many custom scripts for filtering raw data can be found online (e.g., filter_fastq.py github.com/nanoporetech/fastq-filter/blob/master/filter_fastq.py). Users will want to apply the principle of *caveat emptor* when using these scripts, but they can provide invaluable tools.

2.6. Assembly and Assessment

Genome assemblers typically use either short or long read data as input. Short read assemblers have a longer history and many are designed with smaller bacterial or viral genomes in mind. However, because of their longer history, several of the programs that can handle larger genomes have also had extensive work to reduce the amount of computational resources they need such as ABySS 2.0 (www.bcgsc.ca/platform/bioinfo/software/abyss/releases/2.0.0) [97] and SOAPdenovo2 (github.com/aquaskyline/SOAPdenovo2) [98]. In our experience, two genome assemblers that use long read data that are relatively easy to install and use with strong documentation and community support are CANU (canu.readthedocs.io/en/latest) [99] and FALCON (pb-falcon.readthedocs.io/en/latest) [100]. CANU, in particular, appears to be a common choice (Table 1), perhaps because of the clarity of its documentation and recommendations on which parameters (e.g., correctedErrorRate and minOverlapLength) are the most likely to improve the outcome of the assembly. This type of guidance is very helpful as the key parameters for tuning software to a particular species are not always apparent, resulting in an overwhelming number of parameters that could be adjusted. However, when in doubt and lacking documentation, this information can also be gleaned from other users' experience documented in discussion groups for the particular tool. Hybrid assemblers, that use

both short and long read data, such as SPAdes (github.com/ablab/spades) [101], and Platanus-allee (platanus.bio.titech.ac.jp/platanus2 the recent replacement of Plantanus) [102] are available and assembly strategies that merge the results of multiple assemblers have also been used (e.g., [89]).

Once an assembler has completed a draft assembly of the genome, the challenge is determining how "good" the assembly is [19]. The definition of good can depend on the eventual use of the genome and includes parameters such as how contiguous (how many pieces is the genome in) the assembly is, how much of the genome was assembled and whether the assembly contains the expected genes. Often the first tool applied following genome assembly is QUAST (quast.sourceforge.net/quast), which provides a quick summary of the genome including the number of contigs, the total length of the genome as assembled, the N50, and, if the expected genome size is included the NG50 values. This gives an indication of contiguousness and the size of the assembly. BUSCO (busco.ezlab.org) [103,104] is frequently used as a quantitative measure of the completeness of a genome as it indicates whether the shared single copy genes expected in the genome are present—that is how much of the gene space has been captured and assembled. BUSCO indicates how many and which of these are complete and single copy, complete and duplicated, missing or fragmented (Table 1). Finally, BlobTools (blobtools.readme.io/docs) [101] can be used to determine if the assembled sequences are DNA from the expected organism or from contaminating organisms through taxonomic partitioning of the genome. This tool requires the draft genome sequence, a hit file created by BLASTn (blast.ncbi.nlm.nih.gov/Blast.cgi) [105] using the MegaBLAST option [106], a depth file created with a tool such as BWA-MEM (bio-bwa.sourceforge.net) [107], and the raw data used to assemble the genome sequence. After processing this information BlobTools creates a visual indication of which organisms are most closely related to the draft genome (Figure 2). If there is substantial contamination, this information to further filter the raw data for reassembly without the contaminating sequences.

2.7. Polishing

Polishing a genome can lead to significant improvements in the completeness of the genome as assessed by BUSCO and some tools will use short read data to call a consensus SNP, correct indels (insertions and deletions that are common in log read data) and misassembled contigs. Pilon (github.com/broadinstitute/pilon) [108] uses the assembled genome and one or more files containing the alignment of sequencing reads such as mate pairs, paired ends or unpaired sequences to the draft assembly. The program's output includes the files needed for visualizing the changes to the genome using tools such as the Integrative Genomics Viewer (IGV software.broadinstitute.org/software/igv/) [109] and can generate information on the variation with genome sequence. PacBio has developed the tool GenomicConsensus (github.com/PacificBiosciences/GenomicConsensus), which uses mapped PacBio reads to generate a consensus, while Nanopolish (nanopolish.readthedocs.io/en/latest/index.html) has been developed for use with ONT data. In comparison, RACON (github.com/isovic/racon) can be used with either short read or long read data [110].

2.8. Scaffolding

Traditionally, the ordering and orientation of contigs into scaffolds has often relied on the labor intensive and expensive use of fluorescent in situ hybridization of bacterial artificial chromosomes (BACs) and segregating F2 populations that allow for mapping the position of the sequences. More recent methods: Chromosome conformation capture techniques (Hi-C), optical mapping techniques (Bionano) and 10x Genomics Chromium™ Systems can produce data that can be generated and applied to verify the assembly and generate scaffolds with less time and effort [3]. Chromosome conformation capture (3-C) has been a commonly used technique in molecular biology to map chromosomal interactions. It uses a process where genomic DNA is first digested and then ligated in conditions that preserve the 3D organization of the genome to allow the joining of distant sequences that find themselves to be in proximity. Using deep sequencing, the high throughput version of the technique (Hi-C) produces a genome-wide map of proximity contacts between all the different loci. Since the frequency of occurrence of such contacts is based on proximity, with intrachromosome contacts most common and the probability of contacts decreasing with distance, the technique can readily be used for scaffolding contigs [111]. If the analysis of this proximity data is not completed by the provider using proprietary software, once the paired end data has been mapped to assembled contigs, software such as SALSA (github.com/machinegun/SALSA) [112] can use the information to break misassembled contigs and scaffold the genome. FALCON-Phase (github.com/PacificBiosciences/pb-assembly) has also integrated the use of Hi-C data into the FALCON assembly pipeline through a collaboration between PacBio and Phase Genomics (www.phasegenomics.com) [113]. Phase Genomics is a USA based company that can provide kits for HI-C library preparation and bioinformatics support in the use of this data scaffolding of a *de novo* genome with their proprietary software Proximo. Additionally, they provide helpful advice on how to work with Hi-C data generated by their protocols (phasegenomics.github.io/2019/09/19/hic-alignment-and-qc.html). Recently, chromosome level assemblies of black raspberry (*Rubus occidentalis* L.) [114], an ornamental amaranth used by ancient civilizations in South and Central America as a grain crop (*Amaranthus hypochondriacus* L.) [115], and broomcorn millet (*Panicum miliaceum* L.) [59], genomes have been completed using Hi-C data and PacBio data.

Bionano Genomics (San Diego, CA, USA, bionanogenomics.com) contributes to scaffolding by optically mapping specific sequences distributed across the genome. Briefly, high molecular weight DNA is extracted, up to chromosome arm lengths, and labeled at specific sequence motifs for imaging and identification. The DNA molecule is then linearized onto a flowcell where a gradient of micro- and nano-structures gently unwinds and guides DNA into NanoChannels where it is imaged by a high resolution camera. The DNA fragments with similar motif-specific label patterns are assembled together to recreate a whole genome map assembly. This data can be used in a hybrid assembly to scaffold contigs obtained through sequencing of the genome. It can be used to identify regions that are incorrectly assembled or where structural variants can be found. This approach was recently used in the improvement of wheat's hexaploid genome assembly [116] and the large *Sorghum* genome [117].

An alternative approach is used by 10x Genomics Chromium™ System (www.10xgenomics.com). DNA molecules are divided into small sets and provided with an identifying barcode before being sequenced. This provides linked reads that are unlikely to represent the same region from homologous chromosomes. This technique is particularly useful in genomes that are highly heterozygous and/or polyploid because it allows the genome information to be phased, that is the two haplotypes can be distinguished, and it can prevent the collapse of sequence from homologous chromosomes in polyploids. This technique was recently used in the sequencing of the octaploid strawberry genome (*Fragaria* X *ananassa*) [118].

An additional option when a related species with a chromosome-level genome sequence is available, is that this information can be used to create reference based assembly with chromosome-level resolution. However, this method would bias the assembly to more closely resemble that of the relative and will, for example, lack chromosome scale rearrangements. One option for pursuing this route, MeDuSa [119] (github.com/combogenomics/medusa/releases), can use one or more closely related genomes for generating a chromosome-level draft.

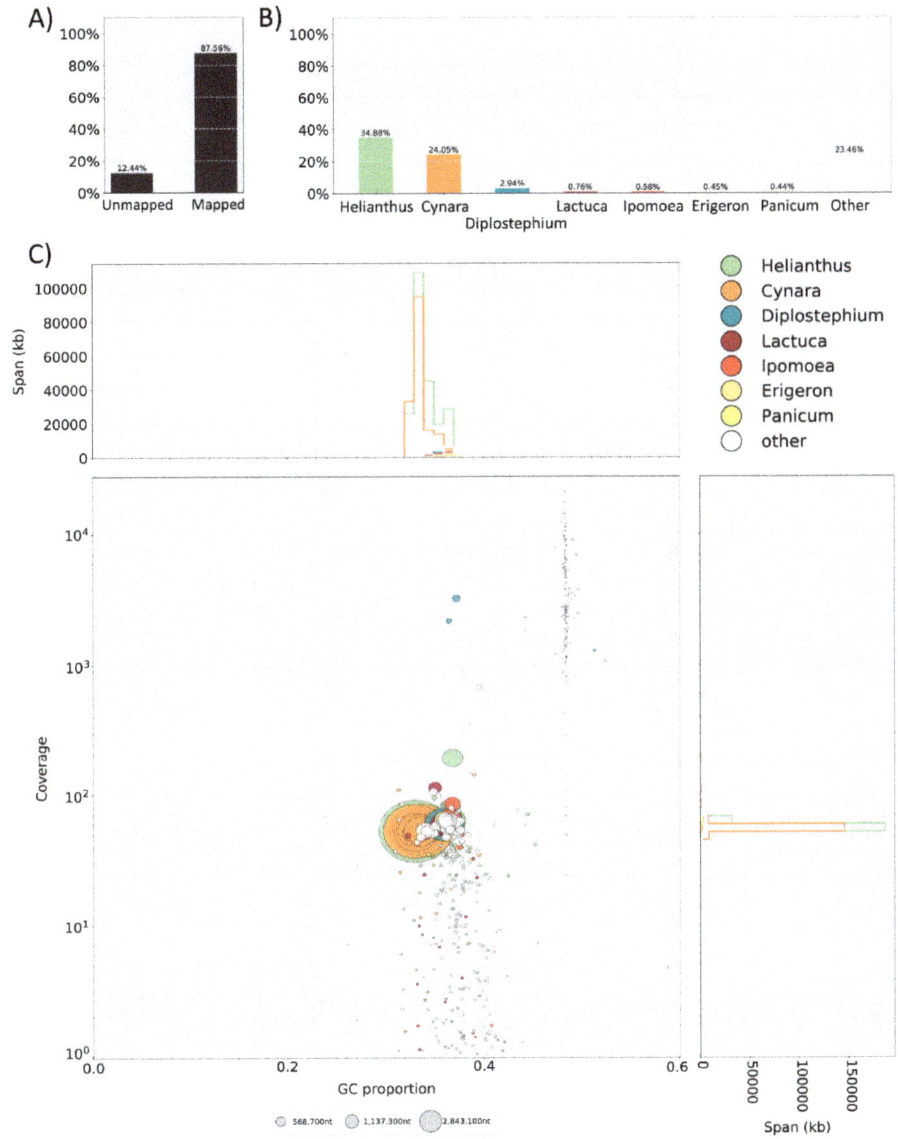

Figure 2. Blobplot generated for *Conzya canadensis* (Asteraceae) draft genome assembly showing the genera with the closest similarity to the sequenced genome (Laforest, Martin, and Page unpublished data). The first panel (**A**) indicates the percentage of reads that were mapped and the second panel (**B**) shows the taxonomic break down of hits at the taxonomic level requested. In this case the majority of hits are from other genera from the Asteraceae. The program generates a text file with more detailed information. The three part third panel (**C**) shows histograms for the proportion of G and C bases in the sequence which typically varies among species (top) and coverage (right) weighted by the cumulative length of sequences in each bin. The main panel has circles colored by taxonomic affiliation positioned on the x-axis by the GC proportion and on the y-axis by coverage within the raw data which gives a sense of the relative concentration of the sequences in the DNA sample.

2.9. Gene Prediction and Annotation

Once a genome sequence of adequate quality has been produced, genes and other genetic elements such as transposons need to be identified. Gene prediction software such as AUGUSTUS (bioinf.uni-greifswald.de/augustus) [120,121] can be used to locate potential coding sequences along the genome sequence. This software has been improved over the years, starting from entirely ab initio gene prediction to include evidence-based discovery using expressed sequence tag (EST) sequences, RNASeq data (by way of hints) and with protein multiple sequence alignments. Repeated elements such as transposable elements (retrotransposons and DNA transposons), tandem or inverted repeats, can be located in the genome with software such as RepeatMasker (www.repeatmasker.org), RepeatFinder (www.cbcb.umd.edu/software/RepeatFinder) [122], or the recently developed Generic Repeat Finder (GRF) [123]. Additionally, there are a host of software packages and resources designed to detect and annotate specific types of transposable elements including SINE_scan (github.com/maohlzj/SINE_Scan) [124] for detected short interspersed nuclear elements (SINEs), the P-Mite database (pmite.hzau.edu.cn) [125] for finding miniature inverted-repeat transposable elements, and HelitronScanner (sourceforge.net/projects/helitronscanner) [126] for detecting helitrons—rolling circles that often capture gene sequences leading to gene duplication.

It is useful to know what the product of identified gene sequences code for and tools have been designed to assign gene ontology—information on a gene's product's molecular function, location and role (GO, geneontology.org) using standardized language. One of the most ubiquitous tools used is the basic local alignment search tool (BLAST) [105] in conjunction with the Genbank [34] databases to assign putative functions through shared identity or similarity of the translated gene product. Blast2Go (www.blast2go.com) [127] is a tool with a subscription fee that can automate this process. Free software packages are also available including the widely used Maker-P (www.yandell-lab.org/software/maker-p.html) [128] as pipeline designed to make the annotation of plant genomes more accessible to new groups and incorporates many of the software packages mentioned above and has extensive documentation and tutorials.

2.10. Examples: Three Recently Sequenced Weed Genomes

Given the wide variety of sequencing strategies and tools that can be employed (or not) at each stage of genome assembly it is unlikely that any two projects have followed the same path to a final assembly. Further, as noted by Del Angel et al. [4], it is important to set goals at the beginning of a project for how contiguous and complete the genome sequence needs to be for the specific project, otherwise the iterative process of analysis and reanalysis with alternative tools can be endless. Given the complexities of genomes (e.g., [129]) and how this complexity is reduced in a genome assembly, it may be helpful to consider a modification of George E. P. Box's aphorism that all genome sequences are wrong, but some are useful. As examples of how these techniques and programs have been applied to weeds, we briefly summarize the methods and outcomes of three recent sequencing projects of two diploids, kochia (*Kochia scoparia* (L.) Schrad. also called *Bassia scoparia* (L.) A.J.Scott), common waterhemp (*Amaranthus tuberculatus* (Moq.) Sauer), and a hexaploid species, barnyard grass (*Echinochloa crus-galli* (L.) Beauv.).

For kochia, a plant with a genome size of approximately [89] 1Gbp (2n = 2x = 18), DNA for sequencing was extracted from a glyphosate susceptible inbred line using a modified CTAB protocol. They sequenced three Illumina libraries, one paired end and two mate-pair libraries using three HiSeq lanes and used 12 PacBio SMRT cells. They then assembled and merged two assembles into a final assembly for analysis. For the first assembly, they used the paired end data and the program Proovread (github.com/BioInf-Wuerzburg/proovread) [130] to correct the PacBio reads, which were then assembled with Canu. For the second assembly, ALLPATHS-LG (software.broadinstitute.org/allpaths-lg/blog) [131] was used to assemble all the Illumina data and scaffolding was completed using the PacBio reads and PBJelly (sourceforge.net/p/pb-jelly/wiki/Home) [132]. They then used the GARM Meta assembler (garm-meta-assem.sourceforge.net) [133] to merge the genomes. This final

711 Mbp assembly consisted of 19,671 scaffolds and had an N50 of 62 kb. Completeness as indicated by BUSCO, using the eudicotyledons odb10 dataset, was estimated at 70.3%. Kochia's sequence was then annotated using the WQ-Maker pipeline transcriptome data from kochia and expressed sequence tags for kochia's family, the Chenopodiaceae, from the National Center for Biotechnology Information (NCBI www.ncbi.nlm.nih.gov). Then then used BLASTN and BLASTP to predict genes and proteins and RepeatMasker to search for repetitive elements.

In the case of common waterhemp, a species with a genome size of approximately 676 Mbp (2n = 2x = 32), DNA from a single female plant was extracted using a modified CTAB protocol and sequenced with both PacBio reads, 15 SMRT cells, and one Illumina HiSeq lane of 150 bp paired end library reads [88]. The long read data provide 87× coverage and was assembled using Canu and then polished with the short read data using Arrow and Pilon. This resulted in a final genome assembly size of 663 Mbp consisting of 2,514 contigs and an N50 of 1.7Mb. The assembly contained 88% of BUSCO's Embryophyta's genes. The program REVEAL (github.com/jasperlinthorst/REVEAL) [134] was then used to produce 16 pseudomolecules using the chromosomal level genome assembly of the cereal crop species *Amaranthus hypochondriacus* L. Both this finished genome and the assembly used to create it were annotated using the MAKER pipeline (yandell-lab.org/software/maker.html) following identification and masking of repetitive elements with RepeatModeler and RepeatMasker.

Barnyard grass has an estimated genome size at 1.4 Gbp based on flow cytometry data and K-mer analysis [54] and a chromosome count of 2n = 6x = 54. DNA was extracted for sequencing from a plant collected from a rice paddy using a CTAB protocol. They sequenced the 48 SMRT cells of PacBio for long read data and both paired end and mate pair Illumina libraries using HiSeq runs. This level of sequencing effort resulted in 171× coverage of the genome. The short read data was assembled with SOAPdenovo2, scaffolded with OPERA-LG (sourceforge.net/p/operasf/wiki/The%20OPERA%20wiki) [135], and then gaps in this assembly were closed with GapCloser from SOAPdenovo2. The long read data was assembled with Canu and used to fill gaps in the short read assembly with PBJelly. The draft genome produced was 1.27 Gbp in length with an N50 of 1.8 Mbp. The authors used BUSCO and determined that 95.5% of the core eukaryotic genes were complete. RepeatModeler and RepeatMasker were used to find and mask repetitive elements. Then they used transcriptome data and three programs to predict genes GeneMark.hmm (exon.gatech.edu/GeneMark) [136], Fgenesh (www.softberry.com) [137], and AUGUSTUS.

3. Current Application: What Are Agricultural Weeds and Where Do They Come From?

Harlan and deWet defined weediness as "an adaptive syndrome which permits a species or variety to thrive and become abundant and difficult to eradicate within areas of human disturbance" [138]. Under this definition, crops are the result of intentional selection for vigor and fertility in the agricultural environment and weeds are the unintentional result [139]. A classic example of this is crop mimicry, where weeds have been selected by agricultural practices such as hand weeding to closely resemble a crop species [140]. This includes species such as false flax (*Camelina sativa* (L.) Crantz), which looks like, has similar time to maturity, and similar seed size to varieties of cultivated flax [141,142], and rice-mimicking varieties of barnyard grass [140]. A more pressing example is the evolution of HR (see Section 4) [143]. This second example illustrates, that as a group, weeds represent multiple independent origins of weediness and numerous examples of rapid adaptive evolution that present an opportunity not only to co-opt these adaptations for crop improvement or guide changes in agricultural practices to slow or thwart this evolution [144], but to provide fundamental insights into evolution [145]. Agricultural weed populations can be selected from populations adapted to natural disturbance regimes or from populations selected for these characteristics as crops, from populations of wild crop relatives, or from hybrids between the two [141,146–148]. Similarly, specific traits that contribute to adaptation to the agricultural environment, including alleles conferring HR, are selected within those populations. These origins and the loci underlying adaptive traits can be elucidated by examining genomic variation with weed populations.

3.1. Detecting the Signatures of Demographic Change and Selection on the Genome

Demographic and selective events change the patterns of variation across the genome, leaving a record of these processes. In weed populations, demographic and selective events may be closely intertwined as artificial selection from weed control measures can drastically change population size and composition. For example, weed populations might undergo rapid declines in population size (bottlenecks) resulting from herbicide application followed by population expansions after the evolution of HR, or the introgression of HR genes from one population into another. These processes can be difficult to disentangle from each other, as well as from patterns related to the variable recombination rate across the genome. However, demographic processes generally leave a signature across the entirety of the genome, while selection leaves a signal localized to the genes that confer higher fitness under the given environmental regime.

Over time, adaptation of a population to its specific environment and associated demographic events lead to divergence in allelic composition across the genome relative to other populations. This divergence leads to population structure and can be used to infer the past history of the sample, with populations sharing more similar allele frequencies more likely to share a recent evolutionary history. When a species exhibits population structure, we can assign individuals to recent common "ancestral populations" that can provide clues to their origin. This is often the basis of human ancestry assignment through home DNA tests, where your genotyping results are compared to the frequency of alleles across the globe to determine which geographic region contains the highest proportion alleles similar to those comprising your genotype [149,150]. Population structure can also provide evidence of hybridization and introgression when individuals show the signal of a mixed affinity to populations or species (admixtures). Again, this is similar to the assignment of percentage affiliation to different groups in human ancestry tests.

Population structure can be estimated at many hierarchical levels, from individual, to subpopulation, and across longer timescales at the phylogenetic level (e.g., STRUCTURE (web.stanford.edu/group/pritchardlab/structure.html) [151], AMOVA [152], and TREEMIX (bitbucket.org/nygcresearch/treemix/wiki/Home) [153]). While these methods aim to cluster individuals into discretely structured groupings, allele frequencies may instead continuously vary across space [154]. This may be especially likely for a recently expanded species due to serial bottlenecks and expansions, or along clines in latitudinal or environmental gradients where there is limited opportunity for long distance dispersal [155,156]. However, methods have been developed to test whether a population is more likely to showing continuous or discrete population structure [157]. In these cases, a model free approach such as principal component analysis may help to clarify population structure [158]. These data can also be used to infer past demographic processes using modelling approaches that allow estimation of parameters including ancestral population size, the number and timing of bottlenecks, time since divergence between populations, ancestral and contemporary levels of gene flow, and contemporary effective population sizes. Demographic modelling has been widely implemented to infer the history of sampled populations including δaδi (bitbucket.org/gutenkunstlab/dadi/src/master/) [159] and FastSimCoal (cmpg.unibe.ch/software/fastsimcoal2/) [160]. With genome-wide data from a population level sample, produced either through a reduced genome representation technique (see Section 4.3) or resequencing (sequencing of a genome of using less coverage and a template draft genome sequence) population structure and demographic history can easily be estimated through these variety of approaches discussed above to provide powerful insights into the source and origins of agricultural weed populations.

While genome wide information provides high resolution data on the distribution of allelic differences among samples due to demography, allelic differences due to selection can be inferred with care using integrative summary statistics and model based approaches. Currently, our understanding is that HR evolution often proceeds through drastic changes in allele frequency at the target gene—conveniently, a single locus of large effect provides the most power for detecting recent signals of selection and differentiating independent events. Three types of signal can be used to recognize selection: changes in allele frequencies (differentiation and diversity), patterns associated with linkage (homozygosity), and the pattern of nucleotide substitutions

First, regions near alleles selected by agricultural practices can be indicated by changes in allele frequencies. When a beneficial allele changes in frequency, becoming highly prevalent or fixed in a population sites nearby, linked to the selected allele due to a low probability of recombination, will show a depletion of genetic variation. The pattern resulting from the fixation of nearby neutral sites along with the selected site is termed a selective sweep [161–164]. An expectation following from this process is that the frequency of alleles under selection is expected to differ among populations experiencing different conditions (e.g., herbicide application or none) and this differentiation between populations is frequently expressed as Wright's fixation index (F_{ST}), though there are a host of related statistics [165,166]. If the F_{ST} of a locus is much larger than at other nearby or neutral loci, this can indicate positive selection.

Second, in addition to differentiation, immediately following selection the frequency of linked alleles will be fixed with new mutations causing new alleles to accrue slowly thereafter. This results in an excess of homozygosity (lack of variant sites) directly after selection. As new alleles will be rare, an excess of rare alleles can indicate positive selection (as well as recent population expansion) and can be quantified by Tajima's D, which compares the number of pair-wise differences between individuals with the total number of segregating polymorphisms [167]. Similarly, Fay and Wu compare the number of pair-wise differences between individuals to the number of individuals that are homozygous for the allele [168].

Third, selection can be detected through a comparison of the rate of nonsynonymous substitutions at a nucleotide (those that alter the amino-acid represented by the codon) to the rate of synonymous substitutions, which are assumed to be silent and neutral. This ratio can indicate selection favoring a change in the structure of a protein (d_N/d_S).

Beyond these summary statistics, many model-based approaches have been developed to distinguish between recent, single genetic origin selective events (hard sweeps) and older or multiple genetic origin selective events (soft sweep) by assessing differences in the magnitude of their signals across the genome (e.g., SweeD (cme.h-its.org/exelixis/web/software/sweed/index.html) [169] and SweepFinder2 (www.personal.psu.edu/mxd60/sf2.html) [170,171]). After assaying within population sweep patterns, one can then compare the extent of convergence in these patterns across populations. A greater or lesser extent of parallel changes in allele frequencies, homozygosity, and diversity in the surrounding sequence provide evidence of shared or independent origins of resistance across populations respectively, and more broadly, may provide the means to identify candidate genes that appear to underlie HR in multiple populations (see Section 4).

While there is great potential to determine the source and number of independent and shared origins of HR from genomic data (e.g., [88]), the task will be more difficult when HR is conferred by many alleles of small effect. With polygenic trait architectures many individuals are needed to have sufficient power to detect the individual small-effect changes, and therefore approaches often rely on taking the sum of allele frequencies weighted by their effect size on the trait [172]. Since these genome-wide association approaches assume allele frequency differences across the genome are all related to selection, one must carefully account for allele frequency changes due to population structure, which has been shown to often be confounded with polygenic signals of selection [173].

3.2. Example: Convergent Adaptation to Glyphosate in Common Waterhemp

Common waterhemp is a problematic, a wind-pollinated, outcrossing, and dioecious weed that occurs throughout the mid-western and eastern United States of America and in Canada from Manitoba to Quebec. It has been hypothesized that weedy agriculture populations result from human-mediated disturbance and mixing of two closely related taxa, *A. tuberculatus* var. *rudis*, a Midwestern native, highly associated with agricultural environments, and *A. tuberculatus* var. *tuberculatus*, a species that occupies a constrained range, and that is limited to riparian environments [174]. Glyphosate resistance was first reported in 2005 in Missouri and one hypothesis is that it may have spread from there across the United States and recently into Ontario. However, considering the strength of selection from herbicides and the highly repetitive nature of HR evolution as suggested from independent glyphosate resistance evolution in multiple *Amaranthus* species [35], it is also possible that glyphosate resistance may have multiple independent origins with *A. tuberculatus*, representing a striking case of convergent evolution.

A recent study used genomic approaches to investigate the history of the species, clarify the origins of agricultural populations, and the evolution of glyphosate resistance [88]. Specifically, Kreiner et al. [88] sequenced the species' genome as described above (see Section 2.6) and then resequenced the genomes of 163 individuals from 19 agricultural populations known to have glyphosate resistance, varying from 13% to 88% of the population, from Missouri, Illinois, and Essex County and Walpole Island within Ontario, as well as ten individuals from a native, non-agricultural population in Ontario that lacked glyphosate resistance. This data and the software freebayes (github.com/ekg/freebayes) [174] were used to identify SNPs across the species genome and then to characterize population demographics, diversity, differentiation, and structure. Demographic modeling completed using δaδi supported the hypothesis of recent secondary contact between lineages. Similarly, analysis with STRUCTURE and principal component analysis, indicated that populations were genetically differentiated by geography and hypothesized species ranges, with populations from Missouri and Illinois clustering and corresponding to *A. tuberculatus* var. *rudis* and natural populations from Ontario clustering and corresponding to *A. tuberculatus* var. *tuberculatus*. These analyses also showed resistant populations from Essex county were unlike nearby natural or agricultural populations found in Ontario, but rather clustered with western Missouri populations. This indicates that populations from Essex County likely represent an introduction of seed from Midwestern *A. tuberculatus* var. *rudis* populations, that harbored multiple independent resistance haplotypes. Interestingly, the second group of resistant populations in Ontario, those from Walpole Island, clustered with natural populations in the area, though with signs of some introgression from the var. *rudis* cluster. With information on the evolutionary origins of these populations, Kreiner et al. set out to distinguish whether populations with shared evolutionary origins have independently evolved resistance, or if resistance spread through the expansion of these populations into new agricultural landscapes. The authors investigated the pattern of selection on the chromosome bearing the glyphosate target-site gene, 5-enolpyruvylshikimate-3-phosphate synthase (EPSPS), using Sweepfinder2 and model-free summary statistics such as diversity, homozygosity, and differentiation. This analysis indicated the plants from Walpole showed a stronger pattern of reduced genetic diversity, increased differentiation and increased extended haplotype homozygosity around the EPSPS genes—evidence of a hard selective sweep—distinct from plants from Essex county, Missouri, or Illinois where a soft-sweep following multiple origins throughout the Midwest appears to have occurred. The authors conclude that glyphosate resistance in newly problematic Ontario populations has multiple genetic origins – both through new seed introduction events and selection on a recently arisen mutation in a previously benign population.

4. Current Application: What Genes Underlie Herbicide Resistance?

Understanding the genetic basis of resistance to an herbicide in a plant species is an essential first step in the development of diagnostic markers, understanding the fitness consequences of the mutation, and, more generally, in understanding how herbicide evolution typically occurs. This information is essential for being able to detect, monitor and develop more effective strategies for managing HR. Of the current total of 500 unique combinations of species (256) and herbicide site mode of action, the underlying genetic basis of these resistances is only known for a minority of cases [35]. The majority of known cases involve mutations to the herbicide's target site (TSR), while the specific genetic basis of non-target site resistance (NTSR) is largely unknown [175,176].

Our lack of understanding of the genetic basis of NTSR, is a major gap in our understanding of weed biology and the evolution and spread of HR [175–177]. Non-target site resistance is the most common mechanism contributing to glyphosate and acetyl CoA carboxylase inhibition resistance (ACCase). It is also the most common mechanism for acetolactate synthase (ALS) resistance in grass species [178] and can confer resistance to several herbicide modes of action simultaneously and unpredictably [179]. Non-target site resistance encompasses a diverse and complex set of traits that likely involve the full gamete of potential genetic basis including dominant to semi-dominant alleles with major effects, copy number variation, multiple minor alleles that incrementally contribute to resistance, and changes in epigenetic regulation (reviewed by [180]). Further, NTSR likely involves varied aspects of the fundamental processes within cells from transcription to translation invoking complex stress responses and altering regulatory pathways [177,180,181]. Until this gap in our knowledge is filled in our ability to make diagnostic tests, draw conclusions about the type and prevalence of mutations/variation that contribute to HR or develop strategies to interfere with NTSR pathways is compromised. However, while we rarely know the specific genetic basis of NTSR in a weed species, we have a good understanding of the types of genes are most likely involved.

4.1. Five Superfamilies of Suspects

Five gene superfamilies have members that have been identified as likely involved in NTSR. Evidence for their involvement comes from either their ability to confer herbicide tolerance or resistance in crop species or *Arabidopsis*, on enzyme and transcriptome analyses of herbicide resistant species or investigations of the molecular mechanisms of drug resistance (reviewed by [182,183]). Evidence from transcriptome studies suggests NTSR is often the result of the action of multiple members of a superfamily and multiple superfamilies [184–187]. Each of these families are large, diverse, and widely represented across the tree of life from bacteria to mammals indicating that they are fundamental to how organisms cope with their environments. In this regard, the evolution of HR has selected variants of genes underlying the complex regulatory and enzymatic pathways that organisms have always used to face biotic and abiotic stresses [188]. These gene superfamilies are considered to form part of what has been termed the xenome, the chemical detection, transport and detoxification system of plants [189] and members of the families are spread throughout plant genomes.

4.1.1. Cytochrome P450 Monooxygenases

The cytochrome P450 monooxygenase gene superfamily (CYP) are the largest enzyme family in plants and are known to be involved in HR [190]. This superfamily, which is involved in detoxification and stress responses, were implicated in HR as a result of the analysis of herbicide residues from plants, their induction following the application of safeners (chemicals that increase herbicide tolerance in grain crops), and the observation of increased P450 metabolism levels in HR annual ryegrass (*Lolium rigidium* Gaud.), black grass (*Alopecurus myosuroides* Huds.) and lesser canary grass (*Phalaris minor* Retz.) [177]. However, the number of these genes [191], with 272 in *Arabidopsis thaliana*, for example [192], and issues with purification from plant material meant that the isolation of specific CYP genes conferring HR in plants was preceded by isolation of these genes in bacteria and mammals, which frequently have higher activity than those from plants [193]. As an example, expression of human CYP genes in potato [194] and rice [195–197] confer HR. Indeed, expression of *CYP1A1* in rice resulted in resistance to ten different herbicides from ten different Herbicide Resistance Action Committee (HRAC) groups [195,198], while expression of *CYP2B6* in resistance to thirteen from six HRAC groups [197]. Despite this demonstrated ability of individual CYP genes to confer broad HR, it is likely that multiple CYP genes are involved in NSTR within each plant species [177]. Plant derived CYP genes that have been demonstrated to confer HR have now been isolated in Jerusalem artichoke (*Helianthus tuberosus* L.) [199], soybean (*Glycine max* (L.) Merr.) [200], *Arabidopsis* [201] and ginseng (*Panax ginseng* Mey.) [202]. Within weeds, two CYP genes have been determined to be associated with ALS resistance in rice barnyardgrass (*Echinochloa phyllopogon* (Staf).) Koso-Pol.) and overexpression of these genes in *Arabidopsis* resulted in resistance to group B herbicides bensulfuron-methyl and penoxsulam [203] and group F4 clomazone [204]. The isolation of CYP genes responsible for HR from other weed species will likely occur in the near future as chemical inhibition of P450 indicate that these genes are involved in HR for flixweed (*Descurainia sophia* L.) [205], water hemp (*Amaranthus tuberculatus* (Moq.) Sauer var. *rudis* (Sauer) Costea & Tardif) [206], and large crabgrass (*Digitaria sanguinalis* L. Scop.) [207] in addition to the grass species mentioned above. Additionally, consistent expansion of CYP copy number across all 69 annotated CYP genes in *Amaranthus tuberculatus* agricultural populations relative to natural populations has been recently found [88].

4.1.2. Glutathione S-Transferases

Glutathione S-transferases (GSTs) are enzymes that play a strong role in plant secondary metabolism and stress response [208–210]. For example, GSTs have been identified as playing a role in salt tolerance [182], copper tolerance [211] and fungal disease resistance [212]. They were first identified in mammals in the 1960s because of their role in drug metabolism and their presence in plants was identified soon after as contributing to atrazine resistance in maize (*Zea mays* L.) [213]. As a result, the role of GSTs for herbicide detoxification in maize have been extensively studied [214] and several of the genes encoding these enzymes have been used to engineer HR. For example, *GST1* [215] expressed in tobacco (*Nicotiana tabacum* L.) [216], resulted in resistance to alachor (group K3) and *GST27*, when expressed in wheat (*Triticum aestivum* L.), resulted in atrazine (group C1) and oxyfluorfen (group E) resistance [217]. Similarly, overexpression of a GSTs from soybean, *GmGSTU4*, in tobacco results in a significant increase in alachor tolerance [218]. Within weeds, two glutathione-S-transferase genes have been identified as being involved in resistance to ACCase and ALS inhibitors in black grass [219]. Indeed, although multiple loci are believed to be involved in NTSR HR for black grass [220], expression of *AmGSTF1* in *Arabidopsis* resulted in resistance to atrazine, alachor, and chlorotoluron (group C2) [185]. Expression analysis suggests that GSTs are involved in HR for a number of other weed species including junglerice (*Echinochloa colona* (L.) Link.) [221], Palmer amaranth (*Amaranthus palmeri* S. Wats.) [222], annual ryegrass [184,223] and sunflower (*Helianthus annuus* L.) [224]. However, as with the CYP genes, the number of GSTs in a plants species makes pinpointing the specific gene or genes responsible for HR challenging. For example, there may be 42 in maize [225] and 54 functional GSTs have been identified in *Arabidopsis* [226].

4.1.3. ATP-Binding Cassette Transporters

ATP-binding cassette (ABC) transporters are a group of proteins that mediate cross membrane transport (reviewed by [227,228]). With more than 80 members they are the largest protein family in *Escherichia coli*. Approximately 130 and 150 members have been located within the *Arabidopsis* [229] and the tomato (*Solanum lycopersicum* (L.) H. Karst.) [230] genomes, respectively. These transporters are understood to be involved in the transport of auxin and glyphosate and may, therefore, play a role when reduced translocation or sequestration of these herbicides is involved in HR [177,231]. In horseweed (*Conzya canadensis* (L.) Cronq.) glyphosate application caused increased expression level in at least seven ABC transporter genes [232] and a transcriptome study on the closely related hairy fleabane (*Conzya bonariensis* (L.) Cronq.) indicated that there were 19 ABC transporter genes in addition to 22 other candidates including GSTs and glycotransferases (see below). Additional evidence of the role of this group is that overexpression of the ABC transporter gene *AtPgp1* in *Arabidopsis* resulted in resistance to dicamba (group O) and oryzalin (group K1) [233] and tobacco overexpressing *pqrA* from the bacterium *Ochrobactrum anthropi* show higher resistance to paraquat (group D) [234].

4.1.4. MFS Transporters

The major facilitator superfamily (MFS) are also transporter proteins. As with the ABC transporters, there are approximately 70 members of the family within the genome of *Escherichia coli* [235] with perhaps 200 in *Arabidopsis* [236]. Like the ABC transporters members of the MFS family have been identified as being upregulated following exposure to auxinic herbicides [237] and the *TPO1* gene from yeast is a member of this group and its homolog from *Arabidopsis*, *At5g13750*, are able to confer resistance to 2,4-D when overexpressed in yeast [238]. However, it does not appear that studies examining the consequences of over expression of this type of gene in plants have been completed.

4.1.5. Glycosyltransferases

Glycosyltransferases (GTs), enzymes that add carbohydrates to molecules, are involved in the detoxification of herbicides in addition to many other roles within plant cells [239,240]. They are numerous in plant genomes with one particular family within this superfamily, the UDP-glucose dependent glycosyltransferases (UGTs), having 107 functional members in *Arabidopsis* [241]. Like CYP and GSTs genes, they are induced by the application of safeners and have been detected in transcriptome studies following herbicide application [189] and enzymes from this group from a wide variety of organisms have been demonstrated to have activity against atrazine and fluorodifen (group F1) [240]. However, unlike the other superfamilies discussed here, we did not find any examples of genes from this family being used to produce HR organisms. Instead, much of the work focused on these enzymes is examining the potential of these enzymes in phytoremediation of organic pollutants [189,242,243]. For example, a gene in *Arabidopsis* (*UGT72B1*) encodes an enzyme that detoxifies 3,4-dichloroaniline (DCA) and 2,4,5-trichlorophenol (TCP) [244].

4.2. A Role for Genomic Approaches

Due to the complexity, diversity, and number of genes that could underlie NTSR; identification of resistance-conferring mutations is a significant challenge even when one has a lead on the potential genetic basis from the above insights [180]. Clearly, significant progress is being made through the application of RNA sequencing to identify the genes being expressed following herbicide application, expression analysis of those genes using quantitative PCR, and transformation of model organisms such as *Arabidopsis* and tobacco to verify the function of the genes. Additional genomic information for weeds is an asset for this type of investigation and can allow comparative genetic approaches and searches with tools such as BLAST [106] to identify and classify members of the multigene families discussed above as has been done in model organisms and crops (e.g., [225]). This can allow for systematic testing of the activity each enzyme (e.g., [241,245]). However, there are undoubtedly more genes and gene families involved in NTSR (e.g., [246]). As with unravelling the demographic history and structure of populations discussed, one method of identifying these genes is to examine the signature of the strong artificial selection pressure of herbicide application across the genome (see Section 3.2). Additionally, a physical map combined with the tools of genetics (e.g., linkage mapping, genome-wide association studies) can inform on small to large effect genomic loci involved in HR.

4.3. Example: Glyphosate NTSR in Morning Glory

A recent *tour de force* investigating glyphosate resistance in morning glory (*Ipomoea purpurea* (L.) Roth.) provides a clear example of how genomics and detection of the signature of selection can be applied to understanding the basis of non-target site resistance. In this work, Van Etten and colleagues [187] generated genome wide DNA markers to examine population structure, the possibility of multiple origins of HR in the species, and to provide an indication of where selection was acting in the genomes. They then sequenced the species' genome and re-sequenced targets within the exome, the regions of the genome that are the parts of a gene that encode the final RNA transcripts, in regions showing selection. This data was used to assemble multiple lines of evidence to identify the candidate genes underlying glyphosate resistance.

To provide information of population differentiation and structure, examine the evidence for HR genes being introduced to populations via gene flow versus the HR arising multiple times, and to search for signatures of selection Van Etten et al. [185] used a reduced genome representation technique (nextRAD). This approach identified single nucleotide polymorphisms (SNPs) across the species' genome for ten individuals from each of four high and four low survival populations. This approach is a variant of restriction site associated DNA sequencing (RADseq), which in general, use restriction enzymes (often a pair) to selectively amplify regions adjacent to restriction sites across a species' genome [247,248]. The number of markers can be manipulated through the length of the restriction enzyme's recognition site allowing for the density of the markers to be manipulated depending on the project's goal and species genome size. As no sequence data is required before hand, this type of data can be generated for species whether or not they have genome sequences available. For each individual, enough Illumina sequencing data needs to be completed to result in approximately 30× coverage for each amplified region. Then programs such as STACKS (catchenlab.life.illinois.edu/stacks) [249–251] or TASSEL (bitbucket.org/tasseladmin/tassel-5-source/wiki/Home) [252] can be used to either group reads by similarity, if a sequenced genome is unavailable, or to align the reads to a draft genome sequence to locate polymorphic (variable) SNPs. These SNPs can then be analyzed with a plethora of packages in the free statistical programing language R [253] to understand the population biology (reviewed by [254]). This can include calculation of population differentiation (F_{ST}) using hierfstat [255] or StAMPP [256]; the generation and visualization of unweighted pair group method with arithmetic mean (UPGMA) or neighbor joining trees using poppr [257] and phytools [258]; and k-means clustering (adegenet [259]) to further investigate population structure. In the case of glyphosate resistance, in both morning glory [185] and Palmer amaranth (*Amaranthus palmeri* S. Wats.) [260], this approach indicated that gene flow introducing HR alleles has likely been responsible for much of the pattern of resistance and susceptible populations. However, in addition to gene flow, a second origin of glyphosate resistance was also suggested in Palmer amaranth [260].

The population level SNP data generated by Van Etten et al. [185] was then further analyzed with two programs, BayeScan [261], which can identify SNPs that show signs of selection and bayenv2 [262], which indicate SNPs associated with levels of HR. BayeScan (cmpg.unibe.ch/software/BayeScan/) calculates pairwise F_{ST} values between each population sampled and a theoretical population comprised of a common gene pool from all sampled populations. Selection is implied as an explanation, if a locus specific factor improves the logistic regression model for these F_{ST} values that includes population structure [261]. The program bayenv2 (bitbucket.org/tguenther/bayenv2_public/src/default/) looks for correlations between an environmental variable, such as HR level, and SNP frequency using a Bayesian method that estimates the pattern of covariance of allele frequencies, uses this as a null model and then tests each SNP [262]. Putative genes in proximity to the 42 outlier SNPs identified by BayeScan and the 83 SNPs flagged by bayenv2 were then identified by annotation tools such as AUGUSTUS (see above).

Next they sequenced a morning glory (diploid, approximately 978 Mb,1C = 1.0 pg [24], 2n = 30 [24])) individual that they considered to be high homozygous using PacBio reads (11 SMRT Cells) and Illumina short read data (100 bp paired end). They completed two genome assemblies one using only the Illumina data with the program ABYSS (github.com/bcgsc/abyss) [263] and the other using a hybrid approach that combined their long and short read data with the program DBG2OLC (github.com/yechengxi/DBG2OLC) [264]. This later assembly consisted of 17,897 scaffolds, had an N50 of 15,425 and a total length of 1,948 Mbp.

They then used their genome assembly to design probes (baits) to perform target-capture resequencing of these genes, the EPSPS genes, genes previously associated with HR and a randomly selected control group. This targeted exome re-sequencing was then completed for five individuals from each of their eight populations. These re-sequenced contigs were aligned to the chromosome level sequence of Japanese morning glory (*Ipomoea nil* (L.) Roth.) [265] to visualize the pattern of outliers indicating selection and they identified five regions of interest which contained 945 genes—including multiple members of the CYP, GSTs GT, and ABC transporter superfamilies. To determine if the number of members identified in these regions was greater than expectation for these large families, they resampled Japanese morning glory's genome to provide a baseline estimate of the number of that would be expected. This indicated that GT, ABC transporters and CYP genes were each overrepresented in the identified regions. These five regions also showed high genetic differentiation between populations with high and low glyphosate survival.

One approximately 29 kb region aligned to Japanese morning glory's chromosome 10 showed reduced nucleotide diversity in resistant individuals, strong evidence of selection based on Tajima's D and Fay and Wu's H as well as stronger linkage among the SNPs of this region. This region contained a tandemly repeated group of seven GT genes and nine CYP genes. For this region, they determined that the majority of resistant individuals shared high genetic similarity and tests of convergence suggesting that this region contains one or more beneficial genes that were introduced by gene flow and rapidly swept through resistant populations. While none of the non-synonymous SNPs in these genes showed fixation in the high survival populations, this region has a strong likelihood of containing loci that underlie glyphosate resistance in the species and are strong candidates for further functional validation.

5. Future Application: Can We Genetically Alter Weed Population to Make Them Easier to Control?

With a greater understanding of the population biology of weed species and the identification of the DNA sequence changes that underlie HR come opportunities for new control strategies. This is made particularly true by the development of genetic engineering methods involving clustered regularly interspaced short palindromic repeats (CRISPR) technologies. CRISPR tools are both simple and versatile, contributing to their successful spread in all aspects of molecular biology (reviewed in [266]). CRISPR systems are found in bacteria and archaea where they provide acquired immunity against invasive elements like phages. They do so by co-opting small pieces of DNA sequence from the pathogen which they subsequently use to generate guide RNA molecules that "program" an endonuclease (e.g., Cas9) to scan the genome and find its target. The recognition of DNA sequence homologous to the guide triggers the cleavage of the DNA strand that leads to mutations and potential inactivation of the targeted element.

In a landmark study, the CRISPR system of *Streptococcus pyogenes* was reduced to two components, an endonuclease (Cas9) and a single guide RNA, that could efficiently and specifically cut DNA in vitro [267]. Following this, similar two-component systems were introduced in a plethora of different organisms to engineer mutations in the DNA sequence with outstanding success [268]. Ultimately, the only requirement for this approach is the knowledge of the targeted DNA sequence, making application in weed control theoretically possible [269]. Consequently, while the short answer to the question "Can we genetically alter weed population to make them easier to control?" is *probably*, there are a great number of technical [270], ethical [271] and ecological [272] hurdles and no current examples of this approach being used in weed science. Here we focus on describing and discussing the potential and technical challenges to developing a weed control strategy using the engineering of whole populations. For an example, we reach beyond weed science to the control of insecticide resistant mosquitoes, summarizing the current findings and approaches of the scientists, who are likely to be the first to release gene drive element into the environment to control a pest population.

5.1. *The Potential for Manipulation of Weed Populations*

Ever since the demonstration of the repurposing of a bacterial CRISPR system as a programmable endonuclease [267], there has been speculation about its potential use for pest control or eradication [273]. Indeed, there were early successes in the application of CRISPR-based "gene drive" systems in order to decimate or modify populations of fruitfly (*Drosophila melanogaster* Meigen) and importantly, disease-spreading mosquitoes (*Anopheles stephensi* Liston and *Anopheles gambiae* Giles) (reviewed by [274]). The basis of a gene drive system relies on using a selfish genetic element capable of either copying itself or biasing reproduction towards its own inheritance so that it propagates through a population in a non-Mendelian fashion. This cheating of the classic inheritance rules can compensate for some deleterious consequences and potentially allow a measure of population control. Adding CRISPR components to this paradigm then allowed homing in on specific targets within the genomes making it available to newly sequenced weed plants [269]. Such a system has yet to be created in plants, but the rapid evolution of plant genetic engineering could make it a reality in the not too distant future.

Indeed, in their report "Gene Drives on the Horizon" the National Academy of Sciences considers the potential of this strategy for the control of Palmer amaranth [271].

The overarching goal of such an endeavor is to create a transgenic weed able to introduce a genetic payload into the populations of its species using biased inheritance and resulted in populations that are easier to control because of a vulnerability introduced with the payload. What the ideal payload would be up for debate, but it is likely to include a CRISPR system composed of a gene encoding a programmable endonuclease like the *Streptococcus pyogenes* Rosenbach Cas9 and a single or multiple guide RNA. These guide RNA could be specifically designed to pair with the locus causing HR or, if this basis is unknown, the target locus could be unrelated to the HR allele, with the goal of introducing sensitivity to a new molecule altogether. The recognition of the target triggers catalytic activity and the cutting of the target DNA creating a lesion. Since DNA breaks are highly detrimental, they are quickly repaired by one of the many pathways existing in the host cell. The gene drive system then subverts the DNA repair pathways ensuring its own propagation. This step represents one of the major challenges to this approach, as plant cells are known to heavily favor non-homologous DNA repair pathways that only produce small DNA sequence changes [275] that would fail to propagate the selfish element.

Indeed, the success of gene drive methods in fruitflies and mosquitoes is due in large part to the frequent use of homology-guided DNA repair in insect cells. However, plant somatic cells seldom use homologous recombination and favor non-homologous repair mechanisms [275]. For gene drive elements to spread efficiently in a plant population, this ratio between the two types of repair would have to be altered. This would be critical as non-homologous repair would create alleles resistant to the CRISPR system that would counter efforts to spread the gene drive. This is why the precise insertion of the gene drive element at a chosen location in the weed genome will likely be a *sine qua non* condition to its propagation. Once integrated, the new allele can start competing with natural alleles, which it can target for cleavage and convert using the host cell machinery. Encouragingly, the molecular mechanism called gene targeting, which uses the same homologous host DNA repair pathways as the gene drive approach, is of great interest in plant genetic engineering and has greatly improved the past few years [276]. Gene targeting aims at delivering a DNA sequence of interest at a specific location within the genome and, therefore, has also greatly benefited from advances in CRISPR technologies. Just like gene drive, gene targeting requires the use of homology-guided DNA repair mechanisms instead of non-homologous DNA repair. The difference between the two is that the final goal of gene targeting is a single isolated event, while a gene drive must self-propagate indefinitely, thereby adding to the challenge.

Excitingly, the case of a bacterial transposon that co-opted a CRISPR system as a means to guide its own propagation within the genome was recently discovered [277]. Transposons are themselves selfish elements that have evolved different means to copy themselves to favor their propagation. For example, some transposons encode an enzyme called integrase that can insert a DNA fragment at a target site in a genome. This new molecular tool has enormous potential as a gene drive system being able to circumvent the need to coax the host repair machinery to use homologous repair mechanisms.

5.2. Additional Technical Challenges

There are a number of additional technical limitations in the creation of a useful gene drive system for weed management beyond a need for the target species to use homologous repair mechanisms. As a first hurdle, this approach would be restricted to plants that can be genetically transformed and little effort has been devoted to the development of transformation techniques in weeds. Plant susceptibility to transformation is highly variable and whether or not it is ultimately possible in a species depends on many intrinsic factors [278]. For instance, species with unfused carpels at the extremity of the stigma may be amenable to the convenient floral dip *Agrobacterium* mediated transformation method. However, the great majority of plant species relies on other methods, such as tissue culture with *Agrobacterium tumefaciens* Smith and Townsend or biolistic bombardment, both being much more time and resource consuming. It could, therefore, take a few months to many years to develop a new transformation protocol for a particular plant—a potentially sizable initial investment of resources.

When transformation is possible, the challenge of precisely integrating a given DNA construct remains. At the molecular level, the problem can be broken down into two distinct parts; the mobilization of the homologous repair machinery and the delivery of the DNA template to be copied in the genome. For the first part, it has been reported that expressing the CRISPR system in specialized cells where homology-guided DNA repair occur at higher frequencies can increase gene targeting [276]. We know for instance that cells undergoing meiosis rely on homologous recombination between DNA molecules for orchestrating proper chromosome segregation. One could take advantage of these cell-specific conditions and engineer a system that would only act in a specific cell context as was recently done in mouse female germline [279]. Another interesting avenue is the tethering of repair machinery components to the endonuclease. Indeed, the fusion of Cas9 with different proteins offers many opportunities including influencing downstream DNA repair as it was successfully done in human cells [280]. Such an approach could be tailored to improve the propagation success of a gene drive element. In the second part of the molecular cascade, a DNA template has to be provided for the homologous repair machinery to integrate at the break site. In the case of gene drive, the engineered allele would bear homology to the wild allele and would therefore present itself as a repair template. Interestingly, recent studies have shown increased success in gene targeting when using components of a geminivirus [281–283]. The rationale behind this approach is that viruses can generate multiple extrachromosomal copies of a given DNA sequence thereby increasing the chances of any one fragment being used as template by the repair machinery. This element could be included into a gene drive system to increase its efficiency.

5.3. Evolutionary Consequences and the Need for Integration with Other Management Strategies

Even without the numerous technical impediments to gene drive strategies in weeds, this approach presents enormous ethical, regulatory, and ecological challenges. Theoretically, a gene drive that reduces the fitness of a population or its ability to reproduce could bring a species to extinction, as it was convincingly demonstrated for caged mosquitoes [284]. Setting this as a goal seems unwise and unlikely to gain societal support [272,285,286] or regulatory approval [287], as a result, strategies to re-sensitized populations to an herbicide or create susceptible to a specific compound unlikely to be found beyond the agroecosystem are likely to be more tenable. The advantage of such an approach is that it does not reduce the fitness of the population in the wild per se. Like the use of herbicides, altering weed populations as a management strategy would not be a silver bullet and would require integration into integrative weed management strategies. In part, this would be a consequence of the time needed for alleles to spread through populations as this could take 10 to 20 generations for a gene drive system to saturate a population [288]. In the re-sensitizing approach, this would mean forsaking the use of a given herbicide for many years thereby relying on other control strategies. In this regard, creating a susceptibility to a new molecule would present advantages but great care would need to be taken in choosing such a compound.

A second reason why this strategy would need to be part of an integrated weed management strategy, comes from the lesson we have learned from our reliance on herbicides. Plants are quite able to evolve in response to selection through modification of genetic machinery, the exome (see Section 4), and the biotic challenge represented by a gene drive element will result in selection on similar genetic machinery used to counter similar genetic attacks from viruses or selfish genetic elements. For example, in the case of a CRISPR-based gene drive, any synonymous mutation to the targeted site(s) would severely reduce the efficiency of the endonucleolytic cleavage [289]. This has already been demonstrated in model species such as fruitflies [290]. The emergence of such allele would be expected and could be mitigated by selecting sites where mutation would have high fitness cost would be more likely to provide a robust solution [291]. Since CRISPR genes come from bacteria, there is also a chance the plant cell would silence them using intrinsic mechanisms and a silenced allele could render then organism "immune" to the subsequent use of a CRISPR-based approach. Taken together, all these considerations argue for thorough modelling and confined population studies before such a strategy could be released in the fields as has been laid out in recommendations by the National Academy of Sciences [271].

5.4. Example: Gene Drive in Malaria Vector Mosquitos

While examples of gene drive development in weed species remain for future reviews, significant work has focused on using the technology to control mosquitoes that spread malaria. This is a system with parallel challenges to those faced in weed science including the emergence of multiple-insecticide resistance with both target site and NTSR mechanisms and a lack of new chemical control options [292,293]. Malaria is a serious and prevalent disease with over 200 million cases a year. It is often fatal, particularly in children, and disproportionally affects people living in South America, South Asia and sub-Saharan Africa where access to health care is often limited. The World Health Organization reported that of the 435,000 deaths reported in 2017 from malaria, ninety-two percent occurred in Africa and sixty-two percent occurred in children under five [294]. Malaria can be caused by any one of five *Plasmodium* parasites and can be transferred by several of the 450 species of *Anopheles* mosquitoes [294]. Within the sub-Saharan Africa region, malaria is primarily the result of infection by *Plasmodium falciparum* Welch transferred by female *Anopheles gambiae* mosquitoes [294]. Chemical strategies for controlling populations of these mosquitoes have resulted in the evolution of insecticide resistance with the first cases of pyrethroid resistance reported in Sudan in the 1970s and reports of resistance now available across Africa and in Madagascar [295]. Currently, *A. gambiae* populations in regions such as the Côte d'Ivoire and Burkina Faso, have evolved complete resistance to all approved classes of insecticides [296,297]. In 2015, researchers developed a CRISPR-based gene drive system designed to reduce reproductive capability by disrupting the sequence of a gene likely involved in the development of the embryo's body plan which results in female sterility. When carriers of this this gene were crossed to wild type mosquitoes the gene had a transmission rate of just over 99% and it was able to spread through a caged populations initiated from equal numbers of wild type and transformed individuals [298]. However, nuclease-resistant variants that completely blocked the spread of the gene could be detected as early as the second generation [285]. More recently, in 2018, the researchers improved on these results by disruption of a gene that controls sex differentiation and that has alternative splicing patterns in male and female mosquitoes, a characteristic believed to increase the constraints in the development of resistant variants. One of the two cages, initiated with 12.5% disrupted allele frequency, reached 100% allele frequency at generation 7 and extinction at generation 8, while for the second cage these two points were reached at generation 11 and 12 respectively. Importantly, they did not detect an evidence for the evolution of resistance to this gene drive, though they note that it may not be "resistance-proof" given a wider sample of mutations [284]. This work relied on foundational genomic information from *A. gambiae*'s genome sequence in 2002 [299] as well as detailed knowledge of the genetic basis of fundamental aspects of *A. gambia*'s biology. In July 2019, the researchers initiated small scale releases of genetically modified, sterile males (not equipped with gene drive) in Burkina Faso to produce the data required to meet the ultimate goal of releasing individuals with gene drive to control malaria [300]. The researchers that have developed this technology work with a consortium, Target Malaria (targetmalaria.org), that includes scientists, regulators, and community engagement specialists. They have also worked to understand the ecological risks associated with the unconfined release of this event [301]. This approach to develop the social license and regulatory approval for this type of intervention provides a valuable template for how weed scientists could approach the modification of a weed species for population management.

6. Conclusions

Genomic approaches are extremely powerful tools for understanding biological systems. These tools, while currently underutilized in weed biology, are exciting in their potential to answer key weed science questions and increasingly accessible. Here our goal is to provide a foothold for weed scientists considering this type of research by providing an introduction to the considerations and process of creating a draft genome and illustrating how that genome could be used as a fundamental tool. Draft weed genomes can provide a resource for demographic analyses that examine the result of selection on the genome. This information can shed light on the evolutionary origins of weeds allowing us to identify management practices that could prevent HR evolution. It can identify strengths and weaknesses of weed populations that can be targeted for control, while providing fundamental information on how plants rapidly respond to selection from humans. The changes that selection makes to the genome and revealed by genomic approaches can also provide evidence of which loci are the genetic basis of NTSR. This information will allow us to form strategies to interfere with these HR mechanisms. Finally, the insights we gain from a better understanding of weed species at the population, genomic and genic level using these approaches open the option of altering the genome of weed species to provide us another tool for weed management—a strategy nearing implementation in mice and mosquitoes.

Author Contributions: All authors contributed to the conceptualization, writing and editing of this paper. Funding acquisition was co-led by E.P and M.L.

Funding: This research was funded by Agriculture and Agri-Food Canada (AAFC), "Deciphering complex mechanisms and inheritance patterns of herbicide resistance cases in Canada" grant number J-001751.

Acknowledgments: We thank Tyler Smith, Connie A. Sauder and Beatriz E. Lujan-Toro for comments on the manuscript.

Conflicts of Interest: The authors declare no conflict of interest.

References

1. Van Dijk, E.L.; Jaszczyszyn, Y.; Naquin, D.; Thermes, C. The Third Revolution in Sequencing Technology. *Trends Genet.* **2018**, *34*, 666–681. [CrossRef] [PubMed]
2. Heather, J.M.; Chain, B. The sequence of sequencers: The history of sequencing DNA. *Genomics* **2016**, *107*, 1–8. [CrossRef] [PubMed]
3. Li, F.-W.; Harkess, A. A guide to sequence your favorite plant genomes. *Appl. Plant Sci.* **2018**, *6*, 1–7. [CrossRef] [PubMed]
4. Dominguez Del Angel, V.; Hjerde, E.; Sterck, L.; Capella-Gutierrez, S.; Notredame, C.; Vinnere Pettersson, O.; Amselem, J.; Bouri, L.; Bocs, S.; Klopp, C.; et al. Ten steps to get started in Genome Assembly and Annotation. *F1000Research* **2018**, *7*, 148. [CrossRef]
5. Armstrong, O.; Fiddes, I.T.; Diekhans, M.; Paten, B. Whole-Genome Alignment and Comparative Annotation. *Annu. Rev. Anim. Biosci.* **2019**, *7*, 41–64. [CrossRef]
6. Gillings, M.R.; Paulsen, I.T.; Tetu, S.G. Genomics and the evolution of antibiotic resistance. *Ann. N. Y. Acad. Sci.* **2017**, *1388*, 92–107. [CrossRef]
7. Loman, N.J.; Pallen, M.J. Twenty years of bacterial genome sequencing. *Nat. Rev. Microbiol.* **2015**, *13*, 787–794. [CrossRef]
8. Hatfull, G.F. Bacteriophage genomics. *Curr. Opin. Microbiol.* **2008**, *11*, 447–453. [CrossRef]
9. Holmes, E.C. Viral Evolution in the Genomic Age. *PLoS Biol.* **2007**, *5*, e278. [CrossRef]
10. Gudbjartsson, D.F.; Helgason, H.; Gudjonsson, S.A.; Zink, F.; Oddson, A.; Gylfason, A.; Besenbacher, S.; Magnusson, G.; Halldorsson, B.V.; Hjartarson, E.; et al. Large-scale whole-genome sequencing of the Icelandic population. *Nat. Genet.* **2015**, *47*, 435–444. [CrossRef]
11. Stranger, B.E.; Nica, A.C.; Forrest, M.S.; Dimas, A.; Bird, C.P.; Beazley, C.; Ingle, C.E.; Dunning, M.; Flicek, P.; Koller, D.; et al. Population genomics of human gene expression. *Nat. Genet.* **2007**, *39*, 1217–1224. [CrossRef] [PubMed]

12. Altshuler, D.L.; Durbin, R.M.; Abecasis, G.R.; Bentley, D.R.; Chakravarti, A.; Clark, A.G.; Collins, F.S.; De La Vega, F.M.; Donnelly, P.; Egholm, M.; et al. A map of human genome variation from population-scale sequencing. *Nature* **2010**, *467*, 1061–1073.
13. Li, J.Z.; Absher, D.M.; Tang, H.; Southwick, A.M.; Casto, A.M.; Ramachandran, S.; Cann, H.M.; Barsh, G.S.; Feldman, M.; Cavalli-Sforza, L.L.; et al. Worldwide human relationships inferred from genome-wide patterns of variation. *Science* **2008**, *319*, 1100–1104. [CrossRef] [PubMed]
14. Ravet, K.; Patterson, E.L.; Krähmer, H.; Hamouzová, K.; Fan, L.; Jasieniuk, M.; Lawton-Rauh, A.; Malone, J.M.; McElroy, J.S.; Merotto, A.; et al. The power and potential of genomics in weed biology and management. *Pest Manag. Sci.* **2018**, *74*, 2216–2225. [CrossRef] [PubMed]
15. Basu, C.; Halfhill, M.D.; Mueller, T.C.; Stewart, C.N. Weed genomics: New tools to understand weed biology. *Trends Plant Sci.* **2004**, *9*, 391–398. [CrossRef] [PubMed]
16. Venter, J.C.; Adams, M.D.; Myers, E.W.; Li, P.W.; Mural, R.J.; Sutton, G.G.; Smith, H.O.; Yandell, M.; Evans, C.A.; Holt, R.A.; et al. The sequence of the human genome. *Science* **2001**, *291*, 1304–1351. [CrossRef] [PubMed]
17. The Arabidopsis Genome Initiative Analysis of the genome sequence of the flowering plant *Arabidopsis thaliana*. *Nature* **2000**, *408*, 796–815. [CrossRef]
18. Michael, T.P.; Jackson, S. The First 50 Plant Genomes. *Plant Genome* **2013**, *6*, 1–7. [CrossRef]
19. Veeckman, E.; Ruttink, T.; Vandepoele, K. Are We There Yet? Reliably Estimating the Completeness of Plant Genome Sequences. *Plant Cell* **2016**, *28*, 1759–1768. [CrossRef]
20. Jung, H.; Winefield, C.; Bombarely, A.; Prentis, P.; Waterhouse, P. Tools and Strategies for Long-Read Sequencing and De Novo Assembly of Plant Genomes. *Trends Plant Sci.* **2019**, *8*, 1–25. [CrossRef]
21. Ekblom, R.; Wolf, J.B.W. A field guide to whole-genome sequencing, assembly and annotation. *Evol. Appl.* **2014**, *7*, 1026–1042. [CrossRef] [PubMed]
22. Wajid, B.; Serpedin, E. Do it yourself guide to genome assembly. *Brief. Funct. Genom.* **2016**, *15*, 1–9. [CrossRef] [PubMed]
23. Leitch, I.; Johnston, E.; Pellicer, J.; Hidalgo, O.; Bennett, M. Angiosperm DNA C-Values Database. Available online: https://cvalues.science.kew.org/ (accessed on 28 May 2019).
24. Rice, A.; Glick, L.; Abadi, S.; Einhorn, M.; Kopelman, N.M.; Salman-Minkov, A.; Mayzel, J.; Chay, O.; Mayrose, I. The Chromosome Counts Database (CCDB)—A community resource of plant chromosome numbers. *New Phytol.* **2015**, *206*, 19–26. [CrossRef] [PubMed]
25. Greilhuber, J.; Temsch, E.M.; Loureiro, J.C.M. Nuclear DNA Content Measurement. In *Flow Cytometry with Plant Cells*; Doležel, J., Greilhuber, J., Suda, J., Eds.; Wiley-VCH Verlag GmbH & Co. KGaA: Weinheim, Germany, 2007; pp. 67–101. ISBN 9783527314874.
26. Doležel, J.; Bartoš, J. Plant DNA flow cytometry and estimation of nuclear genome size. *Ann. Bot.* **2005**, *95*, 99–110. [CrossRef] [PubMed]
27. Leitch, I.J.; Bennett, M.D. Genome size and its uses: The impact of flow cytometry. In *Flow Cytometry with Plant Cells: Analysis of Genes, Chromosomes and Genomes*; Doležel, J., Greilhuber, J., Suda, J., Eds.; Wiley-VCH Verlag GmbH & Co. KGaA: Weinheim, Germany, 2007; pp. 153–176.
28. Galbraith, D.W.; Harkins, K.R.; Maddox, J.M.; Ayres, N.M.; Sharma, D.P.; Firoozabady, E. Rapid flow cytometric analysis of the cell cycle in intact plant tissues. *Science* **1983**, *220*, 1049–1051. [CrossRef] [PubMed]
29. Smith, T.W.; Kron, P.; Martin, S.L. flowPloidy: An R package for genome size and ploidy assessment of flow cytometry data. *Appl. Plant Sci.* **2018**, *6*, e01164. [CrossRef] [PubMed]
30. Doležel, J.; Bartoš, J.; Voglmayr, H.; Greilhuber, J. Nuclear DNA content and genome size of trout and human. *Cytometry* **2003**, *51A*, 127–128. [CrossRef] [PubMed]
31. Barow, M.; Meister, A. Endopolyploidy in seed plants is differently correlated to systematics, organ, life strategy and genome size. *Plant Cell Environ.* **2003**, *26*, 571–584. [CrossRef]
32. Barow, M.; Jovtchev, G. Endopolyploidy in Plants and its Analysis by Flow Cytometry. In *Flow Cytometry with Plant Cells*; Doležel, J., Greilhuber, J., Suda, J., Eds.; WILEY-VCH Verlag GmbH & Co. KGaA: Weinheim, Germany, 2007; pp. 349–372. ISBN 9783527314874.
33. Doležel, J.; Kubaláková, M.; Suchánková, P.; Kovářová, P.; Bartoš, J.; Šimková, H. Chromosome analysis and sorting. In *Flow Cytometry with Plant Cells*; Doležel, J., Greilhuber, J., Suda, J., Eds.; WILEY-VCH Verlag GmbH & Co.: Weinheim, Germany, 2007; pp. 373–404. ISBN 9783527314874.

34. Clark, K.; Karsch-Mizrachi, I.; Lipman, D.J.; Ostell, J.; Sayers, E.W. GenBank. *Nucleic Acids Res.* **2016**, *44*, D67–D72. [CrossRef]
35. Heap, I. The International Survey of Herbicide Resistant Weeds. Available online: www.weedscience.org (accessed on 4 January 2018).
36. United State Department of Agriculture Federal Noxious Weeds. Available online: https://plants.usda.gov/java/noxious (accessed on 25 July 2019).
37. Australian Government Weeds of National Significance. Available online: https://www.environment.gov.au/biodiversity/invasive/weeds/weeds/lists/wons.html (accessed on 25 July 2019).
38. Weber, E.; Gut, D. A survey of weeds that are increasingly spreading in Europe. In *Agronomy for Sustainable Development*; Springer Verlag/EDP Sciences/INRA: Berlin/Heidelberg, Germany, 2005; pp. 109–121.
39. Minister of Agriculture and Agri-Food Canada (AAFC). *Weed Seeds Order*; Minister of Agriculture and Agri-Food (AAFC), Canada: Ottawa, ON, Canada, 2016.
40. Straub, S.C.K.; Cronn, R.C.; Edwards, C.; Fishbein, M.; Liston, A. Horizontal transfer of DNA from the mitochondrial to the plastid genome and its subsequent evolution in milkweeds (*Apocynaceae*). *Genome Biol. Evol.* **2013**, *5*, 1872–1885. [CrossRef]
41. Byrne, S.L.; Erthmann, P.Ø.; Agerbirk, N.; Bak, S.; Hauser, T.P.; Nagy, I.; Paina, C.; Asp, T. The genome sequence of *Barbarea vulgaris* facilitates the study of ecological biochemistry. *Sci. Rep.* **2017**, *7*, 1–14. [CrossRef] [PubMed]
42. Bettgenhaeuser, J.; Corke, F.M.K.; Opanowicz, M.; Green, P.; Hernández-Pinzón, I.; Doonan, J.H.; Moscou, M.J. Natural Variation in *Brachypodium* Links Vernalization and Flowering Time Loci as Major Flowering Determinants. *Plant Physiol.* **2017**, *173*, 256–268. [CrossRef] [PubMed]
43. Cai, C.; Wang, X.; Liu, B.; Wu, J.; Liang, J.; Cui, Y.; Cheng, F.; Wang, X. *Brassica rapa* Genome 2.0: A Reference Upgrade through Sequence Re-assembly and Gene Re-annotation. *Mol. Plant* **2017**, *10*, 649–651. [CrossRef] [PubMed]
44. Van Bakel, H.; Stout, J.M.; Cote, A.G.; Tallon, C.M.; Sharpe, A.G.; Hughes, T.R.; Page, J.E. The draft genome and transcriptome of *Cannabis sativa*. *Genome Biol.* **2011**, *12*, R102. [CrossRef] [PubMed]
45. Kasianov, A.S.; Klepikova, A.V.; Kulakovskiy, I.V.; Gerasimov, E.S.; Fedotova, A.V.; Besedina, E.G.; Kondrashov, A.S.; Logacheva, M.D.; Penin, A.A. High-quality genome assembly of *Capsella bursa-pastoris* reveals asymmetry of regulatory elements at early stages of polyploid genome evolution. *Plant J.* **2017**, *91*, 278–291. [CrossRef] [PubMed]
46. Ye, G.; Zhang, H.; Chen, B.; Nie, S.; Liu, H.; Gao, W.; Wang, H.; Gao, Y.; Gu, L. *De novo* genome assembly of the stress tolerant forest species *Casuarina equisetifolia* provides insight into secondary growth. *Plant J.* **2019**, *97*, 779–794. [CrossRef]
47. Griesmann, M.; Chang, Y.; Liu, X.; Song, Y.; Haberer, G.; Crook, M.B.; Billault-Penneteau, B.; Lauressergues, D.; Keller, J.; Imanishi, L.; et al. Phylogenomics reveals multiple losses of nitrogen-fixing root nodule symbiosis. *Science* **2018**, *361*, eaat1743. [CrossRef]
48. Wang, L.; He, F.; Huang, Y.; He, J.; Yang, S.; Zeng, J.; Deng, C.; Jiang, X.; Fang, Y.; Wen, S.; et al. Genome of Wild Mandarin and Domestication History of Mandarin. *Mol. Plant* **2018**, *11*, 1024–1037. [CrossRef]
49. Peng, Y.; Lai, Z.; Lane, T.; Nageswara-Rao, M.; Okada, M.; Jasieniuk, M.; O'Geen, H.; Kim, R.W.; Sammons, R.D.; Rieseberg, L.H.; et al. *De novo* genome assembly of the economically important weed horseweed using integrated data from multiple sequencing platforms. *Plant Physiol.* **2014**, *166*, 1241–1254. [CrossRef]
50. Sarkar, D.; Mahato, A.K.; Satya, P.; Kundu, A.; Singh, S.; Jayaswal, P.K.; Singh, A.; Bahadur, K.; Pattnaik, S.; Singh, N.; et al. The draft genome of *Corchorus olitorius* cv. JRO-524 (Navin). *Genom. Data* **2017**, *12*, 151–154. [CrossRef]
51. Garcia-Mas, J.; Benjak, A.; Sanseverino, W.; Bourgeois, M.; Mir, G.; Gonzalez, V.M.; Henaff, E.; Camara, F.; Cozzuto, L.; Lowy, E.; et al. The genome of melon (*Cucumis melo* L.). *Proc. Natl. Acad. Sci. USA* **2012**, *109*, 11872–11877. [CrossRef] [PubMed]
52. Scaglione, D.; Reyes-Chin-Wo, S.; Acquadro, A.; Froenicke, L.; Portis, E.; Beitel, C.; Tirone, M.; Mauro, R.; Lo Monaco, A.; Mauromicale, G.; et al. The genome sequence of the outbreeding globe artichoke constructed de novo incorporating a phase-aware low-pass sequencing strategy of F1 progeny. *Sci. Rep.* **2016**, *6*, 1–17.

53. Iorizzo, M.; Ellison, S.; Senalik, D.; Zeng, P.; Satapoomin, P.; Huang, J.; Bowman, M.; Iovene, M.; Sanseverino, W.; Cavagnaro, P.; et al. A high-quality carrot genome assembly provides new insights into carotenoid accumulation and asterid genome evolution. *Nat. Genet.* **2016**, *48*, 657–666. [CrossRef] [PubMed]
54. Guo, L.; Qiu, J.; Ye, C.; Jin, G.; Mao, L.; Zhang, H.; Yang, X.; Peng, Q.; Wang, Y.; Jia, L.; et al. *Echinochloa crus-galli* genome analysis provides insight into its adaptation and invasiveness as a weed. *Nat. Commun.* **2017**, *8*, 1–10. [CrossRef] [PubMed]
55. Badouin, H.; Gouzy, J.; Grassa, C.J.; Murat, F.; Staton, S.E.; Cottret, L.; Lelandais-Brière, C.; Owens, G.L.; Carrère, S.; Mayjonade, B.; et al. The sunflower genome provides insights into oil metabolism, flowering and Asterid evolution. *Nature* **2017**, *546*, 148–152. [CrossRef] [PubMed]
56. Wu, S.; Lau, K.H.; Cao, Q.; Hamilton, J.P.; Sun, H.; Zhou, C.; Eserman, L.; Gemenet, D.C.; Olukolu, B.A.; Wang, H.; et al. Genome sequences of two diploid wild relatives of cultivated sweetpotato reveal targets for genetic improvement. *Nat. Commun.* **2018**, *9*, 1–12. [CrossRef] [PubMed]
57. Honig, J.A.; Zelzion, E.; Wagner, N.E.; Kubik, C.; Averello, V.; Vaiciunas, J.; Bhattacharya, D.; Bonos, S.A.; Meyer, W.A. Microsatellite identification in perennial ryegrass using next-generation sequencing. *Crop Sci.* **2017**, *57*, S-331–S-340. [CrossRef]
58. Vining, K.J.; Johnson, S.R.; Ahkami, A.; Lange, I.; Parrish, A.N.; Trapp, S.C.; Croteau, R.B.; Straub, S.C.K.; Pandelova, I.; Lange, B.M. Draft Genome Sequence of *Mentha longifolia* and Development of Resources for Mint Cultivar Improvement. *Mol. Plant* **2017**, *10*, 323–339. [CrossRef]
59. Zou, C.; Li, L.; Miki, D.; Li, D.; Tang, Q.; Xiao, L.; Rajput, S.; Deng, P.; Peng, L.; Jia, W.; et al. The genome of broomcorn millet. *Nat. Commun.* **2019**, *10*, 491–500. [CrossRef]
60. Guo, L.; Guo, L.; Winzer, T.; Yang, X.; Li, Y.; Ning, Z.; He, Z.; Teodor, R.; Lu, Y.; Tim, A.; et al. The opium poppy genome and morphinan production. *Science* **2018**, *362*, 343–347. [CrossRef]
61. Liu, Y.-J.; Wang, X.-R.; Zeng, Q.-Y. *De novo* assembly of white poplar genome and genetic diversity of white poplar population in Irtysh River basin in China. *Sci. China Life Sci.* **2019**, *62*, 609–618. [CrossRef] [PubMed]
62. Moghe, G.D.; Hufnagel, D.E.; Tang, H.B.; Xiao, Y.L.; Dworkin, I.; Town, C.D.; Conner, J.K.; Shiu, S.H. Consequences of Whole-Genome Triplication as Revealed by Comparative Genomic Analyses of the Wild Radish Raphanus raphanistrum and Three Other Brassicaceae Species. *Plant Cell* **2014**, *26*, 1925–1937. [CrossRef] [PubMed]
63. Xiaohui, Z.; Zhen, Y.; Shiyong, M.; Yang, Q.; Xinhua, Y.; Xiaohua, C.; Feng, C.; Zhangyan, W.; Yuyan, S.; Yi, J.; et al. A *de novo* Genome of a Chinese Radish Cultivar. *Hortic. Plant J.* **2015**, *1*, 155–164.
64. Nakamura, N.; Hirakawa, H.; Sato, S.; Otagaki, S.; Matsumoto, S.; Tabata, S.; Tanaka, Y. Genome structure of *Rosa multiflora*, a wild ancestor of cultivated roses. *DNA Res.* **2018**, *25*, 113–121. [CrossRef] [PubMed]
65. Zhang, J.; Zhang, X.; Tang, H.; Zhang, Q.; Hua, X.; Ma, X.; Zhu, F.; Jones, T.; Zhu, X.; Bowers, J.; et al. Allele-defined genome of the autopolyploid sugarcane *Saccharum spontaneum* L. *Nat. Genet.* **2018**, *50*, 1565–1573. [CrossRef] [PubMed]
66. Bauer, E.; Schmutzer, T.; Barilar, I.; Mascher, M.; Gundlach, H.; Martis, M.M.; Twardziok, S.O.; Hackauf, B.; Gordillo, A.; Wilde, P.; et al. Towards a whole-genome sequence for rye (*Secale cereale* L.). *Plant J.* **2017**, *89*, 853–869. [CrossRef] [PubMed]
67. Giolai, M.; Paajanen, P.; Verweij, W.; Witek, K.; Jones, J.D.G.; Clark, M.D. Comparative analysis of targeted long read sequencing approaches for characterization of a plant's immune receptor repertoire. *BMC Genom.* **2017**, *18*, 1–15. [CrossRef] [PubMed]
68. Paterson, A.H.; Bowers, J.E.; Bruggmann, R.; Dubchak, I.; Grimwood, J.; Gundlach, H.; Haberer, G.; Hellsten, U.; Mitros, T.; Poliakov, A.; et al. The *Sorghum bicolor* genome and the diversification of grasses. *Nature* **2009**, *457*, 551–556. [CrossRef]
69. Dorn, K.M.; Fankhauser, J.D.; Wyse, D.L.; Marks, M.D. A draft genome of field pennycress (*Thlaspi arvense*) provides tools for the domestication of a new winter biofuel crop. *DNA Res.* **2015**, *22*, 121–131. [CrossRef]
70. Creber, H.M.C.; Davies, M.S.; Francis, D.; Walker, H.D. Variation in DNA C value in natural populations of *Dactylis glomerata* L. *New Phytol.* **1994**, *128*, 555–561. [CrossRef]
71. Beck, J. Meiotic Chromosome Counting in Flowering Plants Part 1 [Video File]. Available online: www.youtube.com/watch?v=iXqni6knH5A&t (accessed on 30 May 2019).
72. Beck, J. Meiotic Chromosome Counting in Flowering Plants Part 2 [Video File]. Available online: www.youtube.com/watch?v=xVV4qBfSQLs&t (accessed on 30 May 2019).

73. Kato, A. Air drying method using nitrous oxide for chromosome counting in maize. *Biotech. Histochem.* **1999**, *74*, 160–166. [CrossRef] [PubMed]
74. Kato, A.; Lamb, J.C.; Albert, P.S.; Danilova, T.; Han, F.; Gao, Z.; Findley, S.; Birchler, J.A. Chromosome Painting for Plant Biotechnology. In *Plant Chromosome Engineering. Methods in Molecular Biology (Methods and Protocols)*; Birchler, J.A., Ed.; Humana Press: Totowa, NJ, USA, 2011; Volume 701, pp. 67–96. ISBN 9781617379574.
75. Mandáková, T.; Lysak, M.A. Chromosome Preparation for Cytogenetic Analyses in *Arabidopsis*. *Curr. Protoc. Plant Biol.* **2016**, *1*, 43–51.
76. Chikhi, R.; Medvedev, P. Informed and automated k-mer size selection for genome assembly. *Bioinformatics* **2014**, *30*, 31–37. [CrossRef] [PubMed]
77. Marçais, G.; Kingsford, C. A fast, lock-free approach for efficient parallel counting of occurrences of k-mers. *Bioinformatics* **2011**, *27*, 764–770. [CrossRef] [PubMed]
78. Novák, P.; Neumann, P.; Macas, J. Graph-based clustering and characterization of repetitive sequences in next-generation sequencing data. *BMC Bioinform.* **2010**, *11*, 378–390. [CrossRef] [PubMed]
79. Novák, P.; Neumann, P.; Pech, J.; Steinhaisl, J.; MacAs, J. RepeatExplorer: A Galaxy-based web server for genome-wide characterization of eukaryotic repetitive elements from next-generation sequence reads. *Bioinformatics* **2013**, *29*, 792–793. [CrossRef] [PubMed]
80. Martin, S.L.; Smith, T.; James, T.; Shalabi, F.; Kron, P.; Sauder, C.A. An update to the Canadian range and abundance of Camelina spp. (Brassicaceae) east of the Rocky Mountains. *Botany* **2017**, *95*, 405–417. [CrossRef]
81. Roessler, K.; Muyle, A.; Diez, C.M.; Gaut, G.R.J.; Bousios, A.; Stitzer, M.C.; Seymour, D.K.; Doebley, J.F.; Liu, Q.; Gaut, B.S. The Genomics of Selfing in Maize (*Zea mays* ssp. *mays*): Catching Purging in the Act. *bioRxiv* **2019**, 594812. [CrossRef]
82. Palmer, C.E.D.; Keller, W.A. Overview of Haploidy. In *Biotechnology in Agriculture and Forestry Haploids in Crop Improvement II Vol.56*; Palmer, C.E., Keller, W.A., Kasha, K.J., Eds.; Springer: Berlin/Heidelberg, Germany, 2005; Volume 56, pp. 3–9.
83. Forster, B.P.; Thomas, W.T.B. Doubled haploids in genetics and plant breeding. In *Plant Breeding Reviews Vol. 25*; Janick, J., Ed.; John Wiley & Sons, Inc.: Hoboken, NJ, USA, 2005; Volume 25, pp. 57–88. ISBN 9780471666936.
84. Dunwell, J.M. Haploids in flowering plants: Origins and exploitation. *Plant Biotechnol. J.* **2010**, *8*, 377–424. [CrossRef] [PubMed]
85. Kyriakidou, M.; Tai, H.H.; Anglin, N.L.; Ellis, D.; Strömvik, M.V. Current Strategies of Polyploid Plant Genome Sequence Assembly. *Front. Plant Sci.* **2018**, *9*, 1660–1675. [CrossRef]
86. Carter, R.; Bryson, C.T.; Darbyshire, S.J. Preparation and Use of Voucher Specimens for Documenting Research in Weed Science. *Weed Technol.* **2007**, *21*, 1101–1108. [CrossRef]
87. Hussing, C.; Kampmann, M.L.; Mogensen, H.S.; Børsting, C.; Morling, N. Comparison of techniques for quantification of next-generation sequencing libraries. *Forensic Sci. Int. Genet. Suppl. Ser.* **2015**, *5*, e276–e278. [CrossRef]
88. Kreiner, J.M.; Giacomini, D.; Bemm, F.; Waithaka, B.; Regalado, J.; Lanz, C.; Hildebrandt, J.; Sikkema, P.H.; Tranel, P.J.; Weigel, D.; et al. Multiple modes of convergent adaptation in the spread of glyphosate-resistant *Amaranthus tuberculatus*. *bioRxiv* **2018**, 1–17. [CrossRef]
89. Patterson, E.L.; Saski, C.A.; Sloan, D.B.; Tranel, P.J.; Westra, P.; Gaines, T.A. The draft genome of *Kochia scoparia* and the mechanism of glyphosate resistance via transposon-mediated EPSPS tandem gene duplication. *bioRxiv* **2019**. [CrossRef] [PubMed]
90. Doyle, J.J.; Doyle, J.L. A rapid procedure for DNA purification from small quantities of fresh leaf tissue. *Phytochem. Bull.* **1987**, *19*, 11–15.
91. Healey, A.; Furtado, A.; Cooper, T.; Henry, R.J. Protocol: A simple method for extracting next-generation sequencing quality genomic DNA from recalcitrant plant species. *Plant Methods* **2014**, *10*, 21–29. [CrossRef] [PubMed]
92. Lander, E.S.; Waterman, M.S. Genomic mapping by fingerprinting random clones: A mathematical analysis. *Genomics* **1988**, *2*, 231–239. [CrossRef]
93. Soorni, A.; Haak, D.; Zaitlin, D.; Bombarely, A. Organelle_PBA, a pipeline for assembling chloroplast and mitochondrial genomes from PacBio DNA sequencing data. *BMC Genom.* **2017**, *18*, 49–57. [CrossRef]
94. De Coster, W.; D'Hert, S.; Schultz, D.T.; Cruts, M.; Van Broeckhoven, C. NanoPack: Visualizing and processing long-read sequencing data. *Bioinformatics* **2018**, *34*, 2666–2669. [CrossRef]

95. Salmela, L.; Rivals, E. LoRDEC: Accurate and efficient long read error correction. *Bioinformatics* **2014**, *30*, 3506–3514. [CrossRef]
96. Bolger, A.M.; Lohse, M.; Usadel, B. Trimmomatic: A flexible trimmer for Illumina sequence data. *Bioinformatics* **2014**, *30*, 2114–2120. [CrossRef] [PubMed]
97. Luo, R.; Liu, B.; Xie, Y.; Li, Z.; Huang, W.; Yuan, J.; He, G.; Chen, Y.; Pan, Q.; Liu, Y.; et al. SOAPdenovo2: An empirically improved memory-efficient short-read de novo assembler. *Gigascience* **2012**, *1*, 18–24. [CrossRef] [PubMed]
98. Jackman, S.D.; Vandervalk, B.P.; Mohamadi, H.; Chu, J.; Yeo, S.; Hammond, S.A.; Jahesh, G.; Khan, H.; Coombe, L.; Warren, R.L.; et al. ABySS 2.0: Resource-efficient assembly of large genomes using a Bloom filter. *Genome Res.* **2017**, *27*, 768–777. [CrossRef] [PubMed]
99. Koren, S.; Walenz, B.P.; Berlin, K.; Miller, J.R.; Phillippy, A.M. Canu: Scalable and accurate long-read assembly via adaptive k-mer weighting and repeat separation. *bioRxiv* **2016**, *27*, 722–736. [CrossRef] [PubMed]
100. Chin, C.S.; Peluso, P.; Sedlazeck, F.J.; Nattestad, M.; Concepcion, G.T.; Clum, A.; Dunn, C.; O'Malley, R.; Figueroa-Balderas, R.; Morales-Cruz, A.; et al. Phased diploid genome assembly with single-molecule real-time sequencing. *Nat. Methods* **2016**, *13*, 1050–1054. [CrossRef] [PubMed]
101. Laetsch, D.R.; Blaxter, M.L. BlobTools: Interrogation of genome assemblies. *F1000Research* **2017**, *6*, 1287. [CrossRef]
102. Kajitani, R.; Yoshimura, D.; Okuno, M.; Minakuchi, Y.; Kagoshima, H.; Fujiyama, A.; Kubokawa, K.; Kohara, Y.; Toyoda, A.; Itoh, T. Platanus-allee is a de novo haplotype assembler enabling a comprehensive access to divergent heterozygous regions. *Nat. Commun.* **2019**, *10*, 1–15. [CrossRef]
103. Waterhouse, R.M.; Seppey, M.; Simao, F.A.; Manni, M.; Ioannidis, P.; Klioutchnikov, G.; Kriventseva, E.V.; Zdobnov, E.M. BUSCO applications from quality assessments to gene prediction and phylogenomics. *Mol. Biol. Evol.* **2017**, *35*, 543–548. [CrossRef]
104. Simão, F.A.; Waterhouse, R.M.; Ioannidis, P.; Kriventseva, E.V.; Zdobnov, E.M. BUSCO: Assessing genome assembly and annotation completeness with single-copy orthologs. *Bioinformatics* **2015**, *31*, 3210–3212. [CrossRef]
105. Altschul, S.F.; Gish, W.; Miller, W.; Myers, E.W.; Lipman, D.J. Basic Local Alignment Search Tool. *J. Mol. Biol.* **1990**, *215*, 403–410. [CrossRef]
106. Morgulis, A.; Coulouris, G.; Raytselis, Y.; Madden, T.L.; Agarwala, R.; Schäffer, A.A. Database indexing for production MegaBLAST searches. *Bioinformatics* **2008**, *24*, 1757–1764. [CrossRef] [PubMed]
107. Alonge, M.; Soyk, S.; Ramakrishnan, S.; Wang, X.; Goodwin, S.; Sedlazeck, F.J.; Lippman, Z.B.; Schatz, M.C. Fast and accurate reference-guided scaffolding of draft genomes. *bioRxiv* **2019**, 519637. [CrossRef]
108. Walker, B.J.; Abeel, T.; Shea, T.; Priest, M.; Abouelliel, A.; Sakthikumar, S.; Cuomo, C.A.; Zeng, Q.; Wortman, J.; Young, S.K.; et al. Pilon: An integrated tool for comprehensive microbial variant detection and genome assembly improvement. *PLoS ONE* **2014**, *9*, e112963. [CrossRef] [PubMed]
109. Robinson, J.T.; Thorvaldsdóttir, H.; Winckler, W.; Guttman, M.; Lander, E.S.; Getz, G.; Mesirov, J.P. Integrative Genome Viewer. *Nat. Biotechnol.* **2011**, *29*, 24–26. [CrossRef]
110. Vaser, R.; Nagarajan, N.; Sović, I.; Šikic, M. Fast and accurate *de novo* genome assembly from long uncorrected reads. *Genome Res.* **2017**, *27*, 1–10. [CrossRef] [PubMed]
111. Oddes, S.; Zelig, A.; Kaplan, N. Three invariant Hi-C interaction patterns: Applications to genome assembly. *bioRxiv* **2018**, *142*, 89–99. [CrossRef] [PubMed]
112. Ghurye, J.; Pop, M.; Koren, S.; Bickhart, D.; Chin, C.S. Scaffolding of long read assemblies using long range contact information. *BMC Genom.* **2017**, *18*, 527–538. [CrossRef] [PubMed]
113. Kronenberg, Z.N.; Rhie, A.; Koren, S.; Concepcion, G.T.; Peluso, P.; Munson, K.M.; Hiendleder, S.; Fedrigo, O.; Jarvis, E.D.; Adam, M.; et al. Extended haplotype phasing of de novo genome assemblies with FALCON-Phase. *bioRxiv* **2018**, 1–27. [CrossRef]
114. Jibran, R.; Dzierzon, H.; Bassil, N.; Bushakra, J.M.; Edger, P.P.; Sullivan, S.; Finn, C.E.; Dossett, M.; Vining, K.J.; Vanburen, R.; et al. Chromosome-scale scaffolding of the black raspberry (*Rubus occidentalis* L.) genome based on chromatin interaction data. *Hortic. Res.* **2018**, *5*, 8–19. [CrossRef]
115. Lightfoot, D.J.; Jarvis, D.E.; Ramaraj, T.; Lee, R.; Jellen, E.N.; Maughan, P.J. Single-molecule sequencing and Hi-C-based proximity-guided assembly of amaranth (*Amaranthus hypochondriacus*) chromosomes provide insights into genome evolution. *BMC Biol.* **2017**, *15*, 74. [CrossRef]

116. Appels, R.; Eversole, K.; Feuillet, C.; Keller, B.; Rogers, J.; Stein, N.; Ronen, G. International Wheat Genome Sequencing Consortium Shifting the limits in wheat research and breeding using a fully annotated reference genome. *Science* **2018**, *361*, eaar7191. [PubMed]
117. Deschamps, S.; Zhang, Y.; Llaca, V.; Ye, L.; Sanyal, A.; King, M.; May, G.; Lin, H. A chromosome-scale assembly of the sorghum genome using nanopore sequencing and optical mapping. *Nat. Commun.* **2018**, *9*, 4844. [CrossRef] [PubMed]
118. Edger, P.P.; Poorten, T.J.; VanBuren, R.; Hardigan, M.A.; Colle, M.; McKain, M.R.; Smith, R.D.; Teresi, S.J.; Nelson, A.D.L.; Wai, C.M.; et al. Origin and evolution of the octoploid strawberry genome. *Nat. Genet.* **2019**, *51*, 541–547. [CrossRef] [PubMed]
119. Bosi, E.; Donati, B.; Galardini, M.; Brunetti, S.; Sagot, M.-F.; Lió, P.; Crescenzi, P.; Fani, R.; Fondi, M. MeDuSa: A multi-draft based scaffolder. *Bioinformatics* **2015**, *31*, 2443–2451. [CrossRef] [PubMed]
120. Keller, O.; Kollmar, M.; Stanke, M.; Waack, S. A novel hybrid gene prediction method employing protein multiple sequence alignments. *Bioinformatics* **2011**, *27*, 757–763. [CrossRef] [PubMed]
121. Stanke, M.; Steinkamp, R.; Waack, S.; Morgenstern, B. AUGUSTUS: A web server for gene finding in eukaryotes. *Nucleic Acids Res.* **2004**, *32*, W309–W312. [CrossRef] [PubMed]
122. Volfovsky, N.; Hass, B.J.; Salzberg, S.L. A clustering method for repeat analysis in DNA sequences. *Genome Biol.* **2001**, *2*, 0027.1–0027.11. [CrossRef] [PubMed]
123. Shi, J.; Liang, C. Generic Repeat Finder: A high-sensitivity tool for genome-wide de novo repeat detection. *Plant Physiol.* **2019**, *180*, 1803–1815. [CrossRef] [PubMed]
124. Mao, H.; Wang, H. SINE-scan: An efficient tool to discover short interspersed nuclear elements (SINEs) in large-scale genomic datasets. *Bioinformatics* **2017**, *33*, 743–745. [PubMed]
125. Chen, J.; Hu, Q.; Zhang, Y.; Lu, C.; Kuang, H. P-MITE: A database for plant miniature inverted-repeat transposable elements. *Nucleic Acids Res.* **2014**, *42*, 1176–1181. [CrossRef] [PubMed]
126. Xiong, W.; He, L.; Lai, J.; Dooner, H.K.; Du, C. HelitronScanner uncovers a large overlooked cache of Helitron transposons in many plant genomes. *Proc. Natl. Acad. Sci. USA* **2014**, *111*, 10263–10268. [CrossRef] [PubMed]
127. Götz, S.; García-Gómez, J.M.; Terol, J.; Williams, T.D.; Nagaraj, S.H.; Nueda, M.J.; Robles, M.; Talón, M.; Dopazo, J.; Conesa, A. High-throughput functional annotation and data mining with the Blast2GO suite. *Nucleic Acids Res.* **2008**, *36*, 3420–3435. [CrossRef] [PubMed]
128. Campbell, M.S.; Law, M.; Holt, C.; Stein, J.C.; Moghe, G.D.; Hufnagel, D.E.; Lei, J.; Achawanantakun, R.; Jiao, D.; Lawrence, C.J.; et al. MAKER-P: A Tool Kit for the Rapid Creation, Management, and Quality Control of Plant Genome Annotations. *Plant Physiol.* **2014**, *164*, 513–524. [CrossRef] [PubMed]
129. Audano, P.A.; Sulovari, A.; Graves-Lindsay, T.A.; Cantsilieris, S.; Sorensen, M.; Welch, A.M.E.; Dougherty, M.L.; Nelson, B.J.; Shah, A.; Dutcher, S.K.; et al. Characterizing the Major Structural Variant Alleles of the Human Genome. *Cell* **2019**, *176*, 663–675. [CrossRef] [PubMed]
130. Hackl, T.; Hedrich, R.; Schultz, J.; Förster, F. Proovread: Large-scale high-accuracy PacBio correction through iterative short read consensus. *Bioinformatics* **2014**, *30*, 3004–3011. [CrossRef] [PubMed]
131. Gnerre, S.; Maccallum, I.; Przybylski, D.; Ribeiro, F.J.; Burton, J.N.; Walker, B.J.; Sharpe, T.; Hall, G.; Shea, T.P.; Sykes, S.; et al. High-quality draft assemblies of mammalian genomes from massively parallel sequence data. *Proc. Natl. Acad. Sci. USA* **2011**, *108*, 1513–1518. [CrossRef] [PubMed]
132. English, A.C.; Richards, S.; Han, Y.; Wang, M.; Vee, V.; Qu, J.; Qin, X.; Muzny, D.M.; Reid, J.G.; Worley, K.C.; et al. Mind the Gap: Upgrading Genomes with Pacific Biosciences RS Long-Read Sequencing Technology. *PLoS ONE* **2012**, *7*, e47768. [CrossRef] [PubMed]
133. Mayela Soto-Jimenez, L.; Estrada, K.; Sanchez-Flores, A. GARM: Genome assembly, reconciliation and merging. *Curr. Top. Med. Chem.* **2014**, *14*, 418–424. [CrossRef] [PubMed]
134. Linthorst, J.; Hulsman, M.; Holstege, H.; Reinders, M. Scalable multi whole-genome alignment using recursive exact matching. *bioRxiv* **2015**, 022715. [CrossRef]
135. Gao, S.; Bertrand, D.; Chia, B.K.H.; Nagarajan, N. OPERA-LG: Efficient and exact scaffolding of large, repeat-rich eukaryotic genomes with performance guarantees. *Genome Biol.* **2016**, *17*, 102. [CrossRef]
136. Lukashin, A.V.; Borodovsky, M. GeneMark.hmm: New solutions for gene finding. *Nucleic Acids Res.* **1998**, *26*, 1107–1115. [CrossRef] [PubMed]
137. Solovyev, V.; Kosarev, P.; Seledsov, I.; Vorobyev, D. Automatic annotation of eukaryotic genes, pseudogenes and promoters. *Genome Biol.* **2006**, *7*, S10. [CrossRef] [PubMed]

138. Petit, S.; Boursault, A.; Le Guilloux, M.; Munier-Jolain, N.; Reboud, X. Weeds in agricultural landscapes. A review. *Agron. Sustain. Dev.* **2011**, *31*, 309–317. [CrossRef]
139. De Wet, J.M.J.; Harlan, J.R. Weeds and Domesticates: Evolution in the man-made habitat. *Econ. Bot.* **1975**, *29*, 99–107. [CrossRef]
140. Barrett, S.H. Crop mimicry in weeds. *Econ. Bot.* **1983**, *37*, 255–282. [CrossRef]
141. Harlan, J.R.; de Wet, J.M.J. Some thoughts about weeds. *Econ. Bot.* **1965**, *19*, 16–24. [CrossRef]
142. Tedin, O. Vererbung, Variation Und Syste-Matik in Der Gattung Camelina (German with English Summary). *Hereditas* **1925**, *6*, 275–386. [CrossRef]
143. Powles, S.B.; Yu, Q. Evolution in action: Plants resistant to herbicides. *Annu. Rev. Plant Biol.* **2010**, *61*, 317–347. [CrossRef]
144. Neve, P.; Vila-Aiub, M.; Roux, F. Evolutionary-thinking in agricultural weed management. *New Phytol.* **2009**, *184*, 783–793. [CrossRef]
145. Baker, H.G. The evolution of weeds. *Annu. Rev. Ecol. Syst.* **1974**, *5*, 1–24. [CrossRef]
146. Warwick, S.I.; Stewart, C.N., Jr. Crops come from wild plants: How domestication, transgenes, and linkage together shape ferality. In *Crop Ferality and Volunteerism*; CRC Press: Boca Raton, FL, USA, 2005; pp. 9–30. ISBN 9783540773405.
147. Ellstrand, N.C.; Heredia, S.M.; Leak-Garcia, J.A.; Heraty, J.M.; Burger, J.C.; Yao, L.; Nohzadeh-Malakshah, S.; Ridley, C.E. Crops gone wild: Evolution of weeds and invasives from domesticated ancestors. *Evol. Appl.* **2010**, *3*, 494–504. [CrossRef] [PubMed]
148. Vigueira, C.C.; Olsen, K.M.; Caicedo, A.L. The red queen in the corn: Agricultural weeds as models of rapid adaptive evolution. *Heredity* **2013**, *110*, 303–311. [CrossRef] [PubMed]
149. Royal, C.D.; Novembre, J.; Fullerton, S.M.; Goldstein, D.B.; Long, J.C.; Bamshad, M.J.; Clark, A.G. Inferring Genetic Ancestry: Opportunities, Challenges, and Implications. *Am. J. Hum. Genet.* **2010**, *86*, 661–673. [CrossRef] [PubMed]
150. Kidd, K.K.; Speed, W.C.; Pakstis, A.J.; Furtado, M.R.; Fang, R.; Madbouly, A.; Maiers, M.; Middha, M.; Friedlaender, F.R.; Kidd, J.R. Progress toward an efficient panel of SNPs for ancestry inference. *Forensic Sci. Int.* **2014**, *10*, 23–32. [CrossRef] [PubMed]
151. Pritchard, J.K.; Stephens, M.; Donnelly, P. Inference of population structure using multilocus genotype data. *Genetics* **2000**, *155*, 945–959. [PubMed]
152. Excoffier, L.; Smouse, P.E.; Quattro, J.M. Analysis of Molecular Variance Inferred From Metric Distances Among DNA Haplotypes: Application. *Genetics* **1992**, *131*, 479–491. [PubMed]
153. Pickrell, J.K.; Pritchard, J.K. Inference of Population Splits and Mixtures from Genome-Wide Allele Frequency Data. *PLoS Genet.* **2012**, *8*, e1002967. [CrossRef] [PubMed]
154. Slatkin, M. Isolation by distance in equilibrium and non-equilibrium populations. *Evolution* **1993**, *47*, 264–279. [CrossRef] [PubMed]
155. Vekemans, X.; Hardy, O.J. New insights from fine-scale spatial genetic structure analyses in plant populations. *Mol. Ecol.* **2004**, *13*, 921–935. [CrossRef] [PubMed]
156. Aguillon, S.M.; Fitzpatrick, J.W.; Bowman, R.; Schoech, S.J.; Clark, A.G.; Coop, G.; Chen, N. Deconstructing isolation-by-distance: The genomic consequences of limited dispersal. *PLoS Genet.* **2017**, *13*, e1006911. [CrossRef] [PubMed]
157. Bradburd, G.S.; Coop, G.M.; Ralph, P.L. Inferring continuous and discrete population genetic structure across space. *Genetics* **2018**, *210*, 33–52. [CrossRef] [PubMed]
158. Lawson, D.J.; van Dorp, L.; Falush, D. A tutorial on how not to over-interpret STRUCTURE and ADMIXTURE bar plots. *Nat. Commun.* **2018**, *9*, 3258. [CrossRef] [PubMed]
159. Gutenkunst, R.N.; Hernandez, R.D.; Williamson, S.H.; Bustamante, C.D. Inferring the Joint Demographic History of Multiple Populations from Multidimensional SNP Frequency Data. *PLoS Genet.* **2009**, *5*, e1000695. [CrossRef] [PubMed]
160. Excoffier, L.; Dupanloup, I.; Huerta-Sánchez, E.; Sousa, V.C.; Foll, M. Robust Demographic Inference from Genomic and SNP Data. *PLoS Genet.* **2013**, *9*, e1003905. [CrossRef] [PubMed]
161. Smith, J.M.; Haigh, J. The hitch-hiking effect of a favourable gene. *Genet. Res. Camb.* **1974**, *23*, 23–35. [CrossRef]
162. Przeworski, M. The Signature of Positive Selection at Randomly Chosen Loci. *Genetics* **2002**, *160*, 1179–1189. [PubMed]

163. Fay, J.C.; Wu, C.-I. Hitchhiking Under Positive Darwinian Selection. *Genetics* **2000**, *155*, 1405–1413. [PubMed]
164. Hermisson, J.; Pennings, P.S. Soft Sweeps: Molecular Population Genetics of Adaptation From Standing Genetic Variation. *Genetics* **2005**, *169*, 2335–2352. [CrossRef]
165. Meirmans, P.G.; Hedrick, P.W. Assessing population structure: F_{ST} and related measures. *Mol. Ecol. Resour.* **2011**, *11*, 5–18. [CrossRef]
166. Wright, S. Genetical Structure of Populations. *Nature* **1950**, 247–249. [CrossRef]
167. Tajima, F. Statistical method for testing the neutral mutation hypothesis by DNA polymorphism. *Genetics* **1989**, *123*, 585–595. [PubMed]
168. Fay, J.C.; Wu, C.-I. Sequence divergence, functional constraint, and selection in protein evolution. *Annu. Rev. Genomics Hum. Genet.* **2003**, *4*, 213–235. [CrossRef] [PubMed]
169. Pavlidis, P.; Živković, D.; Stamatakis, A.; Alachiotis, N. SweeD: Likelihood-Based Detection of Selective Sweeps in Thousands of Genomes. *Mol. Biol. Evol.* **2013**, *30*, 2224–2234. [CrossRef] [PubMed]
170. Degiorgio, M.; Huber, C.D.; Hubisz, M.J.; Hellmann, I.; Nielsen, R. SweepFinder2: Increased sensitivity, robustness and flexibility. *Bioinformatics* **2016**, *32*, 1895–1897. [CrossRef]
171. Nielsen, R.; Williamson, S.; Kim, Y.; Hubisz, M.J.; Clark, A.G.; Bustamante, C. Genomic scans for selective sweeps using SNP data. *Genome Res.* **2005**, *15*, 1566–1575. [CrossRef] [PubMed]
172. Berg, J.J.; Coop, G. A Population Genetic Signal of Polygenic Adaptation. *PLoS Genet.* **2014**, *10*, e1004412. [CrossRef] [PubMed]
173. Berg, J.J.; Harpak, A.; Sinnott-armstrong, N.; Joergensen, A.M.; Mostafavi, H.; Field, Y.; Boyle, E.A.; Zhang, X.; Racimo, F.; Pritchard, J.K.; et al. Reduced signal for polygenic adaptation of height in UK Biobank. *Elife* **2019**, *8*, e39725. [CrossRef] [PubMed]
174. Mosyakin, S.L.; Robertson, K.R. Amaranthus. In *Flora of North America North of Mexico*; Flora of North America Editorial Committee: New York, NY, USA; Oxford, UK, 2004.
175. Délye, C.; Menchari, Y.; Michel, S.; Cadet, E.; Le Corre, V. A new insight into arable weed adaptive evolution: Mutations endowing herbicide resistance also affect germination dynamics and seedling emergence. *Ann. Bot.* **2013**, *111*, 681–691. [CrossRef]
176. Baucom, R.S. Evolutionary and ecological insights from herbicide-resistant weeds: What have we learned about plant adaptation, and what is left to uncover? *New Phytol.* **2019**, *223*, 68–82. [CrossRef]
177. Yuan, J.S.; Tranel, P.J.; Stewart, C.N. Non-target-site herbicide resistance: A family business. *Trends Plant Sci.* **2007**, *12*, 6–13. [CrossRef]
178. Délye, C.; Jasieniuk, M.; Le Corre, V. Deciphering the evolution of herbicide resistance in weeds. *Trends Genet.* **2013**, *29*, 649–658. [CrossRef] [PubMed]
179. Beckie, H.J.; Tardif, F.J. Herbicide cross resistance in weeds. *Crop Prot.* **2012**, *35*, 15–28. [CrossRef]
180. Délye, C. Unravelling the genetic bases of non-target-site-based resistance (NTSR) to herbicides: A major challenge for weed science in the forthcoming decade. *Pest Manag. Sci.* **2013**, *62*, 176–187. [CrossRef] [PubMed]
181. Ghanizadeh, H.; Harrington, K.C. Non-target Site Mechanisms of Resistance to Herbicides. *CRC. Crit. Rev. Plant Sci.* **2017**, *36*, 24–34. [CrossRef]
182. Yang, Q.; Liu, Y.J.; Zeng, Q.Y. Overexpression of three orthologous glutathione S-transferases from *Populus* increased salt and drought resistance in *Arabidopsis*. *Biochem. Syst. Ecol.* **2019**, *83*, 57–61. [CrossRef]
183. Conte, S.S.; Lloyd, A.M. Exploring multiple drug and herbicide resistance in plants-Spotlight on transporter proteins. *Plant Sci.* **2011**, *180*, 196–203. [CrossRef] [PubMed]
184. Cummins, I.; Wortley, D.J.; Sabbadin, F.; He, Z.; Coxon, C.R.; Straker, H.E.; Sellars, J.D.; Knight, K.; Edwards, L.; Hughes, D.; et al. Key role for a glutathione transferase in multiple-herbicide resistance in grass weeds. *Proc. Natl. Acad. Sci. USA* **2013**, *110*, 5812–5817. [CrossRef] [PubMed]
185. Van Etten, M.; Lee, K.M.; Chang, S.-M.; Baucom, R.S. Parallel and nonparallel genomic responses contribute to herbicide resistance in *Ipomoea purpurea*, a common agricultural weed. *bioRxiv* **2019**. [CrossRef]
186. Salas-Perez, R.A.; Saski, C.A.; Noorai, R.E.; Srivastava, S.K.; Lawton-Rauh, A.L.; Nichols, R.L.; Roma-Burgos, N. RNA-Seq transcriptome analysis of *Amaranthus palmeri* with differential tolerance to glufosinate herbicide. *PLoS ONE* **2018**, *13*, e0195488. [CrossRef]
187. Bai, S.; Liu, W.; Wang, H.; Zhao, N.; Jia, S.; Zou, N.; Guo, W.; Wang, J. Enhanced herbicide metabolism and metabolic resistance genes identified in tribenuron-methyl resistant *Myosoton aquaticum* L. *J. Agric. Food Chem.* **2018**, *66*, 9850–9857. [CrossRef]

188. Kreuz, K.; Tommasini, R.; Martinoia, E. Old Enzymes for a New Job Herbicide Detoxification in Plants. *Plant Physiol.* **1996**, *111*, 349–353. [CrossRef] [PubMed]
189. Edwards, R.; Del Buono, D.; Fordham, M.; Skipsey, M.; Brazier, M.; Dixon, D.P.; Cummins, I. Differential induction of glutathione transferases and glucosyltransferases in wheat, maize and Arabidopsis thaliana by herbicide safeners. *Z. Nat.* **2005**, *60*, 307–316. [CrossRef] [PubMed]
190. Schuler, M.A.; Werck-Reichhart, D. Functional Genomics of P450s. *Annu. Rev. Plant Biol.* **2003**, *54*, 629–667. [CrossRef] [PubMed]
191. Nelson, D.R.; Schuler, M.A.; Paquette, S.M.; Werck-Reichhart, D.; Bak, S. Comparative Genomics of Rice and Arabidopsis. Analysis of 727 Cytochrome P450 Genes and Pseudogenes from a Monocot and a Dicot. *Plant Physiol.* **2004**, *135*, 756–772. [CrossRef] [PubMed]
192. Ehlting, J.; Provart, N.J.; Werck-Reichhart, D. Functional annotation of the Arabidopsis P450 superfamily based on large-scale co-expression analysis. *Biochem. Soc. Trans.* **2006**, *34*, 1192–1198. [CrossRef] [PubMed]
193. Werck-Reichhart, D.; Hehn, A.; Diderjean, L. Cytochromes P450 for engineering herbicide tolerance. *Plant Cell* **2004**, *5*, 116–123. [CrossRef]
194. Inui, H.; Ueyama, Y.; Shiota, N.; Ohkawa, Y.; Ohkawa, H. Herbicide Metabolism and Cross-Tolerance in Transgenic Potato Plants Expressing Human CYP1A1. *Pestic. Biochem. Physiol.* **1999**, *64*, 33–46. [CrossRef]
195. Kawahigashi, H.; Hirose, S.; Ohkawa, H.; Ohkawa, Y. Herbicide resistance of transgenic rice plants expressing human CYP1A1. *Biotechnol. Adv.* **2007**, *25*, 75–84. [CrossRef] [PubMed]
196. Kawahigashi, H.; Hirose, S.; Ohkawa, H.; Ohkawa, Y. Phytoremediation of the herbicides atrazine and metolachlor by transgenic rice plants expressing human *CYP1A1*, *CYP2B6*, and *CYP2C19*. *J. Agric. Food Chem.* **2006**, *54*, 2985–2991. [CrossRef]
197. Hirose, S.; Kawahigashi, H.; Ozawa, K.; Shiota, N.; Inui, H.; Ohkawa, H.; Ohkawa, Y. Transgenic rice containing human CYP2B6 detoxifies various classes of herbicides. *J. Agric. Food Chem.* **2005**, *53*, 3461–3467. [CrossRef]
198. Herbicide Resistance Action Committee. Available online: https://www.hracglobal.com/ (accessed on 26 January 2019).
199. Robineau, T.; Batard, Y.; Nedelkina, S.; Cabello-Hurtado, F.; LeRet, M.; Sorokine, O.; Didierjean, L.; Werck-Reichhart, D. The Chemically Inducible Plant Cytochrome P450 CYP76B1 Actively Metabolizes Phenylureas and Other Xenobiotics. *Plant Physiol.* **2002**, *118*, 1049–1056. [CrossRef] [PubMed]
200. Siminszky, B.; Corbin, F.T.; Ward, E.R.; Fleischmann, T.J.; Dewey, R.E. Expression of a soybean cytochrome P450 monooxygenase cDNA in yeast and tobacco enhances the metabolism of phenylurea herbicides. *Proc. Natl. Acad. Sci. USA* **1999**, *96*, 1750–1755. [CrossRef] [PubMed]
201. Höfer, R.; Boachon, B.; Renault, H.; Gavira, C.; Miesch, L.; Iglesias, J.; Ginglinger, J.-F.; Allouche, L.; Miesch, M.; Grec, S.; et al. Dual Function of the Cytochrome P450 CYP76 Family from *Arabidopsis thaliana* in the Metabolism of Monoterpenols and Phenylurea Herbicides. *Plant Physiol.* **2014**, *166*, 1149–1161. [CrossRef] [PubMed]
202. Khanom, S.; Jang, J.; Lee, O.R. Overexpression of ginseng cytochrome P450 CYP736A12 alters plant growth and confers phenylurea herbicide tolerance in *Arabidopsis*. *J. Ginseng Res.* **2019**. [CrossRef]
203. Iwakami, S.; Endo, M.; Saika, H.; Okuno, J.; Nakamura, N.; Yokoyama, M.; Watanabe, H.; Toki, S.; Uchino, A.; Inamura, T. Cytochrome P450 CYP81A12 and CYP81A21 Are Associated with Resistance to Two Acetolactate Synthase Inhibitors in *Echinochloa phyllopogon*. *Plant Physiol.* **2014**, *165*, 618–629. [CrossRef] [PubMed]
204. Guo, F.; Iwakami, S.; Yamaguchi, T.; Uchino, A.; Sunohara, Y.; Matsumoto, H. Role of CYP81A cytochrome P450s in clomazone metabolism in *Echinochloa phyllopogon*. *Plant Sci.* **2019**, *283*, 321–328. [CrossRef] [PubMed]
205. Yang, Q.; Li, J.; Shen, J.; Xu, Y.; Liu, H.; Deng, W.; Li, X.; Zheng, M. Metabolic Resistance to Acetolactate Synthase Inhibiting Herbicide Tribenuron-Methyl in *Descurainia sophia* L. Mediated by Cytochrome P450 Enzymes. *J. Agric. Food Chem.* **2018**, *66*, 4319–4327. [CrossRef] [PubMed]
206. Oliveira, M.C.; Gaines, T.A.; Dayan, F.E.; Patterson, E.L.; Jhala, A.J.; Knezevic, S.Z. Reversing resistance to tembotrione in an *Amaranthus tuberculatus* (var. *rudis*) population from Nebraska, USA with cytochrome P450 inhibitors. *Pest Manag. Sci.* **2018**, *74*, 2296–2305. [CrossRef]
207. Hidayat, I.; Preston, C. Cross-resistance to imazethapyr in a fluazifop-P-butyl-resistant population of *Digitaria sanguinalis*. *Pestic. Biochem. Physiol.* **2001**, *71*, 190–195. [CrossRef]
208. Marrs, K.A. The Functions and Regulation of Glutathione S-Transferases in Plants. *Annu. Rev. Plant Physiol. Plant Mol. Biol.* **1996**, *47*, 127–158. [CrossRef]

209. Stavridou, E.; Voulgari, G.; Bosmali, I.; Chronopoulou, E.G.; Lo Cicero, L.; Lo Piero, A.R.; Labrou, N.E.; Tsaftaris, A.; Nianiou-Obeidat, I.; Madesis, P. Plant Adaptation to Stress Conditions: The Case of Glutathione S-Transferases (GSTs). In *Biotic and Abiotic Stress Tolerance in Plants*; Vats, S., Ed.; Springer Nature Singapore Pte Ltd.: Berlin/Heidelberg, Germany, 2018; pp. 173–202. ISBN 978-981-10-9028-8.
210. Labrou, N.E.; Papageorgiou, A.C.; Pavli, O.; Flemetakis, E. Plant GSTome: Structure and functional role in xenome network and plant stress response. *Curr. Opin. Biotechnol.* **2015**, *32*, 186–194. [CrossRef] [PubMed]
211. Li, L.; Hou, M.; Cao, L.; Xia, Y.; Shen, Z.; Hu, Z. Glutathione S-transferases modulate Cu tolerance in *Oryza sativa*. *Environ. Exp. Bot.* **2018**, *155*, 313–320. [CrossRef]
212. Li, Z.-K.; Chen, B.; Li, X.-X.; Wang, J.-P.; Zhang, Y.; Wang, X.-F.; Yan, Y.-Y.; Ke, H.F.; Yang, J.; Wu, J.-H.; et al. A newly identified cluster of glutathione S-transferase genes provides *Verticillium* wilt resistance in cotton. *Plant J.* **2019**, *98*, 213–227. [CrossRef] [PubMed]
213. Dixon, D.P.; Lapthorn, A.; Edwards, R. Protein family review Plant glutathione transferases. *Genome Biol.* **2002**, *3*, 1–10. [CrossRef] [PubMed]
214. Dixon, D.; Cole, D.J.; Edwards, R. Characterisation of multiple glutathione transferases containing the GST I subunit with activities toward herbicide substrates in maize (*Zea mays*). *Pestic. Sci.* **1997**, *50*, 72–82. [CrossRef]
215. Grove, G.; Zarlengo, R.P.; Timmerman, K.P.; Li, N.Q.; Tam, M.F.; Tu, C.-P.D. Characterization and heterospecific expression of cDNA clones of genes in the maize GSH S-transferase multigene family. *Nucleic Acids Res.* **1988**, *16*, 425–438. [CrossRef] [PubMed]
216. Karavangeli, M.; Labrou, N.E.; Clonis, Y.D.; Tsaftaris, A. Development of transgenic tobacco plants overexpressing maize glutathione S-transferase I for chloroacetanilide herbicides phytoremediation. *Biomol. Eng.* **2005**, *22*, 121–128. [CrossRef]
217. Milligan, A.S.; Daly, A.; Parry, M.A.J.; Lazzeri, P.A.; Jepson, I. The expression of a maize glutathione S-transferase gene in transgenic wheat confers herbicide tolerance, both in planta and in vitro. *Mol. Breed.* **2001**, *7*, 301–315. [CrossRef]
218. Benekos, K.; Kissoudis, C.; Nianiou-Obeidat, I.; Labrou, N.; Madesis, P.; Kalamaki, M.; Makris, A.; Tsaftaris, A. Overexpression of a specific soybean GmGSTU4 isoenzyme improves diphenyl ether and chloroacetanilide herbicide tolerance of transgenic tobacco plants. *J. Biotechnol.* **2010**, *150*, 195–201. [CrossRef]
219. Cummins, I.; Cole, D.J.; Edwards, R. A role for glutathione transferases functioning as glutathione peroxidases in resistance to multiple herbicides in black-grass. *Plant J.* **1999**, *18*, 285–292. [CrossRef]
220. Petit, C.; Duhieu, B.; Boucansaud, K.; Délye, C. Complex genetic control of non-target-site-based resistance to herbicides inhibiting acetyl-coenzyme A carboxylase and acetolactate-synthase in *Alopecurus myosuroides* Huds. *Plant Sci.* **2010**, *178*, 501–509. [CrossRef]
221. Wright, A.A.; Rodriguez-Carres, M.; Sasidharan, R.; Koski, L.; Peterson, D.G.; Nandula, V.K.; Ray, J.D.; Bond, J.A.; Shaw, D.R. Multiple Herbicide–Resistant Junglerice (*Echinochloa colona*): Identification of Genes Potentially Involved in Resistance through Differential Gene Expression Analysis. *Weed Sci.* **2018**, *66*, 347–354. [CrossRef]
222. Nakka, S.; Godar, A.S.; Thompson, C.R.; Peterson, D.E.; Jugulam, M. Rapid detoxification via glutathione S-transferase (GST) conjugation confers a high level of atrazine resistance in Palmer amaranth (*Amaranthus palmeri*). *Pest Manag. Sci.* **2017**, *73*, 2236–2243. [CrossRef] [PubMed]
223. Dücker, R.; Zöllner, P.; Lümmen, P.; Ries, S.; Collavo, A.; Beffa, R. Glutathione transferase plays a major role in flufenacet resistance of ryegrass (*Lolium* spp.) field populations. *Pest Manag. Sci.* **2019**. [CrossRef]
224. Balabanova, D.; Remans, T.; Vassilev, A.; Cuypers, A.; Vangronsveld, J. Possible involvement of glutathione S-transferases in imazamox detoxification in an imidazolinone-resistant sunflower hybrid. *J. Plant Physiol.* **2018**, *221*, 62–65. [CrossRef] [PubMed]
225. Sharma, R.; Draicchio, F.; Bull, H.; Herzig, P.; Maurer, A.; Pillen, K.; Thomas, W.T.B.; Flavell, A.J. Genome-wide association of yield traits in a nested association mapping population of barley reveals new gene diversity for future breeding. *J. Exp. Bot.* **2018**, *69*, 3811–3822. [CrossRef] [PubMed]
226. Krajewski, M.P.; Kanawati, B.; Fekete, A.; Kowalski, N.; Schmitt-Kopplin, P.; Grill, E. Analysis of *Arabidopsis* glutathione-transferases in yeast. *Phytochemistry* **2013**, *91*, 198–207. [CrossRef] [PubMed]
227. Theodoulou, F.L. Plant ABC transporters. *Biochim. Biophys. Acta Biomembr.* **2000**, *1465*, 79–103. [CrossRef]
228. Hwang, J.U.; Song, W.Y.; Hong, D.; Ko, D.; Yamaoka, Y.; Jang, S.; Yim, S.; Lee, E.; Khare, D.; Kim, K.; et al. Plant ABC Transporters Enable Many Unique Aspects of a Terrestrial Plant's Lifestyle. *Mol. Plant* **2016**, *9*, 338–355. [CrossRef] [PubMed]

229. Sánchez-Fernández, R.; Davies, T.G.E.; Coleman, J.O.D.; Rea, P.A. The *Arabidopsis thaliana* ABC Protein Superfamily, a Complete Inventory. *J. Biol. Chem.* **2001**, *276*, 30231–30244. [CrossRef] [PubMed]
230. Ofori, P.A.; Mizuno, A.; Suzuki, M.; Martinoia, E.; Reuscher, S.; Aoki, K.; Shibata, D.; Otagaki, S.; Matsumoto, S.; Shiratake, K. Genome-wide analysis of atp binding cassette (ABC) transporters in tomato. *PLoS ONE* **2018**, *13*, e0200854. [CrossRef] [PubMed]
231. Busi, R.; Goggin, D.E.; Heap, I.M.; Horak, M.J.; Jugulam, M.; Masters, R.A.; Napier, R.M.; Riar, D.S.; Satchivi, N.M.; Torra, J.; et al. Weed resistance to synthetic auxin herbicides. *Pest Manag. Sci.* **2018**, *74*, 2265–2276. [CrossRef] [PubMed]
232. Peng, Y.; Abercrombie, L.L.G.; Yuan, J.S.; Riggins, C.W.; Sammons, R.D.; Tranel, P.J.; Stewart, C.N., Jr.; Characterization of the horseweed (*Conyza canadensis*) transcriptome using GS-FLX 454 pyrosequencing and its application for expression analysis of candidate non-target herbicide resistance genes. *Pest Manag. Sci.* **2010**, *66*, 1053–1062. [CrossRef] [PubMed]
233. Windsor, B.; Roux, S.J.; Lloyd, A. Multiherbicide tolerance conferred by AtPgp1 and apyrase overexpression in *Arabidopsis thaliana*. *Nat. Biotechnol.* **2003**, *21*, 428–433. [CrossRef] [PubMed]
234. Jo, J.; Won, S.H.; Son, D.; Lee, B.H. Paraquat resistance of transgenic tobacco plants over-expressing the *Ochrobactrum anthropi pqrA* gene. *Biotechnol. Lett.* **2004**, *26*, 1391–1396. [CrossRef] [PubMed]
235. Quistgaard, E.M.; Löw, C.; Guettou, F.; Nordlund, P. Understanding transport by the major facilitator superfamily (MFS): Structures pave the way. *Nat. Rev. Mol. Cell Biol.* **2016**, *17*, 123–132. [CrossRef]
236. Ward, J.M. Identification of novel families of membrane proteins from the model plant *Arabidopsis thaliana*. *Bioinformatics* **2001**, *17*, 560–563. [CrossRef]
237. Teixeira, M.C.; Duque, P.; Sá-Correia, I. Environmental genomics: Mechanistic insights into toxicity of and resistance to the herbicide 2,4-D. *Trends Biotechnol.* **2007**, *25*, 363–370. [CrossRef]
238. Cabrito, T.R.; Teixeira, M.C.; Duarte, A.A.; Duque, P.; Sá-Correia, I. Heterologous expression of a Tpo1 homolog from *Arabidopsis thaliana* confers resistance to the herbicide 2,4-D and other chemical stresses in yeast. *Appl. Microbiol. Biotechnol.* **2009**, *84*, 927–936. [CrossRef]
239. Tiwari, P.; Sangwan, R.S.; Sangwan, N.S. Plant secondary metabolism linked glycosyltransferases: An update on expanding knowledge and scopes. *Biotechnol. Adv.* **2016**, *34*, 714–739. [CrossRef]
240. Pflugmacher, S.; Schröder, P.; Sandermann Jr, H. Taxonomic distribution of plant glutathione S-transferases acting on xenobiotics. *Phytochemistry* **2000**, *54*, 267–273. [CrossRef]
241. Brazier-Hicks, M.; Offen, W.A.; Gershater, M.C.; Revett, T.J.; Lim, E.-K.; Bowles, D.J.; Davies, G.J.; Edwards, R. Characterization and engineering of the bifunctional *N*- and *O*-glucosyltransferase involved in xenobiotic metabolism in plants. *Proc. Natl. Acad. Sci. USA* **2007**, *104*, 20238–20243. [CrossRef] [PubMed]
242. Loutre, C.; Dixon, D.P.; Brazier, M.; Slater, M.; Cole, D.J.; Edwards, R. Isolation of a glucosyltransferase from *Arabidopsis thaliana* active in the metabolism of the persistent pollutant 3,4-dichloroaniline. *Plant J.* **2003**, *34*, 485–493. [CrossRef] [PubMed]
243. Wetzel, A.; Sandermann, H. Plant biochemistry of xenobiotics: Isolation and characterization of a soybean *O*-glucosyltransferase of DDT metabolism. *Arch. Biochem. Biophys.* **1994**, *314*, 323–328. [CrossRef] [PubMed]
244. Brazier-Hicks, M.; Edwards, R. Functional importance of the family 1 glucosyltransferase UGT72B1 in the metabolism of xenobiotics in *Arabidopsis thaliana*. *Plant J.* **2005**, *42*, 556–566. [CrossRef] [PubMed]
245. Meβner, B.; Thulke, O.; Schäffner, A.R. *Arabidopsis* glucosyltransferases with activities toward both endogenous and xenobiotic substrates. *Planta* **2003**, *217*, 138–146.
246. Cha, J.-Y.; Lee, D.-Y.; Ali, I.; Jeong, S.Y.; Shin, B.; Ji, H.; Kim, J.S.; Kim, M.-G.; Kim, W.-Y. *Arabidopsis* GIGANTEA negatively regulates chloroplast biogenesis and resistance to herbicide butafenacil. *Plant Cell Rep.* **2019**, *38*, 793–801. [CrossRef]
247. Etter, P.D.; Bassham, S.; Hohenlohe, P.A.; Johnson, E.; Cresko, W.A. SNP Discovery and genotyping for evolutionary genetics using RAD Sequencing. In *Molecular Methods for Evolutionary Genetics*; Orgogozo, V., Rockman, M.V., Eds.; Methods in Molecular Biology; Humana Press: Totowa, NJ, USA, 2011; Volume 772, pp. 1–9. ISBN 978-1-61779-227-4.
248. Peterson, B.K.; Weber, J.N.; Kay, E.H.; Fisher, H.S.; Hoekstra, H.E. Double digest RADseq: An inexpensive method for de novo SNP discovery and genotyping in model and non-model species. *PLoS ONE* **2012**, *7*, e37135. [CrossRef]
249. Catchen, J.M.; Hohenlohe, P.A.; Bassham, S.; Amores, A.; Cresko, W.A. Stacks: An analysis tool set for population genomics. *Mol. Ecol.* **2013**, *22*, 3124–3140. [CrossRef]

250. Rochette, N.C.; Catchen, J.M. Deriving genotypes from RAD-seq short-read data using Stacks. *Nat. Protoc.* **2017**, *12*, 2640–2659. [CrossRef]
251. Paris, J.R.; Stevens, J.R.; Catchen, J.M. Lost in parameter space: A road map for stacks. *Methods Ecol. Evol.* **2017**, *8*, 1360–1373. [CrossRef]
252. Bradbury, P.J.; Zhang, Z.; Kroon, D.E.; Casstevens, T.M.; Ramdoss, Y.; Buckler, E.S. TASSEL: Software for association mapping of complex traits in diverse samples. *Bioinformatics* **2007**, *23*, 2633–2635. [CrossRef] [PubMed]
253. R Core Team, R. *A Language and Environment for Statistical Computing*; R Core Team R: Vienna, Austria, 2017.
254. Paradis, E.; Gosselin, T.; Goudet, J.; Jombart, T.; Schliep, K. Linking genomics and population genetics with R. *Mol. Ecol. Resour.* **2017**, *17*, 54–66. [CrossRef] [PubMed]
255. Goudet, J.; Jombart, T. *Hierfstat: Estimation and Tests of Hierarchical F-Statistics*; CRAN: 2015; p. 58. Available online: https://cran.r-project.org/web/packages/hierfstat/index.html (accessed on 30 May 2019).
256. Pembleton, L.W.; Cogan, N.O.I.; Forster, J.W. StAMPP: An R package for calculation of genetic differentiation and structure of mixed-ploidy level populations. *Mol. Ecol. Resour.* **2013**, *13*, 946–952. [CrossRef] [PubMed]
257. Kamvar, Z.N.; Tabima, J.F.; Grünwald, N.J. Poppr: An R package for genetic analysis of populations with clonal, partially clonal, and/or sexual reproduction v2.6.1. *PeerJ* **2018**, *2*, e281. [CrossRef]
258. Revell, L.J. phytools: Phylogenetic tools for comparative biology (and other things). *Methods Ecol. Evol.* **2012**, *3*, 217–223. [CrossRef]
259. Jombart, T.; Ahmed, I. Adegenet 1.3-1: New tools for the analysis of genome-wide SNP data. *Bioinformatics* **2011**, *27*, 1403. [CrossRef] [PubMed]
260. Küpper, A.; Manmathan, H.K.; Giacomini, D.; Patterson, E.L.; Mccloskey, W.B.; Gaines, T.A. Population Genetic Structure in Glyphosate-Resistant and -Susceptible Palmer Amaranth (*Amaranthus palmeri*) Populations Using Genotyping-by-sequencing (GBS). *Front. Plant Sci.* **2018**, *9*, 29. [CrossRef]
261. Foll, M.; Gaggiotti, O. A genome-scan method to identify selected loci appropriate for both dominant and codominant markers: A Bayesian perspective. *Genetics* **2008**, *180*, 977–993. [CrossRef]
262. Günther, T.; Coop, G. Robust identification of local adaptation from allele frequencies. *Genetics* **2013**, *195*, 205–220. [CrossRef]
263. Simpson, J.T.; Wong, K.; Jackman, S.D.; Schein, J.E.; Jones, S.J.M.; Birol, I. ABySS: A parallel assembler for short read sequence data. *Genome Res.* **2009**, *19*, 1117–1123. [CrossRef] [PubMed]
264. Ye, C.; Hill, C.M.; Wu, S.; Ruan, J.; Ma, Z. (Sam) DBG2OLC: Efficient assembly of large genomes using long erroneous reads of the third generation sequencing technologies. *Sci. Rep.* **2016**, *6*, 1–9.
265. Hoshino, A.; Jayakumar, V.; Nitasaka, E.; Toyoda, A.; Noguchi, H.; Itoh, T.; Shin, T.; Minakuchi, Y.; Koda, Y.; Nagano, A.J.; et al. Genome sequence and analysis of the Japanese morning glory *Ipomoea nil*. *Nat. Commun.* **2016**, *7*, 1–10. [CrossRef] [PubMed]
266. Pickar-Oliver, A.; Gersbach, C.A. The next generation of CRISPR–Cas technologies and applications. *Nat. Rev. Mol. Cell Biol.* **2019**. [CrossRef] [PubMed]
267. Jinek, M.; Chylinski, K.; Fonfara, I.; Hauer, M.; Doudna, J.A.; Charpentier, E. A Programmable Dual-RNA –Guided DNA Endonuclease in Adaptive Bacterial Immunity. *Science* **2012**, *337*, 816–822. [CrossRef] [PubMed]
268. Doudna, J.A.; Charpentier, E. The new frontier of genome engineering with CRISPR-Cas9. *Science* **2014**, *346*, 1258096. [CrossRef] [PubMed]
269. Neve, P. Gene drive systems: Do they have a place in agricultural weed management? *Pest Manag. Sci.* **2018**, *74*, 2671–2679. [CrossRef] [PubMed]
270. Bull, J.J.; Malik, H.S. The gene drive bubble: New realities. *PLOS Genet.* **2017**, *13*, e1006850. [CrossRef] [PubMed]
271. Committee on Gene Drive Research in Non-Human Organisms: Recommendations for Responsible Conduct; Board on Life Sciences; Division on Earth and Life Studies; National Academies of Sciences, Engineering, and Medicine. *Gene Drives on the Horizon: Advancing Science, Navigating Uncertainty, and Aligning Research with Public Values*; The National Academies Press: Washington, DC, USA, 2016; ISBN 9780309437875.
272. Courtier-Orgogozo, V.; Morizot, B.; Boëte, C. Agricultural pest control with CRISPR-based gene drive: Time for public debate. *EMBO Rep.* **2017**, *18*, 878–880. [CrossRef]
273. Esvelt, K.M.; Smidler, A.L.; Catteruccia, F.; Church, G.M. Concerning RNA-guided gene drives for the alteration of wild populations. *Elife* **2014**, e03401. [CrossRef]

274. Champer, J.; Buchman, A.; Akbari, O.S. Cheating evolution: Engineering gene drives to manipulate the fate of wild populations. *Nat. Rev. Genet.* **2016**, *17*, 146–159. [CrossRef] [PubMed]
275. Knoll, A.; Fauser, F.; Puchta, H. DNA recombination in somatic plant cells: Mechanisms and evolutionary consequences. *Chromosom. Res.* **2014**, *22*, 191–201. [CrossRef] [PubMed]
276. Huang, T.K.; Puchta, H. CRISPR/Cas-mediated gene targeting in plants: Finally a turn for the better for homologous recombination. *Plant Cell Rep.* **2019**, *38*, 443–453. [CrossRef] [PubMed]
277. Klompe, S.E.; Vo, P.L.H.; Halpin-Healy, T.S.; Sternberg, S.H. Transposon-encoded CRISPR–Cas systems direct RNA-guided DNA integration. *Nature* **2019**, *571*, 219–225. [CrossRef] [PubMed]
278. Que, Q.; Chilton, M.D.M.; Elumalai, S.; Zhong, H.; Dong, S.; Shi, L. Repurposing macromolecule delivery tools for plant genetic modification in the era of precision genome engineering. *Methods Mol. Biol.* **2019**, *1864*, 3–18. [PubMed]
279. Grunwald, H.A.; Gantz, V.M.; Poplawski, G.; Xu, X.-R.S.; Bier, E.; Cooper, K.L. Super-Mendelian inheritance mediated by CRISPR–Cas9 in the female mouse germline. *Nature* **2019**, *566*, 105–109. [CrossRef]
280. Charpentier, M.; Khedher, A.H.; Menoret, S.; Brion, A.; Lamribet, K.; Dardillac, E.; Boix, C.; Perrouault, L.; Tesson, L.; Geny, S.; et al. CtIP fusion to Cas9 enhances transgene integration by homology-dependent repair. *Nat. Commun.* **2018**, *9*, 1–11. [CrossRef]
281. Wang, M.; Lu, Y.; Botella, J.R.; Mao, Y.; Hua, K.; Zhu, J. Gene Targeting by Homology-Directed Repair in Rice Using a Geminivirus-Based CRISPR/Cas9 System. *Mol. Plant* **2017**, *10*, 1007–1010. [CrossRef]
282. Dahan-Meir, T.; Filler-Hayut, S.; Melamed-Bessudo, C.; Bocobza, S.; Czosnek, H.; Aharoni, A.; Levy, A.A. Efficient in planta gene targeting in tomato using geminiviral replicons and the CRISPR/Cas9 system. *Plant J.* **2018**, *95*, 5–16. [CrossRef]
283. Gil-Humanes, J.; Wang, Y.; Liang, Z.; Shan, Q.; Ozuna, C.V.; Sánchez-León, S.; Baltes, N.J.; Starker, C.; Barro, F.; Gao, C.; et al. High-efficiency gene targeting in hexaploid wheat using DNA replicons and CRISPR/Cas9. *Plant J.* **2017**, *89*, 1251–1262. [CrossRef]
284. Kyrou, K.; Hammond, A.M.; Galizi, R.; Kranjc, N.; Burt, A.; Beaghton, A.K.; Nolan, T.; Crisanti, A. A CRISPR-Cas9 gene drive targeting doublesex causes complete population suppression in caged *Anopheles gambiae* mosquitoes. *Nat. Biotechnol.* **2018**, *36*, 1062–1066. [CrossRef] [PubMed]
285. Webber, B.L.; Raghu, S.; Edwards, O.R. Opinion: Is CRISPR-based gene drive a biocontrol silver bullet or global conservation threat? *Proc. Natl. Acad. Sci. USA* **2015**, *112*, 10565–10567. [CrossRef] [PubMed]
286. Baltzegar, J.; Cavin Barnes, J.; Elsensohn, J.E.; Gutzmann, N.; Jones, M.S.; King, S.; Sudweeks, J. Anticipating complexity in the deployment of gene drive insects in agriculture. *J. Responsible Innov.* **2018**, *5*, S81–S97. [CrossRef]
287. Oye, K.A.; Esvelt, K.; Appleton, E.; Catteruccia, F.; Church, G.; Kuiken, T.; Lightfoot, S.B.-Y.; McNamara, J.; Smidler, A.; Collins, J.P. Regulating gene drives. *Science* **2014**, *345*, 626–628. [CrossRef] [PubMed]
288. Unckless, R.L.; Messer, P.W.; Connallon, T.; Clark, A.G. Modeling the manipulation of natural populations by the mutagenic chain reaction. *Genetics* **2015**, *201*, 425–431. [CrossRef] [PubMed]
289. Unckless, R.L.; Clark, A.G.; Messer, P.W. Evolution of resistance against CRISPR/Cas9 gene drive. *Genetics* **2017**, *205*, 827–841. [CrossRef] [PubMed]
290. Champer, J.; Reeves, R.; Oh, S.Y.; Liu, C.; Liu, J.; Clark, A.G.; Messer, P.W. Novel CRISPR/Cas9 gene drive constructs reveal insights into mechanisms of resistance allele formation and drive efficiency in genetically diverse populations. *PLoS Genet.* **2017**, *13*, e1006796. [CrossRef] [PubMed]
291. Noble, C.; Olejarz, J.; Esvelt, K.M.; Church, G.M.; Nowak, M.A. Evolutionary dynamics of CRISPR gene drives. *Sci. Adv.* **2017**, *3*, e1601964. [CrossRef] [PubMed]
292. Hemingway, J.; Hawkes, N.J.; McCarroll, L.; Ranson, H. The molecular basis of insecticide resistance in mosquitoes. *Insect Biochem. Mol. Biol.* **2004**, *34*, 653–665. [CrossRef] [PubMed]
293. Ranson, H.; Lissenden, N. Insecticide Resistance in African *Anopheles* Mosquitoes: A Worsening Situation that Needs Urgent Action to Maintain Malaria Control. *Trends Parasitol.* **2016**, *32*, 187–196. [CrossRef] [PubMed]
294. Cohuet, A.; Harris, C.; Robert, V.; Fontenille, D. Evolutionary forces on *Anopheles*: What makes a malaria vector? *Trends Parasitol.* **2010**, *26*, 130–136. [CrossRef] [PubMed]
295. Hemingway, J.; Ranson, H.; Magill, A.; Kolaczinski, J.; Fornadel, C.; Gimnig, J.; Coetzee, M.; Simard, F.; Roch, D.K.; Hinzoumbe, C.K.; et al. Averting a malaria disaster: Will insecticide resistance derail malaria control? *Lancet* **2016**, *387*, 1785–1788. [CrossRef]

296. Namountougou, M.; Simard, F.; Baldet, T.; Diabaté, A.; Ouédraogo, J.B.; Martin, T.; Dabiré, R.K. Multiple Insecticide Resistance in *Anopheles gambiae* s.l. Populations from Burkina Faso, West Africa. *PLoS ONE* **2012**, *7*, e48412. [CrossRef] [PubMed]
297. Edi, C.V.A.; Koudou, B.G.; Jones, C.M.; Weetman, D.; Ranson, H. Multiple-Insecticide Resistance in *Anopheles gambiae* Mosquitoes, Southern Côte d'Ivoire. *Emerg. Infect. Dis.* **2012**, *18*, 1508–1511. [CrossRef] [PubMed]
298. Hammond, A.; Galizi, R.; Kyrou, K.; Simoni, A.; Siniscalchi, C.; Katsanos, D.; Gribble, M.; Baker, D.; Marois, E.; Russell, S.; et al. A CRISPR-Cas9 gene drive system targeting female reproduction in the malaria mosquito vector *Anopheles gambiae*. *Nat. Biotechnol.* **2016**, *34*, 78–83. [CrossRef]
299. Holt, R.A.; Broder, S.; Subramanian, G.M.; Halpern, A.L.; Sutton, G.G.; Charlab, R.; Nusskern, D.R.; Wincker, P.; Clark, A.G.; Ribeiro, J.M.C.; et al. The genome sequence of the malaria mosquito *Anopheles gambiae*. *Science* **2002**, *298*, 129–149. [CrossRef] [PubMed]
300. Scudellari, M. Hijacking evolution. *Nature* **2019**, *571*, 160–162. [CrossRef]
301. Collins, C.M.; Bonds, J.A.S.; Quinlan, M.M.; Mumford, J.D. Effects of the removal or reduction in density of the malaria mosquito, *Anopheles gambiae* s.l., on interacting predators and competitors in local ecosystems. *Med. Vet. Entomol.* **2019**, *33*, 1–15. [CrossRef]

plants

MDPI

Review

Omics Potential in Herbicide-Resistant Weed Management

Eric L. Patterson [1], Christopher Saski [2], Anita Küpper [3], Roland Beffa [3] and Todd A. Gaines [4,*]

1. Department of Plant, Soil and Microbial Sciences, Michigan State University, 1066 Bogue St., East Lansing, MI 48824, USA; patte543@msu.edu
2. Plant and Environmental Sciences Department, Clemson University, 306B Biosystems Research Complex, Clemson, SC 29634, USA; saski@clemson.edu
3. Bayer AG, CropScience Division, Weed Control Research, Building H872, 65926 Frankfurt, Germany; anita.kuepper@bayer.com (A.K.); roland.beffa@bayer.com (R.B.)
4. Department of Bioagricultural Sciences and Pest Management, Colorado State University, 1177 Campus Delivery, Fort Collins, CO 80523, USA

* Correspondence: todd.gaines@colostate.edu; Tel.: +1-970-491-6824

Received: 14 November 2019; Accepted: 12 December 2019; Published: 14 December 2019

Abstract: The rapid development of omics technologies has drastically altered the way biologists conduct research. Basic plant biology and genomics have incorporated these technologies, while some challenges remain for use in applied biology. Weed science, on the whole, is still learning how to integrate omics technologies into the discipline; however, omics techniques are more frequently being implemented in new and creative ways to address basic questions in weed biology as well as the more practical questions of improving weed management. This has been especially true in the subdiscipline of herbicide resistance where important questions are the evolution and genetic basis of herbicide resistance. This review examines the advantages, challenges, potential solutions, and outlook for omics technologies in the discipline of weed science, with examples of how omics technologies will impact herbicide resistance studies and ultimately improve management of herbicide-resistant populations.

Keywords: weed genomics; herbicide resistance database; herbicide resistance diagnostics; precision herbicide resistance management; functional genomics; weed biology; weed evolution; integrated pest management

1. Introduction

Reference genome assemblies have enabled many advances in our understanding of gene function and the linkages between the genome and phenome. Modern plant biology has become quantitative, systems-oriented, and predictable. The fields of genomics, transcriptomics, proteomics, and metabolomics—collectively referred to as 'omics'—describe the component parts of the biological system that lead to the presentation of traits. Profound developments have been realized in model plant and crop species where the genome and associated omics systems have led to new biological understanding and application [1]; however, the question remains—how can omics and associated systems-scale biology contribute to our understanding of herbicide resistance and ultimately help improve weed management? Fundamentally, this is a question of how omics discoveries can translate into applied outcomes and innovations. Within weed science, genomics and transcriptomics have been the most utilized of the various omics techniques and are the focus of this review. Proteomics and metabolomics are also emerging as potential areas of research for herbicide resistance [2–4]; however, the full potential of omics techniques has not yet been realized [5].

Several weed genomes have been completed to various levels of assembly completeness (Figure 1). Plans are in progress to rapidly and substantially expand the availability of weed genomics resources [6].

While the costs for sequencing are on a continuous decline and computational capacity is increasing, major challenges remain to fully realize the potential of omics and their contribution to improved weed management. In this review, we present what omics studies have already contributed to herbicide resistance and weed management, explore the challenges for omics in weeds, identify translational aspects of model systems, discuss the trajectory and impact of integrating omics in weed science, and propose a road map for where the discipline should go in the future to harness the power of omics for improved herbicide resistance management.

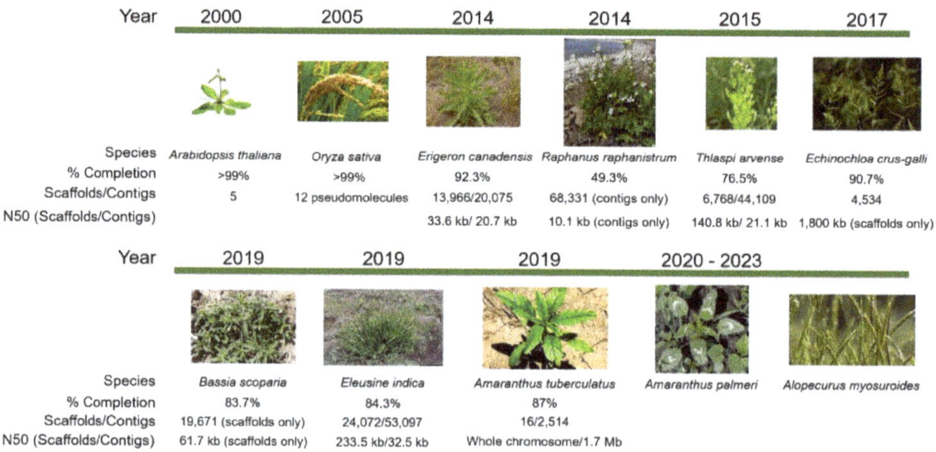

Figure 1. Timeline of weed genome assembly in comparison to the model plants *Arabidopsis thaliana* and rice. The first weed genome assembled to chromosome-level scaffolds is *Amaranthus tuberculatus* [7], for which scaffolding was completed by aligning with a related crop genome, *Amaranthus hypochondriacus* [8]. Other weeds with assembled genomes in various stages of completeness include *Erigeron canadensis* [9], *Thlaspi arvense* [10], *Echinochloa crus-galli* [11], *Bassia scoparia* [12], and *Eleusine indica* [13]. Assemblies for *Amaranthus palmeri* and *Alopecurus myosuroides* are in progress. Image sources: Arabidopsis, https://www.eurekalert.org/multimedia/pub/159783.php; field pennycress, https://www.agweb.com/article/pennycress-gets-in-the-middle-chris-bennett; horseweed, https://oregonstate.edu/dept/nursery-weeds/weedspeciespage/horseweed/horseweed_habit.html; wild radish, http://science.halleyhosting.com/nature/plants/4petal/must/raphanus/raphanistrum.html; barnyardgrass, http://swbiodiversity.org/seinet/taxa/index.php?taxon=2915&taxauthid=1; kochia, photo courtesy of Phil Westra, CSU; goosegrass, https://www.invasive.org/browse/detail.cfm?imgnum=5387295; Palmer amaranth, https://www.mda.state.mn.us/plants/pestmanagement/weedcontrol/noxiouslist/palmeramaranth; waterhemp, https://agfaxweedsolutions.com/2019/02/11/waterhemp-scores-again-new-resistance-found/; blackgrass, https://www.fwi.co.uk/arable/crop-management/weed-management/blackgrass/how-to-use-integrated-methods-to-control-blackgrass; rice, http://aaasjournal.org/rice-fields-chemical-physical-properties-implications-breeding-strategies/rice-plant/.

2. Challenges Specific to Weed Science

Omics research in weed science faces several challenges, some specific to weed science and some generic to the entire field of omics research. Several of these will be addressed with new discoveries and technologies that are currently being developed, while others may need a concerted effort by the weed science community to address. We will lay out several of these challenges and some possible solutions that may arise to meet them.

2.1. Managing Omics Datasets

The size and complexity of omics datasets being generated necessitates excellent database resources including large data storage, data backups, easy access, and data manipulation tools both in weed science and omics research at large. Several toolkits for genome databases have been developed and successfully implemented with support from both private and public sectors. For example, Tripal was developed with support from various academic and government funding agencies and is freely available for download [14]. Tripal was designed to streamline and simplify the process of omics database generation and organization, even in an online format [14]. Tripal also allows for the integration and use of several important bioinformatics tools such as BLAST, InterPro, gene function enrichment analysis, etc., an approach employed by several plant genome groups such as the Cucurbit Genomics Database [15] and the Genome Database for Rosaceae [16]. Other database services for omics can be licensed from the private sector, e.g., CropPedia by KeyGene (https://www.croppedia.com/).

Aside from establishing a contemporary platform for data housing and manipulation, deciphering a complex, quantitative phenotype still remains a challenge. Data from the genome, epigenome, transcriptome, proteome, and metabolome can now be collected from the same plant, and even single cells in some cases. A primary goal is to understand the latent relationships among the omics datasets to derive a comprehensive understanding of the underlying biology. In the example above, taking a holistic approach (e.g., collection of different omics datasets) offers power and resolution in comprehensively understanding the cellular and molecular components (and their interactions) [17]; however, integrating discrete experimental results is still difficult because of the inherent differences in the data [18]. Furthermore, there are limitations in omics technologies that are confounded by the complex nature of living systems [19]. As data integration techniques and strategies continue to advance, holistic interpretation of systems data will improve our biological understanding of complex phenotypes.

2.2. Genome Annotation

Another significant challenge facing the entire genome community is efficient and accurate annotation of reference genome assemblies and eventual pan genomes. Homology-based gene annotation pipelines, such as Maker [20] and Blast2GO [21], rely heavily on well-annotated, phylogenetically close relatives to the species of interest for gene model evidence. These tools perform even better with the availability of transcriptome datasets that are representative of key tissue sources selected across the developmental life cycle. Many weed species of interest do not reside close enough to a genomically-enabled neighbor species to be useful in homology-based gene annotation. Frequently, the closest species to weeds with sequenced genomes reside in distant plant families or even orders. Typical gene annotation strategies include the use of several popular prediction algorithms, such as SNAP [22], Augustus, GenesFH, GeneMark, Glimmer, and others. These algorithms can be trained with species specific data, manual curation, and consensus predictions extracted with programs such as EVidenceModeler [23]. In any case, weed species are often described as having exceptional genomes, with dynamic genomic plasticity and unique genetic content that can rapidly adapt and endow extreme phenotypes [7,12,24,25].

Another profound gap exists between computational prediction of genes and gene function, and validation of gene expression and its role and interaction with components of the omics system. A primary goal in weed science is to develop modern tools that leverage omics datasets to enable the study and verification of gene function *in situ*. A current constraint in closing this gap is the lack of curated and well-maintained germplasm banks and the lack of gene editing and transformation protocols. The lack of such tools prevents functional studies on genes and gene families, limiting the ability to fully harness the power of genomics for weedy traits.

2.3. Diversity of Evolutionary Strategies in Weeds

Weed genomics has other challenges that are specific. One of the biggest challenges is the number of species being studied globally. It seems impossible to select a weed species (or even a handful of weed species) that represent the diversity in weed science (not only the phylogenetic diversity but also the diversity of weed management problems). Currently, the priority species for genomics research are those that have the largest economic impact; species such as *Amaranthus palmeri, Alopecurus myosuroides, Echinochloa crus-galli,* and *Lolium* spp. However, these species are not always the most tractable for basic biology research. For instance, *Amaranthus palmeri* is dioecious, confounding the development of specific populations for population-level genetic analysis; or the fact that *Alopecurus myosuroides* has an exceptionally large, repeat-rich genome (~3.5 Gb) with high amounts of heterozygosity; or polyploid genomes like *Echinochloa crus-galli*. A 'model weed' approach could be used to deeply investigate fundamental questions about the great diversity of weedy traits and variation in evolutionary strategies found in weeds [26–28], while new resources may be developed for specific applications to compare results across multiple weed species.

One proposed explanation for the way in which some weed species continue to be dynamic in the face of elastic environmental pressures (avoiding genetic bottlenecks) is through maintenance or generation of genetic diversity [29]. Genetic diversity is critical for adaptation, and is perhaps, a key component in understanding the origins of traits and speciation; however, distinguishing genetic diversity from environmentally-induced phenotypic variability and linking phenotypes to genes poses several challenges. First and foremost is the ability to find, maintain, and accurately characterize lines with quantifiable heritability for traits of interest. Without consistent, well-characterized phenotypes, finding the genes through traditional methods (test crosses, genome-wide association studies (GWAS), QTL-seq, etc.) becomes a much more difficult task. Secondly, highly homogenous lines are desired as the starting point for genome assembly projects. Highly heterogeneous genomes are much more difficult to assemble and typically result in lower contiguity and completeness with a higher degree of inaccuracy [12]. To compensate, extra sequencing and haplotype phasing is typical in the assembly process, requiring additional time and expense. Furthermore, genetic studies that take advantage of segregating populations comprised of recombinant inbred lines (RILs) [30] offer high degrees of resolution and discrete QTL windows. For weed species that are obligate outcrossers (e.g., dioecious *Amaranthus* spp., self-incompatible *Lolium* spp.), the development of homozygous populations is not possible, leaving mapping resolution to be defined by half-sibling segregating populations and/or GWAS approaches where population structure confounds mapping resolution.

Weed scientists have ambitious goals to study complex traits in weeds, such as abiotic stress, seed germination, and non-target site resistance (NTSR) [6,31,32]. A major challenge is that these traits will be investigated across multiple weed species representing diverse plant families. No one model weed can represent the full range of life history traits and biology present across weeds. An interesting example of a (generally, but not always) highly quantitative complex trait is NTSR to herbicides. The study of NTSR is further complicated by the many combinations of weed species and registered herbicides for which multiple resistance mechanisms are possible [31]. The genetic basis and inheritance of herbicide resistance can be complex [33,34], such as NTSR mechanisms that are quantitative between populations and between individuals in a given population [35–39]. Elucidating the basis of NTSR to one herbicide in one species is not necessarily extensible to other herbicides and other species. In addition, NTSR mechanisms can endow cross-resistance to multiple herbicides with different sites of action [40–42], and single plants can contain multiple different NTSR mechanisms [36,43]. This further confounds the use of omics strategies to disentangle the underlying genetic mechanisms. Likewise, weed species in general display an interesting disposition of resilience for complex abiotic traits that are of agronomic importance such as drought, heat, salt, and cold resilience, as well as seed longevity and many others. Dissecting these traits on a molecular basis can prove to be difficult without modern omics approaches.

3. Addressing Challenges by Looking at Other Disciplines

Many research disciplines that work in model systems have already begun to fully exploit the decreased costs of next generation sequencing (NGS). Looking at the diverse ways researchers working in model systems are using omics technologies in their respective fields can provide established tools and templates to address the unmet needs of the weed science community.

3.1. *Method Standardization for Utilizing NGS in Weed Science*

With more and more researchers utilizing NGS, the need for quality and methods standards that enable comparisons between studies becomes paramount. First and foremost, weed scientists need access to reference lines used in NGS studies, especially since the high usage of herbicides worldwide has made it more difficult to obtain purely susceptible populations. For some species, reference susceptible lines are in common use such as the 'Roth' line of *Alopecurus myosuroides* maintained by Rothamsted Research Institute, which has never had herbicide exposure in the past 150 years [44], the *Lolium rigidum* line VLR1 from Victoria, Australia [45], and the susceptible *Bassia scoparia* line 7710 from Colorado [12,46]. Likewise, references exist for resistant weed populations like the established *A. myosuroides* "Peldon" [44], the *L. rigidum* lines VLR69 [47] and SLR31 [42], and the first *Amaranthus palmeri* population reported to be resistant to glyphosate [48,49]. Distribution of these reference lines is currently only on an ad hoc basis by contacting the labs that maintain them. The greater challenge is to capture the diversity of resistance mechanisms and combinations and maintain their availability over a long period for future studies. Reference lines should be stored with institutions like the USDA National Laboratory for Genetic Resource Preservation (USDA-NLGRP), which already houses a broad germplasm collection, or an arrangement similar to the NSF-funded Sequence-Indexed Library of Insertion Mutations for *A. thaliana* [50] that propagates seed for distribution to the community. Easily accessible reference lines can then be used for sequencing projects, dose response experiments for herbicide sensitivity or fitness penalty studies, population genetics studies, as control groups, or to test gene function. In the future, we hope to see homozygous recombinant inbred lines (RIL) or multi-parent mapping populations in weed science for the identification of more complex quantitative trait loci (QTL). For situations in which homozygous lines may be difficult to produce (e.g., self-incompatible species, dioecious species, multiple resistance mechanisms), we encourage the production and availability of multiple reference populations.

Due to the high demand from researchers working on model systems, NGS data analysis can be performed through several publicly available platforms, for example the NSF-funded CyVerse with data storage and bioinformatic tools through the Discovery Environment web interface [51], or the Galaxy project [52]. Furthermore, NSF-funded labs have produced easy to use online tools like the Genome Sequence Annotation Server (GenSAS) that provides a pipeline for *de novo* gene prediction and whole genome structural and functional annotation [53]. More tools that are weed science specific may need to be developed or adapted from other existing tools; for instance, a database of consistent annotations and gene ontologies. The Antibiotic Resistance Ontology (ARO) service [54] and the Cytochrome P450 homepage [55] have shown how important proper annotations are to provide consistent vocabulary for genes, which form much of the foundation of genomic bioinformatics.

Currently, omics techniques used for weeds are limited in scope, usually to a pair of samples and just a few individuals per population. In the future, more NGS studies will be available for meta-analyses that can provide insights into more complex evolutionary questions and the basic mechanisms driving complex traits like metabolic herbicide resistance. Additionally, we may soon be able to perform whole genome sequencing from many individuals of a single species for genome-wide association studies (GWAS) and pangenome analysis, which will provide key information about genetic variability and evolutionary history of individuals and populations. Similar to human genotypic ancestry services, the more individual genome information is available for weeds, the better genetic relatedness, movement patterns, and invasion biology can be understood.

3.2. Improving Herbicide Resistance Diagnostics with Omics

We predict that improved resistance diagnostics in combination with field history data will allow for field-tailored precision weed control recommendations that avoid unnecessary one-size-fits-all treatments and improve risk prediction tools. Improved diagnostics and precision mapping might also support or refute zero tolerance approaches in the case of new and agriculturally troublesome herbicide resistance mutations.

Currently, the International Survey of Herbicide Resistant Weeds [56] is the main database for herbicide resistant weeds and provides an extensive collection of new resistance reports and genomic DNA sequences that encode for herbicide targets in various weed species. The antibiotic resistance field maintains the Resistance Map, which provides interactive data on antibiotic use and resistance patterns worldwide and predicts resistance trends [57]. The Comprehensive Antibiotic Resistance Database (CARD) collects antibiotic resistance genes and associated proteins and takes the idea a step further to also provide information on antibiotics, resistance mechanisms, antibiotic targets, associated phenotypes, and tools to analyze molecular sequences. It also predicts putative antibiotic resistance genes from unannotated but assembled contigs and their prevalence from sequenced genomes [54,58]. These data are also essential to weed scientists to ask questions such as when and where are the first cases of resistance, how widespread are they, by what mechanism of resistance it is conferred, what is the agricultural relevance for the grower, and how are herbicides being used on a global scale? We foresee the need for weed resources such as weedscience.org expanding in scope to include more reporting of resistance mechanism (e.g., target-site resistance, TSR, and NTSR) and being partially modeled based on resources developed by microbiologists.

3.3. Improved Gene Function Validation for Herbicide Resistance Mechanisms

The increase in sequencing efforts to investigate mechanisms of resistance has led to an increase in the identification of candidate driver genes that correlate with resistant phenotypes. Unfortunately, many studies do not continue to functionally validate these candidate genes and the actual cause for resistance remains undetermined. In successful validation studies, researchers have utilized *Agrobacterium tumefaciens*-mediated transformation of candidate genes in model plant systems, such as *Arabidopsis thaliana* [59–62], tobacco (*Nicotiana benthamiana*) [63], rice calli (*Oryza sativa*) [64,65], transgenic rice [66,67], or budding yeast (*Saccharomyces cerevisiae*) [68]. However, most weed science studies fall short of functionally validating identified genes due to lack of investment to date in stable plant transformation methods for weeds. Plant transformation is an area where method development is urgently needed.

Plant gene function can be investigated using transient expression systems to either knock out or overexpress a candidate gene variant. Currently, there are several alternative techniques available in non-model species for the investigation of gene function by RNA interference (RNAi) such as virus-induced gene silencing (VIGS) [69,70]. Relevant to herbicide resistance, this technique has been recently utilized to silence CYP749A16 in trifloxysulfuron-tolerant cotton [71] and to silence a GST gene cluster in *Verticillium* wilt-resistant cotton [63]. Plants can also be inoculated with modified virus alone, resulting in transcription of anti-sense RNA and subsequent target mRNA cleavage, such as the barley stripe mosaic virus system used in cereals [72]. Alternative techniques to suppress target mRNA by direct topical applications of anti-sense silencing oligos have been developed such as small interfering RNAs (siRNAs) in a complex with a protein carrier [73], high-pressure spraying of double-stranded RNA (dsRNA) [74], or through simple application of long dsRNA [75]. In contrast to reverse genetics approaches that knock out gene function by anti-sense transcript silencing, gain of function due to candidate gene variants can be assessed with transient expression in plants using promoter-targeted RNA-directed DNA methylation (in cases where DNA methylation can affect gene transcription) [76] and transient infection with *Agrobacterium* to express a candidate gene [77].

Most gene function studies in model systems have used alternative transfer DNA (t-DNA) or transposon insertional mutagenesis to create mutant plants (gene knock-outs) for phenotype screening where plants with interesting phenotypes were further characterized for the affected gene(s). These techniques require the production and maintenance of a large amount of germplasm as well as huge resource input. This is only feasible when a large community is working on a single species (e.g., *Arabidopsis*). For weed scientists it may be more viable to take a targeted approach for gene knockouts using gene editing techniques like zinc finger nucleases (ZFNs) [78,79] or transcription activator-like effector nucleases (TALEN) [80]. Additionally, gene editing using clustered regularly interspaced short palindromic repeats/CRISPR-associated protein 9 (CRISPR/Cas9) guided by small RNA instead of proteins for sequence-specific DNA cleavage [81] may be the quickest way to achieve targeted gene editing. CRISPR systems have been shown to work both transiently or stably and with high efficiency and specificity [82]. The weed science community would benefit greatly from implementing these techniques to validate candidate gene function; however, as for other approaches to study gene function, investment in plant transformation methods for weed species is needed to fully enable gene editing in weeds.

4. Using Current and Future Omics Tools to Improve Herbicide Resistant Weed Management

Potential applications of genomics for improving applied weed control have been reviewed [5,6,27,28,83]. A striking example of technology that could advance weed management is the gene drive system [84]. Gene drives that could result in species extinction may be unfeasible for regulatory and/or public acceptance reasons. However, some weedy traits may be excellent gene drive targets to reduce the impacts of weeds. For example, if genomics can identify the basis of extreme allergenicity in weeds (e.g., ragweed species), a gene drive system could target elimination of the allergen from populations. If genomics can identify the basis of seed dormancy, a gene drive system could lead to greater synchronization of germination. Tumbleweeds require the development of an abscission layer at the base of the plant to break off, tumble, and disperse seed. A gene drive system could potentially eliminate the abscission layer trait from a population, reducing spread of the tumbleweed seeds.

Externally-applied gene silencing techniques to manipulate gene expression and potentially reverse herbicide resistance mechanisms are another application of new knowledge gained from genomics [85]. However, major challenges remain to utilize externally-applied gene silencing in plants, specifically difficulties in stability, delivery, and efficacy of gene silencing oligonucleotides [74,75]. The resources from expanded weed genomics efforts will be crucial to design effective gene silencing triggers with maximum specificity to target species and with minimal off-target effects, both in the target organism and in non-target organisms.

Improved understanding of pathogen response pathways in weeds could lead to opportunities for improved biocontrol. For example, pathogens could be engineered to be more virulent on weeds but not on crops [86,87]. Gene drive systems could be combined with bio-control to spread susceptibility to a pathogen within a weed population, potentially enabling long-term suppression of populations without further intervention. Weeds that are alternate hosts for crop pathogens could be targeted with gene drive or gene silencing to eliminate their ability to serve as alternate hosts.

The UK BioBank provides an example from the biomedical science field as to how large-scale availability of genotypic and phenotypic data on thousands of individuals can democratize genomics and make possible the discovery of the genetic basis of many diseases and traits in humans [88]. For the model plant *Arabidopsis*, full genome sequences and phenotyping data exist for more than 1000 lines, along with databases of corresponding RNA-Seq gene expression data and gene knockout mutation phenotypic effects [89]. We envision a similar weed biobank database empowering research on weeds across the world, consisting of reference genomes for multiple species, phenotypic data contributed from collaborators around the world, and genome wide genotype data sets that are publicly available and can be mined to discover the basis of quantitative traits, complex herbicide resistance mechanisms, and other traits of interest in weeds. A weed biobank for GWAS will be complemented by other tools from quantitative genetics, such as utilizing F2 mapping for herbicide resistance traits and abiotic stress tolerance traits. The integration of quantitative genetics with phenotyping including metabolomics, proteomics, and transcriptomics on segregating individuals will initially enable markers associated with traits of interest, and ultimately identify genomic regions and specific genes controlling the traits. In addition, as in cancer therapy [90], genomic diagnostics might help to choose the best herbicide combination(s) to mitigate the evolution of NTSR, in particular metabolic resistance.

5. Where Is Weed Omics Going?

Looking ahead to the next five to 10 years, we see several applications for weed omics. Large scale, high-throughput detection of known resistance mutations is possible using targeted amplicon NGS, bringing down the cost of genotyping and increasing the scope of available information [91]. The precision to identify resistant genotypes at low frequency within field-scale management units will enable improved management recommendations specific for growers and their unique situation of resistance mechanism(s), frequency, crop rotation, soil type, etc. The detection of low frequency resistance will enable early warning systems, both for individual growers and within regions. The use of metadata from digital agriculture will enable integration of field history and geospatial data on weed populations to further inform best practice recommendations for growers.

Like the standards proposed in Section 2 for defining and characterizing the phenotype of herbicide resistance, we envision the same standards to define and report herbicide resistance based on characterized mutations in candidate genes. Currently, resistance is defined according to biological criteria, primarily greenhouse dose responses, which has pros (reliable, not dependent on specific mechanism), but it also has cons, including the cost and time required. Additionally, the current resistance definitions in use consider resistance to be defined only when resistant individuals are at a high frequency in a population. The common term used is biotype, which is not necessarily an accurate term for many of the reports in the resistance database when there may be mixtures of different resistance mechanisms within a population (e.g., TSR and NTSR). We ask to consider whether a few highly resistant individuals within a population of mostly sensitive individuals should have a definition (e.g., early stage resistance), as this initially rare resistance frequency is when active measures can be taken to slow the increase in resistance. Improved diagnostics (faster, cheaper, more individuals tested) will enable early detection of resistant individuals within populations, and corresponding management measures to be prescribed. In-field diagnostics may have utility to provide rapid information for grower decision making, similar to how various rapid PCR techniques can be used to identify plant pathogens in the field [92].

We argue that resistance databases should accept molecular criteria to report known, well-characterized cases of resistance, both for TSR and NTSR (when the genetic basis is known). Improved resistance testing with rapid markers and database tracking is possible with modern molecular biology. Resistance cases are likely underreported in databases because, for example, reporting the next observation of acetolactate synthase (ALS) resistance has little incentive for researchers to conduct laborious and expensive assays, while ALS resistance can be easily diagnosed with molecular markers for target-site mutations. The ease in identification and reporting should help address the current bias in data for prevalence of common herbicide resistance mechanisms, data that will be important for the herbicide discovery industry.

In addition to utilization of molecular markers, resistance databases should be further improved through advances from omics technologies. More technologies should be developed to diagnose known resistance mechanisms, including nucleic acid probes, antibodies, and metabolite screens. More knowledge gaps exist for NTSR, with a few examples of known genetic variants for metabolic resistance characterized to date in weeds [61,64,93], but with many more cases of metabolic resistance to be discovered. With an improved understanding of metabolic resistance genes and pathways, transcriptional and/or protein markers can be screened as a diagnostic panel, in which the presence of defined subsets of markers indicates a sample is positive for metabolic resistance, similar to what is currently performed in cancer diagnostics [94,95]. Such a diagnostic panel has already been shown for weeds, with different sets of transcriptional markers for cytochrome P450s and other NTSR genes able to differentiate metabolic resistant and susceptible *Lolium* field populations collected in France [96,97].

We propose a system to classify metabolic resistance genes, such as cytochrome P450s and GSTs, by their capacity to metabolize the known herbicide structures. To achieve this will require both discovery and validation of genes in these gene families utilizing genomics, as well as cloning these genes into heterologous systems (e.g., yeast, *Arabidopsis*) to determine their metabolic activity on each herbicide of interest. Undertaking this objective will require considerable investment and coordination, due to the high number of cytochrome P450s genes in plants and their sequence and functional divergence across plant families [98]. Collectively, this information will inform management by shedding light on cross-resistance patterns due to metabolism, as well as enable testing of compounds in development and those yet to be discovered for their susceptibility to metabolism by resistant weeds.

6. Summary

The weed omics era is enabling translational research to bridge from basic science to field applications, by linking systems-scale science to applied science for practitioners. The rise of digital farming and dense geospatial data will enable prediction tools for the occurrence and spread of herbicide resistance within fields and across landscapes. This metainformation will improve diagnostics as well as provide greater insight into the factors driving selection for various resistance mechanisms. Machine learning will lead to algorithms to select the best options from chemical and non-chemical control technologies [99]. Weed omics will contribute to better define these prediction tools and associated algorithms. These benefits of weed omics will be more challenging to realize for farms not utilizing the advanced data science approaches necessary for implementation of digital farming.

While there are substantial challenges today to apply omics to weed science, the coming years will see development of new approaches to help overcome these challenges. As the increase in data acquisition continues to coincide with the development of new statistical approaches to systems biology, what seems like insurmountable obstacles now may soon be trivial issues. Whole genome sequencing projects have evolved from large-scale international efforts to routine tasks often undertaken by an individual lab. For example, obtaining a high quality reference genome of a heterozygous plant would not have been possible only a decade ago, and now the International Weed Genomes Consortium has pledged to generate 10 in only a few years [6], in addition to several key species recently completed outside this collaboration [7,12].

Author Contributions: Writing—original draft preparation, E.L.P., C.S., A.K., R.B., T.A.G.; writing—review and editing, E.L.P., C.S., A.K., R.B., T.A.G.

Funding: This research received no external funding.

Conflicts of Interest: The authors declare no conflict of interest.

References

1. Sheth, B.P.; Thaker, V.S. Plant systems biology: Insights, advances and challenges. *Planta* **2014**, *240*, 33–54. [CrossRef] [PubMed]
2. Maroli, A.; Nandula, V.; Duke, S.; Tharayil, N. Stable isotope resolved metabolomics reveals the role of anabolic and catabolic processes in glyphosate-induced amino acid accumulation in *Amaranthus palmeri* biotypes. *J. Agric. Food Chem.* **2016**, *64*, 7040–7048. [CrossRef] [PubMed]
3. Maroli, A.S.; Nandula, V.K.; Dayan, F.E.; Duke, S.O.; Gerard, P.; Tharayil, N. Metabolic profiling and enzyme analyses indicate a potential role of antioxidant systems in complementing glyphosate resistance in an *Amaranthus palmeri* biotype. *J. Agric. Food Chem.* **2015**, *63*, 9199–9209. [CrossRef] [PubMed]
4. Fernández-Escalada, M.; Zulet-González, A.; Gil-Monreal, M.; Zabalza, A.; Ravet, K.; Gaines, T.; Royuela, M. Effects of EPSPS copy number variation (CNV) and glyphosate application on the aromatic and branched chain amino acid synthesis pathways in *Amaranthus palmeri*. *Front. Plant Sci.* **2017**, *8*, 1970. [CrossRef] [PubMed]

5. Maroli, A.S.; Gaines, T.A.; Foley, M.E.; Duke, S.O.; Doğramacı, M.; Anderson, J.V.; Horvath, D.P.; Chao, W.S.; Tharayil, N. Omics in weed science: A perspective from genomics, transcriptomics, and metabolomics approaches. *Weed Sci.* **2018**, *66*, 681–695. [CrossRef]
6. Ravet, K.; Patterson, E.L.; Krähmer, H.; Hamouzová, K.; Fan, L.; Jasieniuk, M.; Lawton-Rauh, A.; Malone, J.M.; McElroy, J.S.; Merotto, A.; et al. The power and potential of genomics in weed biology and management. *Pest Manag. Sci.* **2018**, *74*. [CrossRef]
7. Kreiner, J.M.; Giacomini, D.A.; Bemm, F.; Waithaka, B.; Regalado, J.; Lanz, C.; Hildebrandt, J.; Sikkema, P.H.; Tranel, P.J.; Weigel, D.; et al. Multiple modes of convergent adaptation in the spread of glyphosate-resistant *Amaranthus tuberculatus*. *Proc. Natl. Acad. Sci. USA* **2019**. [CrossRef]
8. Lightfoot, D.; Jarvis, D.E.; Ramaraj, T.; Lee, R.; Jellen, E.; Maughan, P. Single-molecule sequencing and Hi-C-based proximity-guided assembly of amaranth (*Amaranthus hypochondriacus*) chromosomes provide insights into genome evolution. *BMC Biol.* **2017**, *15*, 74. [CrossRef]
9. Peng, Y.; Lai, Z.; Lane, T.; Nageswara-Rao, M.; Okada, M.; Jasieniuk, M.; O'Geen, H.; Kim, R.W.; Sammons, R.D.; Rieseberg, L.H.; et al. *De novo* genome assembly of the economically important weed horseweed using integrated data from multiple sequencing platforms. *Plant Physiol.* **2014**, *166*, 1241–1254. [CrossRef]
10. Dorn, K.M.; Fankhauser, J.D.; Wyse, D.L.; Marks, M.D. A draft genome of field pennycress (*Thlaspi arvense*) provides tools for the domestication of a new winter biofuel crop. *DNA Res.* **2015**, *22*, 121–131. [CrossRef]
11. Guo, L.; Qiu, J.; Ye, C.; Jin, G.; Mao, L.; Zhang, H.; Yang, X.; Peng, Q.; Wang, Y.; Jia, L.; et al. *Echinochloa crus-galli* genome analysis provides insight into its adaptation and invasiveness as a weed. *Nat. Commun.* **2017**, *8*, 1031. [CrossRef] [PubMed]
12. Patterson, E.L.; Saski, C.A.; Sloan, D.B.; Tranel, P.J.; Westra, P.; Gaines, T.A. The draft genome of *Kochia scoparia* and the nechanism of glyphosate resistance via transposon-mediated *EPSPS* tandem gene duplication. *Genome Biol. Evol.* **2019**. [CrossRef] [PubMed]
13. Zhang, H.; Hall, N.; Goertzen, L.R.; Bi, B.; Chen, C.Y.; Peatman, E.; Lowe, E.K.; Patel, J.; McElroy, J.S. Development of a goosegrass (*Eleusine indica*) draft genome and application to weed science research. *Pest Manag. Sci.* **2019**, in press. [CrossRef] [PubMed]
14. Ficklin, S.P.; Sanderson, L.A.; Cheng, C.H.; Staton, M.E.; Lee, T.; Cho, I.H.; Jung, S.; Bett, K.E.; Main, D. Tripal: A construction toolkit for online genome databases. *Database* **2011**, *2011*, bar044. [CrossRef]
15. Zheng, Y.; Wu, S.; Bai, Y.; Sun, H.; Jiao, C.; Guo, S.; Zhao, K.; Blanca, J.; Zhang, Z.; Huang, S. Cucurbit Genomics Database (CuGenDB): A central portal for comparative and functional genomics of cucurbit crops. *Nucleic Acids Res.* **2018**, *47*, D1128–D1136. [CrossRef]
16. Jung, S.; Lee, T.; Cheng, C.-H.; Buble, K.; Zheng, P.; Yu, J.; Humann, J.; Ficklin, S.P.; Gasic, K.; Scott, K. 15 years of GDR: New data and functionality in the Genome Database for Rosaceae. *Nucleic Acids Res.* **2018**, *47*, D1137–D1145. [CrossRef]
17. Wanichthanarak, K.; Fahrmann, J.F.; Grapov, D. Genomic, Proteomic, and Metabolomic Data Integration Strategies. *Biomark. Insights* **2015**, *10*, 1–6. [CrossRef]
18. Pinu, F.R.; Beale, D.J.; Paten, A.M.; Kouremenos, K.; Swarup, S.; Schirra, H.J.; Wishart, D. Systems Biology and Multi-Omics Integration: Viewpoints from the Metabolomics Research Community. *Metabolites* **2019**, *9*, 76. [CrossRef]
19. Gomez-Cabrero, D.; Abugessaisa, I.; Maier, D.; Teschendorff, A.; Merkenschlager, M.; Gisel, A.; Ballestar, E.; Bongcam-Rudloff, E.; Conesa, A.; Tegner, J. Data integration in the era of omics: Current and future challenges. *BMC Syst. Biol.* **2014**, *8* (Suppl. 2), I1. [CrossRef]
20. Cantarel, B.L.; Korf, I.; Robb, S.M.; Parra, G.; Ross, E.; Moore, B.; Holt, C.; Sanchez Alvarado, A.; Yandell, M. MAKER: An easy-to-use annotation pipeline designed for emerging model organism genomes. *Genome Res.* **2008**, *18*, 188–196. [CrossRef]
21. Conesa, A.; Gotz, S.; Garcia-Gomez, J.M.; Terol, J.; Talon, M.; Robles, M. Blast2GO: A universal tool for annotation, visualization and analysis in functional genomics research. *Bioinformatics* **2005**, *21*, 3674–3676. [CrossRef] [PubMed]
22. Stanke, M.; Morgenstern, B. AUGUSTUS: A web server for gene prediction in eukaryotes that allows user-defined constraints. *Nucleic Acids Res.* **2005**, *33*, W465–W467. [CrossRef] [PubMed]

23. Haas, B.J.; Salzberg, S.L.; Zhu, W.; Pertea, M.; Allen, J.E.; Orvis, J.; White, O.; Buell, C.R.; Wortman, J.R. Automated eukaryotic gene structure annotation using EVidenceModeler and the Program to Assemble Spliced Alignments. *Genome Biol.* **2008**, *9*, R7. [CrossRef] [PubMed]
24. Koo, D.-H.; Molin, W.T.; Saski, C.A.; Jiang, J.; Putta, K.; Jugulam, M.; Friebe, B.; Gill, B.S. Extrachromosomal circular DNA-based amplification and transmission of herbicide resistance in crop weed *Amaranthus palmeri*. *Proc. Natl. Acad. Sci. USA* **2018**, *115*, 3332–3337. [CrossRef] [PubMed]
25. Molin, W.T.; Wright, A.A.; Lawton-Rauh, A.; Saski, C.A. The unique genomic landscape surrounding the *EPSPS* gene in glyphosate resistant *Amaranthus palmeri*: A repetitive path to resistance. *BMC Gen.* **2017**, *18*, 91. [CrossRef]
26. Gressel, J. Arabidopsis is not a weed, and mostly not a good model for weed genomics; There is no good model for weed genomics. In *Weedy and Invasive Plant Genomics*; Stewart, C.N., Ed.; John Wiley & Sons: Ames, IA, USA, 2009; pp. 25–32.
27. Stewart, C.N.; Tranel, P.J.; Horvath, D.P.; Anderson, J.V.; Rieseberg, L.H.; Westwood, J.H.; Mallory-Smith, C.A.; Zapiola, M.L.; Dlugosch, K.M. Evolution of weediness and invasiveness: Charting the course for weed genomics. *Weed Sci.* **2009**, *57*, 451–462. [CrossRef]
28. Basu, C.; Halfhill, M.D.; Mueller, T.C.; Stewart, C.N., Jr. Weed genomics: New tools to understand weed biology. *Trends Plant Sci.* **2004**, *9*, 391–398. [CrossRef]
29. Vigueira, C.C.; Olsen, K.M.; Caicedo, A.L. The red queen in the corn: Agricultural weeds as models of rapid adaptive evolution. *Heredity* **2013**, *110*, 303–311. [CrossRef]
30. Jugulam, M.; McLean, M.D.; Chen, S.; Christopher Hall, J. Development of near-isogenic lines and identification of markers linked to auxinic herbicide resistance in wild mustard (*Sinapis arvensis* L.). *Pest Manag. Sci.* **2012**, *68*, 548–556.
31. Délye, C. Unravelling the genetic bases of non-target-site-based resistance (NTSR) to herbicides: A major challenge for weed science in the forthcoming decade. *Pest Manag. Sci.* **2013**, *69*, 176–187. [CrossRef]
32. Westwood, J.H.; Charudattan, R.; Duke, S.O.; Fennimore, S.A.; Marrone, P.; Slaughter, D.C.; Swanton, C.; Zollinger, R. Weed management in 2050: Perspectives on the future of weed science. *Weed Sci.* **2018**, *66*, 275–285. [CrossRef]
33. Baucom, R.S. Evolutionary and ecological insights from herbicide-resistant weeds: What have we learned about plant adaptation, and what is left to uncover? *New Phytol.* **2019**, *223*, 68–82. [CrossRef] [PubMed]
34. Ghanizadeh, H.; Buddenhagen, C.E.; Harrington, K.C.; James, T.K. The genetic inheritance of herbicide resistance in weeds. *Crit. Rev. Plant Sci.* **2019**, *38*, 295–312. [CrossRef]
35. Oliveira, M.C.; Gaines, T.A.; Jhala, A.J.; Knezevic, S.Z. Inheritance of mesotrione resistance in an *Amaranthus tuberculatus* (var. *rudis*) population from Nebraska, USA. *Front. Plant Sci.* **2018**, *9*, 60. [CrossRef]
36. Huffman, J.; Hausman, N.E.; Hager, A.G.; Riechers, D.E.; Tranel, P.J. Genetics and inheritance of nontarget-site resistances to atrazine and mesotrione in a waterhemp (*Amaranthus tuberculatus*) population from Illinois. *Weed Sci.* **2015**, *63*, 799–809. [CrossRef]
37. Busi, R.; Gaines, T.A.; Vila-Aiub, M.M.; Powles, S.B. Inheritance of evolved resistance to a novel herbicide (pyroxasulfone). *Plant Sci.* **2014**, *217*, 127–134. [CrossRef]
38. Busi, R.; Neve, P.; Powles, S. Evolved polygenic herbicide resistance in *Lolium rigidum* by low-dose herbicide selection within standing genetic variation. *Evol. Appl.* **2013**, *6*, 231–242. [CrossRef]
39. Dücker, R.; Zöllner, P.; Lümmen, P.; Ries, S.; Collavo, A.; Beffa, R. Glutathione transferase plays a major role in flufenacet resistance of ryegrass (*Lolium* spp.) field populations. *Pest Manag. Sci.* **2019**, *75*, 3084–3092. [CrossRef]
40. Neve, P.; Powles, S. Recurrent selection with reduced herbicide rates results in the rapid evolution of herbicide resistance in *Lolium rigidum*. *Theor. Appl. Genet.* **2005**, *110*, 1154–1166. [CrossRef]
41. Vila-Aiub, M.M.; Neve, P.; Powles, S. Resistance cost of a cytochrome P450 herbicide metabolism mechanism but not an ACCase target site mutation in a multiple resistant *Lolium rigidum* population. *New Phytol.* **2005**, *167*, 787–796. [CrossRef]
42. Christopher, J.T.; Powles, S.B.; Liljegren, D.R.; Holtum, J.A.M. Cross-resistance to herbicides in annual ryegrass (*Lolium rigidum*). II. Chlorsulfuron resistance involves a wheat-like detoxification system. *Plant Physiol.* **1991**, *95*, 1036–1043. [CrossRef] [PubMed]
43. Shergill, L.S.; Barlow, B.R.; Bish, M.D.; Bradley, K.W. Investigations of 2, 4-D and multiple herbicide resistance in a Missouri waterhemp (*Amaranthus tuberculatus*) population. *Weed Sci.* **2018**, *66*, 386–394. [CrossRef]

44. Hall, L.M.; Moss, S.R.; Powles, S.B. Mechanism of resistance to chlorotoluron in two biotypes of the grass weed *Alopecurus myosuroides*. *Pestic. Biochem. Physiol.* **1995**, *53*, 180–192. [CrossRef]
45. Tardif, F.; Hokum, J.; Powles, S. Occurrence of a herbicide-resistant acetyl-coenzyme A carboxylase mutant in annual ryegrass (*Lolium rigidum*) selected by sethoxydim. *Planta* **1993**, *190*, 176–181. [CrossRef]
46. Preston, C.; Belles, D.S.; Westra, P.H.; Nissen, S.J.; Ward, S.M. Inheritance of resistance to the auxinic herbicide dicamba in kochia (*Kochia scoparia*). *Weed Sci.* **2009**, *57*, 43–47. [CrossRef]
47. Burnet, M.W.; Hart, Q.; Holtum, J.A.; Powles, S.B. Resistance to nine herbicide classes in a population of rigid ryegrass (*Lolium rigidum*). *Weed Sci.* **1994**, *42*, 369–377. [CrossRef]
48. Culpepper, A.S.; Grey, T.L.; Vencill, W.K.; Kichler, J.M.; Webster, T.M.; Brown, S.M.; York, A.C.; Davis, J.W.; Hanna, W.W. Glyphosate-resistant Palmer amaranth (*Amaranthus palmeri*) confirmed in Georgia. *Weed Sci.* **2006**, *54*, 620–626. [CrossRef]
49. Gaines, T.A.; Zhang, W.; Wang, D.; Bukun, B.; Chisholm, S.T.; Shaner, D.L.; Nissen, S.J.; Patzoldt, W.L.; Tranel, P.J.; Culpepper, A.S.; et al. Gene amplification confers glyphosate resistance in *Amaranthus palmeri*. *Proc. Natl. Acad. Sci. USA* **2010**, *107*, 1029–1034. [CrossRef]
50. Alonso, J.M.; Stepanova, A.N.; Leisse, T.J.; Kim, C.J.; Chen, H.; Shinn, P.; Stevenson, D.K.; Zimmerman, J.; Barajas, P.; Cheuk, R. Genome-wide insertional mutagenesis of *Arabidopsis thaliana*. *Science* **2003**, *301*, 653–657. [CrossRef]
51. Merchant, N.; Lyons, E.; Goff, S.; Vaughn, M.; Ware, D.; Micklos, D.; Antin, P. The iPlant collaborative: Cyberinfrastructure for enabling data to discovery for the life sciences. *PLOS Biol.* **2016**, *14*, e1002342. [CrossRef]
52. Afgan, E.; Baker, D.; Van den Beek, M.; Blankenberg, D.; Bouvier, D.; Čech, M.; Chilton, J.; Clements, D.; Coraor, N.; Eberhard, C. The Galaxy platform for accessible, reproducible and collaborative biomedical analyses: 2016 update. *Nucleic Acids Res.* **2016**, *44*, W3–W10. [CrossRef] [PubMed]
53. Humann, J.L.; Lee, T.; Ficklin, S.; Main, D. Structural and functional annotation of eukaryotic genomes with GenSAS. In *Gene Prediction: Methods and Protocols*; Kollmar, M., Ed.; Springer: New York, NY, USA, 2019; pp. 29–51. [CrossRef]
54. McArthur, A.G.; Waglechner, N.; Nizam, F.; Yan, A.; Azad, M.A.; Baylay, A.J.; Bhullar, K.; Canova, M.J.; De Pascale, G.; Ejim, L. The comprehensive antibiotic resistance database. *Antimicrob. Agents Chemother.* **2013**, *57*, 3348–3357. [CrossRef] [PubMed]
55. Nelson, D.R. The cytochrome p450 homepage. *Hum. Genom.* **2009**, *4*, 59. [CrossRef] [PubMed]
56. Heap, I. The International Survey of Herbicide Resistant Weeds. Available online: http://www.weedscience.com (accessed on 2 April 2019).
57. CDDEP. ResistanceMap: Center for Disease Dynamics, Economics & Policy. Available online: https://resistancemap.cddep.org/index.php (accessed on 16 July 2019).
58. Jia, B.; Raphenya, A.R.; Alcock, B.; Waglechner, N.; Guo, P.; Tsang, K.K.; Lago, B.A.; Dave, B.M.; Pereira, S.; Sharma, A.N. CARD 2017: Expansion and model-centric curation of the comprehensive antibiotic resistance database. *Nucleic Acids Res.* **2017**, *45*, D566–D573. [CrossRef]
59. LeClere, S.; Wu, C.; Westra, P.; Sammons, R.D. Cross-resistance to dicamba, 2,4-D, and fluroxypyr in *Kochia scoparia* is endowed by a mutation in an *AUX/IAA* gene. *Proc. Natl. Acad. Sci. USA* **2018**, *115*, E2911–E2920. [CrossRef]
60. Sammons, R.D.; You, J.; Qi, Y.; Flasinski, S.; Kavanaugh, C.; Washam, J.; Ostrander, E.; Wang, D.; Heck, G. Evaluation of glyphosate resistance in *Arabidopsis thaliana* expressing an altered target site EPSPS. *Pest Manag. Sci.* **2018**, *74*, 1174–1183. [CrossRef]
61. Guo, F.; Iwakami, S.; Yamaguchi, T.; Uchino, A.; Sunohara, Y.; Matsumoto, H. Role of CYP81A cytochrome P450s in clomazone metabolism in *Echinochloa phyllopogon*. *Plant Sci.* **2019**, *283*, 321–328. [CrossRef]
62. Saika, H.; Horita, J.; Taguchi-Shiobara, F.; Nonaka, S.; Nishizawa-Yokoi, A.; Iwakami, S.; Hori, K.; Matsumoto, T.; Tanaka, T.; Itoh, T. A novel rice cytochrome P450 gene, CYP72A31, confers tolerance to acetolactate synthase-inhibiting herbicides in rice and Arabidopsis. *Plant Physiol.* **2014**, *166*, 1232–1240. [CrossRef]
63. Li, Z.K.; Chen, B.; Li, X.X.; Wang, J.P.; Zhang, Y.; Wang, X.F.; Yan, Y.Y.; Ke, H.F.; Yang, J.; Wu, J.H. A newly identified cluster of glutathione S-transferase genes provides *Verticillium* wilt resistance in cotton. *Plant J.* **2019**, *98*, 213–227. [CrossRef]

64. Iwakami, S.; Endo, M.; Saika, H.; Okuno, J.; Nakamura, N.; Yokoyama, M.; Watanabe, H.; Toki, S.; Uchino, A.; Inamura, T. Cytochrome P450 CYP81A12 and CYP81A21 are associated with resistance to two acetolactate synthase inhibitors in *Echinochloa phyllopogon*. *Plant Physiol.* **2014**, *165*, 618–629. [CrossRef]
65. Chu, Z.; Chen, J.; Nyporko, A.; Han, H.; Yu, Q.; Powles, S. Novel α-tubulin mutations conferring resistance to dinitroaniline herbicides in *Lolium rigidum*. *Front. Plant Sci.* **2018**, *9*, 97. [CrossRef] [PubMed]
66. Chen, J.; Chu, Z.; Han, H.; Goggin, D.E.; Yu, Q.; Sayer, C.; Powles, S.B. A Val-202-Phe α-tubulin mutation and enhanced metabolism confer dinitroaniline resistance in a single *Lolium rigidum* population. *Pest Manag. Sci.* **2019**. [CrossRef] [PubMed]
67. Pan, L.; Yu, Q.; Han, H.; Mao, L.; Nyporko, A.; Fan, L.; Bai, L.; Powles, S.B. Aldo-keto reductase metabolizes glyphosate and confers glyphosate resistance in *Echinochloa colona*. *Plant Physiol.* **2019**, *181*, 1–16. [CrossRef] [PubMed]
68. Iwakami, S.; Kamidate, Y.; Yamaguchi, T.; Ishizaka, M.; Endo, M.; Suda, H.; Nagai, K.; Sunohara, Y.; Toki, S.; Uchino, A. CYP 81A P450s are involved in concomitant cross-resistance to acetolactate synthase and acetyl-CoA carboxylase herbicides in *Echinochloa phyllopogon*. *New Phytol.* **2019**, *221*, 2112–2122. [CrossRef] [PubMed]
69. Lange, M.; Yellina, A.L.; Orashakova, S.; Becker, A. Virus-induced gene silencing (VIGS) in plants: An overview of target species and the virus-derived vector systems. In *Virus-Induced Gene Silencing: Methods and Protocols*; Becker, A., Ed.; Humana Press: Totawa, NJ, USA, 2013; pp. 1–14.
70. Becker, A.; Lange, M. VIGS–genomics goes functional. *Trends Plant Sci.* **2010**, *15*, 1–4. [CrossRef] [PubMed]
71. Thyssen, G.N.; Naoumkina, M.; McCarty, J.C.; Jenkins, J.N.; Florane, C.; Li, P.; Fang, D.D. The P450 gene CYP749A16 is required for tolerance to the sulfonylurea herbicide trifloxysulfuron sodium in cotton (*Gossypium hirsutum* L.). *BMC Plant Biol.* **2018**, *18*, 186. [CrossRef] [PubMed]
72. Bennypaul, H.S.; Mutti, J.S.; Rustgi, S.; Kumar, N.; Okubara, P.A.; Gill, K.S. Virus-induced gene silencing (VIGS) of genes expressed in root, leaf, and meiotic tissues of wheat. *Funct. Integr. Genom.* **2012**, *12*, 143–156. [CrossRef]
73. Numata, K.; Ohtani, M.; Yoshizumi, T.; Demura, T.; Kodama, Y. Local gene silencing in plants via synthetic ds RNA and carrier peptide. *Plant Biotechnol. J.* **2014**, *12*, 1027–1034. [CrossRef]
74. Dalakouras, A.; Wassenegger, M.; McMillan, J.N.; Cardoza, V.; Maegele, I.; Dadami, E.; Runne, M.; Krczal, G.; Wassenegger, M. Induction of silencing in plants by high-pressure spraying of in vitro-synthesized small RNAs. *Front. Plant Sci.* **2016**, *7*, 1327. [CrossRef]
75. Dubrovina, A.S.; Aleynova, O.A.; Kalachev, A.V.; Suprun, A.R.; Ogneva, Z.V.; Kiselev, K.V. Induction of transgene suppression in plants via external application of synthetic dsRNA. *Int. J. Mol. Sci.* **2019**, *20*, 1585. [CrossRef]
76. Shibuya, K.; Fukushima, S.; Takatsuji, H. RNA-directed DNA methylation induces transcriptional activation in plants. *Proc. Natl. Acad. Sci. USA* **2009**, *106*, 1660–1665. [CrossRef] [PubMed]
77. Yoo, S.-D.; Cho, Y.-H.; Sheen, J. *Arabidopsis* mesophyll protoplasts: A versatile cell system for transient gene expression analysis. *Nat. Protoc.* **2007**, *2*, 1565. [CrossRef] [PubMed]
78. Townsend, J.A.; Wright, D.A.; Winfrey, R.J.; Fu, F.; Maeder, M.L.; Joung, J.K.; Voytas, D.F. High-frequency modification of plant genes using engineered zinc-finger nucleases. *Nature* **2009**, *459*, 442. [CrossRef] [PubMed]
79. Urnov, F.D.; Rebar, E.J.; Holmes, M.C.; Zhang, H.S.; Gregory, P.D. Genome editing with engineered zinc finger nucleases. *Nat. Rev. Genet.* **2010**, *11*, 636–646. [CrossRef] [PubMed]
80. Li, T.; Liu, B.; Spalding, M.H.; Weeks, D.P.; Yang, B. High-efficiency TALEN-based gene editing produces disease-resistant rice. *Nat. Biotechnol.* **2012**, *30*, 390–392. [CrossRef] [PubMed]
81. Jinek, M.; Chylinski, K.; Fonfara, I.; Hauer, M.; Doudna, J.A.; Charpentier, E. A programmable dual-RNA–guided DNA endonuclease in adaptive bacterial immunity. *Science* **2012**, *337*, 816–821. [CrossRef] [PubMed]
82. Feng, Z.; Zhang, B.; Ding, W.; Liu, X.; Yang, D.-L.; Wei, P.; Cao, F.; Zhu, S.; Zhang, F.; Mao, Y. Efficient genome editing in plants using a CRISPR/Cas system. *Cell Res.* **2013**, *23*, 1229–1232. [CrossRef]
83. Tranel, P.J.; Horvath, D.P. Molecular biology and genomics: New tools for weed science. *BioScience* **2009**, *59*, 207–215. [CrossRef]
84. Neve, P. Gene drive systems: Do they have a place in agricultural weed management? *Pest Manag. Sci.* **2018**, *74*, 2671–2679. [CrossRef]

85. Sammons, R.D.; Ivashuta, S.; Liu, H.; Wang, D.; Feng, P.C.C.; Kouranov, A.Y.; Andersen, S.E. Method for Controlling Herbicide-Resistant Plants. U.S. Patent US9121022B2, 1 September 2015.
86. St Leger, R.; Screen, S.; Butt, T.; Jackson, C.; Magan, N. Prospects for strain improvement of fungal pathogens of insects and weeds. In *Fungi as Biocontrol Agents: Progress, Problems and Potential*; CAB International: New York, NY, USA, 2001; pp. 219–237.
87. Amsellem, Z.; Cohen, B.A.; Gressel, J. Engineering hypervirulence in a mycoherbicidal fungus for efficient weed control. *Nat. Biotechnol.* **2002**, *20*, 1035–1039. [CrossRef]
88. Bycroft, C.; Freeman, C.; Petkova, D.; Band, G.; Elliott, L.T.; Sharp, K.; Motyer, A.; Vukcevic, D.; Delaneau, O.; O'Connell, J.; et al. The UK Biobank resource with deep phenotyping and genomic data. *Nature* **2018**, *562*, 203–209. [CrossRef] [PubMed]
89. Togninalli, M.; Seren, Ü.; Freudenthal, J.A.; Monroe, J.G.; Meng, D.; Nordborg, M.; Weigel, D.; Borgwardt, K.; Korte, A.; Grimm, D.G. AraPheno and the AraGWAS Catalog 2020: A major database update including RNA-Seq and knockout mutation data for *Arabidopsis thaliana*. *Nucleic Acids Res.* **2019**. [CrossRef] [PubMed]
90. Berger, M.F.; Mardis, E.R. The emerging clinical relevance of genomics in cancer medicine. *Nat. Rev. Clin. Oncol.* **2018**, *15*, 353–365. [CrossRef] [PubMed]
91. Délye, C.; Michel, S.; Pernin, F.; Gautier, V.; Gislard, M.; Poncet, C.; Le Corre, V. Harnessing the power of Next-Generation Sequencing technologies to the purpose of high-throughput pesticide resistance diagnosis. *Pest Manag. Sci.* **2019**, in press.
92. Donoso, A.; Valenzuela, S. In-field molecular diagnosis of plant pathogens: Recent trends and future perspectives. *Plant Pathol.* **2018**, *67*, 1451–1461. [CrossRef]
93. Cummins, I.; Wortley, D.J.; Sabbadin, F.; He, Z.; Coxon, C.R.; Straker, H.E.; Sellars, J.D.; Knight, K.; Edwards, L.; Hughes, D.; et al. Key role for a glutathione transferase in multiple-herbicide resistance in grass weeds. *Proc. Natl. Acad. Sci. USA* **2013**, *110*, 5812–5817. [CrossRef]
94. Muthu, M.; Nordström, A. Current status and future prospects of clinically exploiting cancer-specific metabolism—Why is tumor metabolism not more extensively translated into clinical targets and biomarkers? *Int. J. Mol. Sci.* **2019**, *20*, 1385. [CrossRef]
95. Wang, L.; Wheeler, D.A. Genomic sequencing for cancer diagnosis and therapy. *Ann. Rev. Med.* **2014**, *65*, 33–48. [CrossRef]
96. Duhoux, A.; Carrère, S.; Gouzy, J.; Bonin, L.; Délye, C. RNA-Seq analysis of rye-grass transcriptomic response to an herbicide inhibiting acetolactate-synthase identifies transcripts linked to non-target-site-based resistance. *Plant Mol. Biol.* **2015**, *87*, 473–487. [CrossRef]
97. Gaines, T.A.; Lorentz, L.; Figge, A.; Herrmann, J.; Maiwald, F.; Ott, M.C.; Han, H.; Busi, R.; Yu, Q.; Powles, S.B.; et al. RNA-Seq transcriptome analysis to identify genes involved in metabolism-based diclofop resistance in *Lolium rigidum*. *Plant J.* **2014**, *78*, 865–876. [CrossRef]
98. Werck-Reichhart, D.; Hehn, A.; Didierjean, L. Cytochromes P450 for engineering herbicide tolerance. *Trends Plant Sci.* **2000**, *5*, 116–123. [CrossRef]
99. Herrmann, J. Analysis of the Spatial and Temporal Dynamics of Herbicide Resistance to ACCase- and ALS-Inhibitors in *Alopecurus myosuroides* Huds. and Their Causes. Ph.D. Thesis, University of Braunschweig, Braunschweig, Germany, 2016.

plants

MDPI

Review

Current Status and Future Prospects in Herbicide Discovery

Franck E. Dayan

Department of Bioagricultural Sciences and Pest Management, 1177 Campus Delivery, Colorado State University, Fort Collins, CO 80523, USA; franck.dayan@colostate.edu; Tel.: +1-662-816-6214

Received: 21 August 2019; Accepted: 9 September 2019; Published: 11 September 2019

Abstract: Herbicides represent about 60% of the pesticides (by volume) used worldwide. The success of herbicides can be attributed in part to a relatively steady discovery of one unique mechanisms of action (MOA) every two years from the early 1950s to the mid-1980s. While this situation changed dramatically after the introduction of glyphosate-resistant crops, evolution of resistance to glyphosate has renewed the agrichemical industry interest in new chemistry interacting with novel target sites. This review analyses recent characterization of new herbicide target sites, the chemical classes developed to inhibit these target sites, and where appropriate the innovative technologies used in these discovery programs.

Keywords: amino acid biosynthesis; lipid biosynthesis; mechanism of action; plastoquinone biosynthesis; pyrimidine biosynthesis; target site

---

1. Introduction

The first synthetic herbicide was discovered in the early 1940s [1] and its efficacy and selectivity caused a paradigm change in agricultural weed management practices. New herbicide mechanisms of action (MOA) were discovered at a relatively steady rate of one unique MOA every two years from the early 1950s to the mid-1980s. Today, herbicides represent about 60% of the pesticides used worldwide, and most large-scale crop production systems rely extensively on synthetic herbicides to manage weeds. This has led to the relatively slow but steady evolution of many herbicide-resistant (HR) biotypes. The introduction of glyphosate-resistant (GR) crops in the last 25 years has compounded the selection pressure imposed by the repeated application of herbicides over larger areas. Managing these HR plants is problematic and the lack of control threatens farm profitability while challenging environmentally beneficial farming practices (e.g., no-till) [2,3]. The emergence and spread of HR weeds will require farmer cooperation to successfully control them [4].

While a tremendous commercial success, GR crops have been detrimental to herbicide discovery programs, causing a lapse in innovative research and development programs and a dearth of new chemistry with novel mechanisms of action (MOA) [5,6]. Factors that have contributed to a reduced investment in herbicide research and development are multifold and have been discussed elsewhere but include [7]:

- The success of GR crops that revolutionized weed management [8].
- The increased cost of R&D programs for production of a single new active ingredient from $184 million in 2000 to nearly $286 million in 2016 [9].
- The increased barriers imposed by toxicological and environmental regulations that must be fulfilled to ensure safety of the products [10].
- The severe attrition of the Agchem industry from more than 100 R&D companies to a few dominating companies [11]. This may in part be due to late stage failures (duPont), expense of liabilities and the depth of intellectual property.

Readers are encouraged to read Gerwick's review on the problems facing the agchem industry and Duke's review on why no new herbicide modes of action have been commercialized in recent years to have a broader analysis of these problems [6,12].

So what are the current status and future prospects in herbicide discovery? Herbicides are small (usually <500 MW) molecules that tend to target plant-specific processes. Generally speaking, herbicides can be grouped into three main categories: 1) herbicides that target biochemical pathways and physiological processes involved with photosynthesis, 2) herbicides that inhibit the formation of biological building blocks (i.e., sugars, amino acids and fatty acids) or their assembly into macromolecules, and 3) herbicides with other modes of action (Table 1). The many different active ingredients can be categorized based on their physicochemical properties [13] or organized around their respective mechanisms of action (Table 1). This low number of mechanisms of action is somewhat surprising, considering the thousands of potential molecular target sites that exist in plants and the hundreds of thousands of molecules screened for herbicidal activity every year.

This review will not cover the known mechanisms of action, and readers interested in the topic are referred to the original reports and several reviews on that topic for more information (e.g., [14,15]). However, a good knowledge of herbicide target sites and their mechanisms of action is crucial to decipher the way new herbicides may exert their action.

Table 1. Classification of mechanisms of action for current and potential new commercial herbicides discussed in this review.

Group	Type	Mechanism/Target
Biochemical pathways and physiological processes involved with photosynthesis	Light reaction	Photosystem II
		Photosystem I
	Carotenoid	Deoxyxylulose-5-phosphate synthase
		Phytoene desaturase
	Plastoquinone	*p*-Hydroxyphenylpyruvate dioxygenase
		Homogentisate solanesyltransferase [1]
		Solanyl diphosphate synthase [1]
	Chlorophylls	Protoporphyrinogen oxidase
	Uncouplers	Oxidative (photo)phosphorylation
Formation of biological building blocks or their assembly into macromolecules	Amino acids	5-enolpyruvylshikimate-3-phosphate (EPSP) Synthase
		Acetolactate synthase
		Glutamine synthetase
		Dihydroxy-acid dehydratase [1]
	Lipids	Acetyl-CoA carboxylase
		Fatty acid thioesterase [1]
		Very long chain fatty acid elongases
	Cell walls	Cellulose synthase and others
	Microtubule assembly	α- and/or β-Tubulin
	Microtubule organization	Microtubule organizing centers
	Folates	Dihydropteroate synthetase
		Dihydrofolate reductase [1]
	Nucleic acids	***DNA gyrase*** [1]
		Dihydroorotate dehydrogenase [1]
Other processes	Protein synthesis	***Peptide deformylase***
	Protein regulation	***Serine-threonine protein phosphatases*** [1]
	Hormone	Synthetic auxins
		Auxin-transport inhibition

[1] bold and italics indicates recently described or potentially new mechanisms or targets.

As mentioned above, no herbicides with truly new molecular targets had been introduced in the past 30 years. Yet, the need for new tools is more dire than ever, especially to combat HR weeds, and in particular those that have evolved resistance to glyphosate [16]. Though there does not appear to be a 'silver bullet' coming down the Agchem pipeline, there has been a recent flurry of reports of new mechanisms of action. So what has changed?

The dominance of glyphosate has been a destabilizing force affecting other agchem companies' decisions to move forward with new chemistry discovered through their own R&D programs. As mentioned before, the potential market shares of new compounds were not sufficient to justify the cost of developing these products. While the Agchem market is still relying on glyphosate for weed control in all the major row crops, the emergence of GR weeds has begun to impact the current usefulness and future prospect of glyphosate. Indeed, farmers are already returning to older chemistry to control GR weeds. In this new environment, companies may be projecting that the time is ripe for introducing new chemistry and new MOA. This seems to be reflected with recent presentations at the 2019 International Union of Pure and Applied Chemistry (IUPAC) congress on plant protection in Ghent, Belgium [17] and the Agrochemical Division of American Chemical Society's program at their 2019 national meeting in San Diego, CA. Additionally, several new startup companies have developed innovative technologies to explore new chemical spaces and/or facilitate the elucidation of their target sites. For example, MoA technology uses in vivo high throughput platforms, proprietary bioinformatics and Artificial Intelligence (AI) tools to discover novel herbicides with novel MOAs. Enko Chem Inc. aims to become a leading innovator in crop protection chemistry by utilizing a target-based discovery platform to produce high quality and novel small molecule starting points and a suite of tools and approaches to develop these into product candidates, and Agrematch develops AI and big-data tools for rational identification of molecules with desired biological activity and high potential to become crop protection products while significantly reducing R&D costs and accelerating time to market.

2. Novel Mechanisms of Action

2.1. Lipid Biosynthesis

Lipid synthesis involves several biochemical pathways leading to the formation of important building blocks for membranes, cuticles and waxes necessary for plant survival. Consequently, it has been the target of several herbicide classes.

Acetyl-coenzyme A carboxylase (ACCase) catalyzes the first committed step to fatty acid synthesis (Figure 1A). Cyclohexanediones (e.g., sethoxydim) and aryloxyphenoxypropionates (e.g., diclofop-methyl) are the two major groups of herbicides targeting this enzyme. Inhibitors of ACCase are grass-selective because two forms of ACCase occur in plants. A prokaryotic form is insensitive to these herbicides and is found only in the plastids of dicotyledonous plants, whereas the herbicide-sensitive eukaryotic form is found in the cytoplasm of all plants and in the plastids of grasses. Non-grass plants are resistant because they can sustain lipid biosynthesis in the presence of such herbicides.

While there are a host of enzymes catalyzing the many subsequent steps in fatty acid synthesis, the only other herbicide target site in this pathway are the very long chain fatty acid elongases (VLCFAE) (Figure 1A). These enzymes are much further down the metabolic pathway, responsible for synthesis of waxes, cutins, and suberins. VLFCAEs are the targets of several classes of herbicides, including the chloroacetanilides (e.g., alachlor), the thiocarbamates (e.g., EPTC), and the oxyacetamides (e.g., flufenacet). Inhibition of VLFCAEs by these herbicides results in decreased growth and leaf curling or twisting [14].

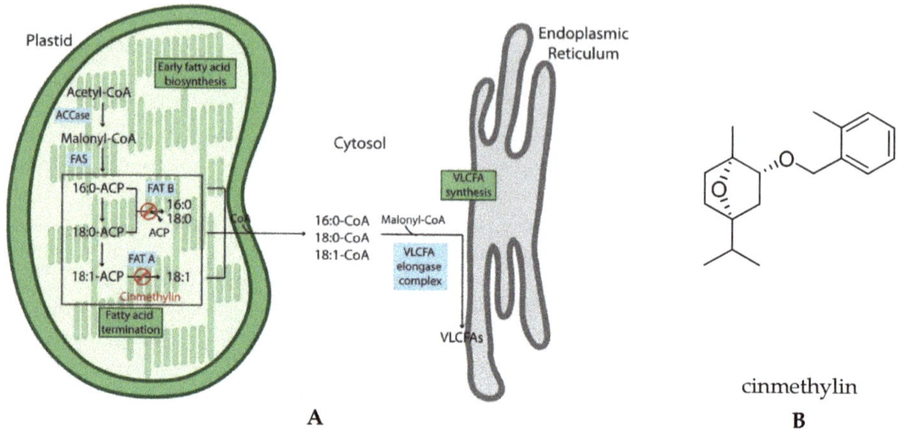

Figure 1. (**A**) Overview of fatty acid biosynthesis and herbicide targets in plant cells. Inhibition of acetyl-CoA carboxylase (ACCase) by haloxyfopmethyl or tepraloxydim disrupts early fatty acid biosynthesis. Cinmethylin prevents the release of both unsaturated and saturated fatty acids from the plastids through inhibition of fatty acid thioesterase (FAT) A and B, respectively. Inhibitors of very-long-chain fatty acid (VLCFA) biosynthesis act at the endoplasmic reticulum. ACP: acyl carrier protein; CoA: Coenzyme A. From [18] with permission. (**B**) Structure of cinmethylin.

Fatty Acid Thioesterase (FAT)

An exciting new development has been the discovery of new target sites in fatty acid synthesis of an old herbicide, cinmethylin (Figure 1B). These are the fatty acid thioesterases (FAT) (Figure 1B) [18]. FATs are plastid localized enzymes that mediate the release of fatty acids from its acyl carrier protein (ACP) which is necessary for FA export out of the chloroplast and transfer to the endoplasmic reticulum as fatty acyl-CoAs.

FATs are inhibited by cinmethylin, a natural product-like benzyl ether derivative of 1,4-cineole that was first described by Shell in 1981 and commercialized in 1989 [19]. Plants treated with cinmethylin have reduced levels of the saturated C14:0 and C16:0 fatty acids, indicating that the herbicide inhibits both classes of FAT proteins [18]. The direct interaction of cinmethylin with FAT proteins was confirmed by fluorescence-based thermal shift assays and co-crystallization of cinmethylin within the FAT enzyme.

2.2. *Plastoquinone Biosynthesis*

Plastoquinone is a lipid (prenyl) quinone with important biological functions. It is best known for its role as an electron acceptor in the light reaction of photosynthesis. Specifically, it accepts electrons from photosystem II and transfers them to the cytochrome b6 complex. Its importance in agrochemistry cannot be overstated. Some of the earliest commercial herbicides (e.g., triazines, ureas, and nitriles) inhibit photosynthesis by competing for the plastoquinone binding site on PSII. Later on, interest in plastoquinone renewed with the discovery of triketone herbicides that inhibit *p*-hydroxyphenylpyruvate dioxygenase (HPPD), a key enzyme in plastoquinone synthesis (Figure 2). These herbicides cause bleaching of photosynthetic tissues because plastoquinone is required for the activity of phytoene desaturase [20], a well-known target of herbicides inhibiting carotenoid biosynthesis (Figure 2). Consequently, industry started focusing on the biosynthesis pathway of plastoquinone in hope of identifying additional herbicide target sites. Its biosynthesis involves the convergence of two pathways (Figure 2). On one hand, the quinone head is derived from tyrosine and involves HPPD to form homogentisate. On the other hand, the lipophilic tail is derived from the 2-C-Methyl-D-erythritol 4-Phosphate (MEP)-derived terpenoid pathway (Figure 2).

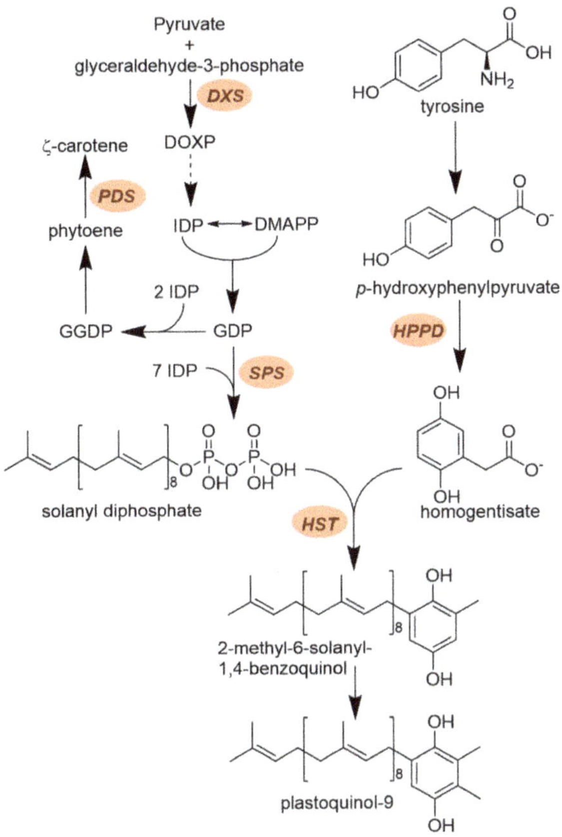

Figure 2. Overview of the relationship between carotenoid and prenyl quinone biosynthesis. Biosynthesis of carotenoids and plastoquinone requires the MEP, terpenoid and homogentisate pathways. Older chemistry such as clomazone inhibits 1-deoxy-D-xylulose 5-phosphate synthase (DXS), the first step in the MEP pathway; a number of chemical classes inhibit carotenoid biosynthesis by targeting phytoene desaturase (PDS); the newer triketone herbicides inhibit p-hydroxyphenylpyruvate dioxygenase (HPPD) involved in homogentisate biosynthesis. The two newest target sites affect solanyl diphosphate synthase (SPS) responsible for the synthesis of the terpenoid tail of plastoquinone or homogentisate solanesyl transferase (HST), the enzyme combining solanyl diphosphate and homogentisate to form a plastoquinone precursor.

2.2.1. Solanyl Diphosphate Synthase (SPS)

The building block of the lipid tail of plastoquinone is solanyl diphosphate. It is obtained by the activity of solanyl diphosphate synthase (SPS) which catalyzes the sequential addition of seven isopentenyl diphosphate to geranyl diphosphate [21]. A collaboration between the herbicide discovery group of Bayer CropScience and Targenomix recently discovered using a systems biology approach that aclonifen (Figure 3) causes bleaching of treated plants by inhibiting SPS. Aclonifen is a relatively old diphenylether herbicide whose MOA was unknown. The binding of aclonifen to SPS was confirmed by crystallography. Phenylalanine residues within the catalytic domain of SPS are involved in the binding of aclonifen. Plants possess three genes encoding SPS. Two of the genes encode for SPS1 and SPS2 proteins that are localized in the chloroplast and involved in plastoquinone synthesis [22]. The other gene encodes for the mitochondrial isoforms (SPS3) involved in ubiquinone synthesis [22]. SPS1 and SPS2 are sensitive to aclonifen, whereas SPS3 is insensitive to the herbicide.

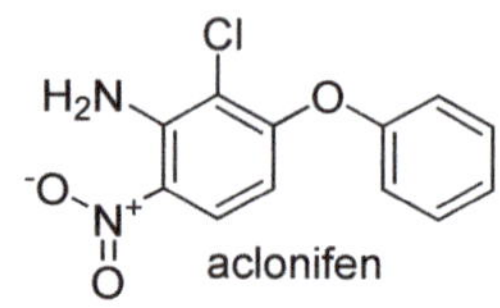

Figure 3. Structure of aclonifen, an inhibitor of chloroplastic solanyl diphosphate synthase (SPS).

2.2.2. Homogentisate Solanesyl Transferase (HST)

As mentioned above, plastoquinone biosynthesis involves the convergence of homogentisate and terpernoid synthesis. This step is catalyzed by homogentisate solanesyl transferase (HST) [23]. HST catalyzes the prenylation and decarboxylation of homogentisate to form 2-methyl-6-solanesyl-1,4-benzoquinol, the first intermediate in plastoquinone-9 biosynthesis. (Figure 2). This enzyme was known to be sensitive to inhibition by haloxydine [23]. Haloxydine acts as a suicide inhibitor mimicking homogentisate binding.

Mitsui Chemical Agro Inc. reported the discovery and development of cyclopyrimorate (Figure 4) as a new bleaching herbicide inhibiting HST. This herbicide was discovered in a program aiming at combining the pharmacophore backbone of credazine and pyridafol. Structure optimization against the weeds *Scirpus juncoides* (sedge) and *Sagittaria trifolia* (threeleaf arrowhead) demonstrated that the cyclopropane ring and the methyl group at the ortho positions (2 and 6, respectively) on the phenyl ring were critical for activity. The hydrophobicity of the moiety at position 2 can modulate activity, with cyclopropane being optimal. Additionally, a hydroxy group at position 4 on the pyridazine ring is important. Extensive biochemical work determined that plants treated with this herbicide have decreased levels of chlorophyll, carotenoids and plastoquinone, and accumulate homogentisate. The effect of the herbicide was strongly reversed by decyl plastoquinone and moderately reversed by homogentisate, suggesting that cyclopyrimorate targeted HST. Further work demonstrated that cyclopyrimorate was a proherbicide that needed to be bioactivated into its des-morpholinocarbonyl cyclopyrimorate (DMC) metabolite (Figure 4) to inhibit HST [24]. In planta metabolism of cyclopyrimorate into DMC releasing the free hydroxy group on position 4 of the pyridazine ring is critical for bioactivation of this herbicide. The I_{50} values for cyclopyrimorate and DMC on HST were 3.9 and 561 μM, respectively. DMC is a competitive inhibitor of HST for homogentisate but uncompetitive toward the prenyl diphosphate (Figure 2).

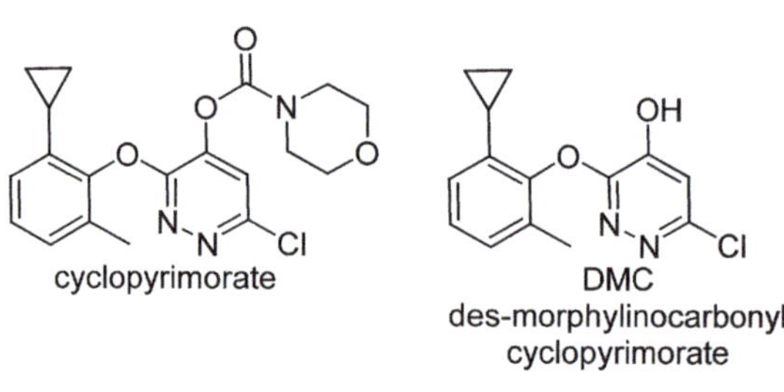

Figure 4. Cyclopyrimorate and its bioactive metabolite des-morpholinocarbonyl cyclopyrimorate (DMC).

Bleaching symptoms are the results of a dramatic decrease in plastoquinone levels in treated plants. Since this target site is downstream enzyme of HPPD in the plastoquinone biosynthesis pathway, activity of cyclopyrimorate is enhanced in tank mix with 4-HPPD inhibitors. This herbicide will be developed for weed management in rice paddies, including acetolactate synthase (ALS) resistant weed species with projected commercialization in 2020.

2.3. *Amino Acid Biosynthesis and Protein Regulation*

2.3.1. Dihydroxy-Acid Dehydratase (DHAD)

A large number of herbicides inhibit branched chain amino acid biosynthesis by targeting acetolactate synthase, the first step committed in this pathway (Figure 5) [14]. In light of the commercial success of this chemistry, industry has searched for chemicals that could inhibit this pathway by other means. A number of inhibitors of acetohydroxy acid isomeroreductase have been discovered but none of them have been developed into commercial products.

An innovative resistant-gene-directed discovery approach led to the discovery of a new herbicide target site in the branched chain amino acid pathway [25]. Aspterric acid is a natural herbicide produced by the soil fungus *Aspergillus terreus* (Figure 5). A research group at University of California Los Angeles analyzed the microbial gene cluster involved in the biosynthesis of this microbial phytotoxin. They discovered that that the cluster also included a paralog form of dihydroxy acid dehydratase (DHAD), the last common enzyme of the branched chain amino acid biosynthesis pathway (Figure 5). Further work demonstrated that aspterric acid targets DHAD, and the DHAD paralog present in the gene cluster was aspterric acid-resistant, its target enzyme [25]. Aspterric acid is a relatively weak phytotoxin which may not rise to a successful commercial herbicide, but it might serve as a structural backbone to elaborate new herbicide classes with improved physicochemical properties. As well, it is not clear whether DHAD is a good target site for herbicide to control weeds under agronomic conditions, and more work must be carried out to validate it as a desirable target site. Nevertheless, one of the advantages of this microbial gene co-clustering analysis is that it can lead to the discovery of genes involved in a phytotoxin biosynthesis, the identification of its target site, and the isolation of a herbicide-resistant form of this target site [26].

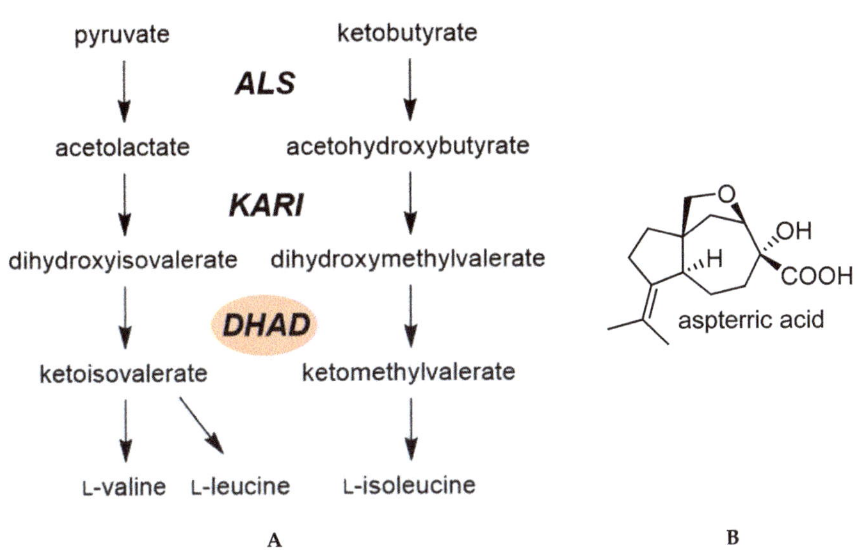

Figure 5. (**A**) Key enzymes involved in branched chain amino acid biosynthesis. DHAD is the most recent putative herbicide target site in this pathway. (**B**) Structure of the microbial metabolite aspterric acid, an inhibitor of DHAD. ALS: acetolactate synthase; KARI: acetohydroxy acid isomeroreductase; DHAD: dihydroxy acid dehydratase.

2.3.2. 3-Dehydroquinate Synthase

The shikimate pathway is one of the central pathways associated in plant metabolism, providing the carbon skeletons for the aromatic amino acids L-tryptophan, L-phenylalanine, and L-tyrosine

(Figure 6) and many important secondary metabolites (e.g., chlorogenic acid, alkaloids, glucosinolates, auxin, tannins, suberin, lignin and lignan, and tocopherols) [27]. It is estimated that at least 30% of all fixed carbon is directed through this pathway to support the flux required to produce these plant components.

Since mammals cannot synthesize these amino acids, this pathway is particularly desirable as a potential target for herbicides. To date, glyphosate is the only herbicide targeting this pathway by acting as an irreversible inhibitor of 5-enolpyruvylshikimate-3-phospate synthase (Figure 6). While glyphosate slowly depletes the pools of aromatic amino acids, its herbicidal activity is associated with a deregulation of the shikimate pathway, leading to accumulation of high levels of shikimate-3-phosphate and shikimate and siphoning of carbon and phosphate from other pathways, disrupting more than just the shikimate pathway [14].

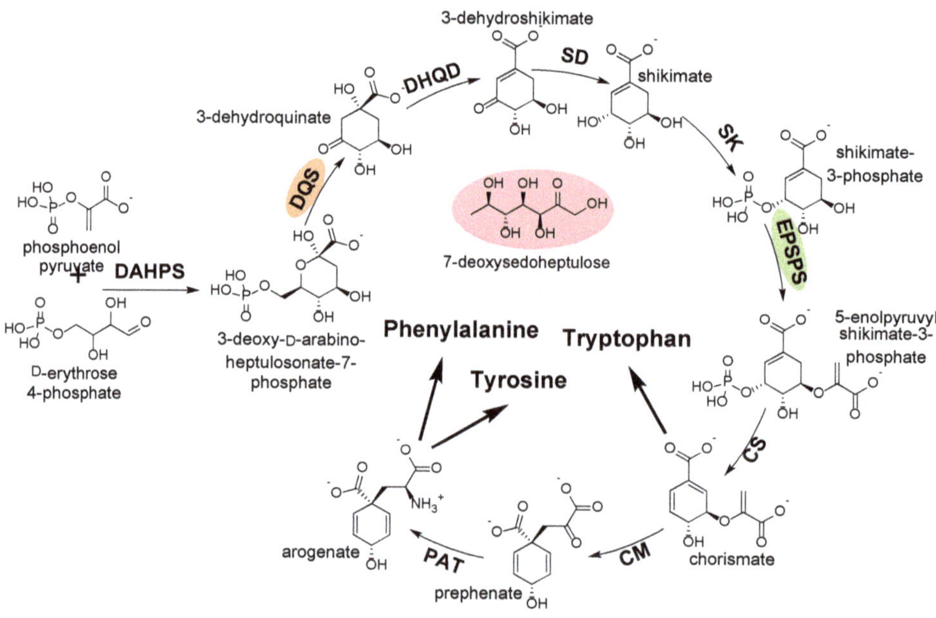

Figure 6. Shikimate pathway and aromatic amino acid biosynthesis showing the metabolites formed at each step catalyzed by the enzymes abbreviated in bold. The target site of glyphosate is EPSPS (green). The target site of the antimetabolite 7-deoxysedoheptulose (red) is DQS (orange). DAHPS: 3-deoxy-D-arabino-heptulosonate-7-phosphate synthase, DQS: 3-dehydroquinate synthase, DHQD/SD: 3-dehydroquinatede hydratase, SK: shikimate kinase, EPSPS: 5-enolpyruvylshikimate 3-phosphate synthase, CS: chorismate synthase, CM: chorismate mutase, PAT: prephenate aminotransferase.

Recently, a group from Tübingen University (Germany) has focused on antimetabolites as novel structural backbones to discover inhibitors affecting new target sites. Antimetabolites are interesting molecules as they inhibit enzymes by mimicking their physiological substrates. Their study identified the rare sugar 7-deoxy-sedoheptulose (7dSh) as an inhibitor of 3-dehydroquinate synthase (Figure 6), a key enzyme of the shikimate pathway. 7dSh is active at low micromolar range [28]. The growth of plants treated with 25 μM 7dSH was inhibited to the same degree as an equivalent amount of glyphosate. However, treatment with 7dSH caused an accumulation of 3-deoxy-D-arabino-heptulosonate-7-phosphate, the substrate of 3-dehydroquinate synthase (Figure 6), whereas glyphosate caused a rapid accumulation of shikimate, a well-known biomarker of 5-enolpyruvylshikimate 3-phosphate synthase inhibition [29]. Not surprisingly, plants treated with 7dSh have lower levels of free aromatic amino acids (tyrosine, phenylalanine, and tryptophan). This apparently deregulates the biosynthesis of other amino acids, resulting in accumulation of the branched chain amino acids (valine, leucine, and isoleucine), as well as arginine. 7dSh did not have any preemergence activity. However, it controlled velvetleaf (*Abutilon theophrasti*) when applied as a postemergence herbicide at a rate of 2 kg ha^{-1}. The addition of an adjuvant was required to obtain this level of activity. On the other hand, 7dSh had no activity on green foxtail (*Setaria viridis*) suggesting selectivity for broadleaf weed control.

2.3.3. Serine/Threonine Protein Phosphatases (PPs)

More than 70% of all proteins have multiple phosphorylation sites and many of these proteins' activities are regulated via phosphorylation. This is achieved by the concerted action of protein kinases and phosphatases, that account for between 2–4% of the protein-encoding genes of most plants [30,31]. The specificity of protein kinases is based on primary sequence recognition, whereas protein phosphatases tend to be non-discriminate. However, studies across many eukaryote systems confirmed that the phosphatases are not involved in generic dephosphorylation but are in fact as highly regulated as their kinase counterparts. Phosphoprotein phosphatases represent a large group of proteins, that include a sub-class called serine/threonine phosphatases (PPs) [32].

As their names imply, protein serine/threonine phosphatases (PPs) remove phosphate groups bound to serine and threonine residues. PPs are categorized into three subclasses—phosphoprotein phosphatases, metal-dependent protein phosphatases, and aspartate-based phosphatases. The PPs in plants belongs to the phosphoprotein phosphatase sub-class [32].

PPs are the target of endothall (Figure 7), an old herbicide that was first commercialized in the 1950s. Endothall induces severe growth inhibition [15]. Endothall is a structural analog of cantharidin, a natural product from the blister beetle (*Epicauta* spp.) and the Spanish fly (*Lytta vesicatoria*) (Figure 7). Both of these molecules cause similar symptoms on plants [33].

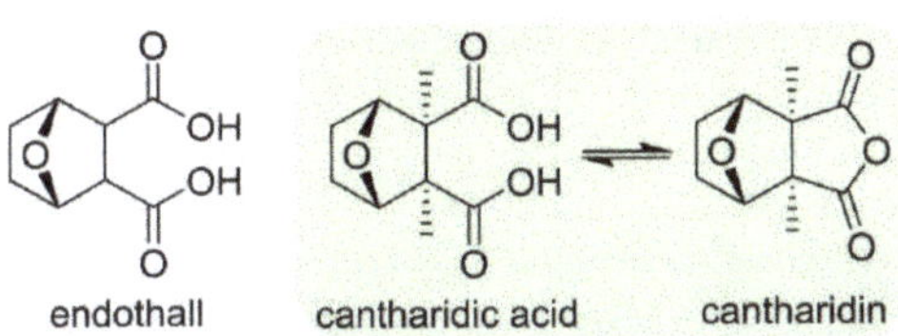

Figure 7. Structure of endothall, cantharidin and its dicarboxylic acid analog.

Endothall and cantharidin both inhibit plant serine/threonine protein phosphatases in a time-dependent manner, suggesting that these compounds act as slow, irreversible inactivators of the serine/threonine protein phosphatase activities [33]. The catalytic domain of all PP highly conserved across animals, plants and protozoans. Inhibitors, such as cantharidin and endothall, bind to a hydrophobic pocket of the PP active site. Endothall is a very effective herbicide to manage weeds in aquatic environments [34].

2.4. *Pyruvate Dehydrogenase Complex (PDHc)*

Pyruvate dehydrogenase complex (PDHc) catalyzes the oxidative decarboxylation of pyruvate to form acetate and its subsequent acetylation of coenzyme A (CoA) to produce acetyl-CoA [35]. As such it is a critically important for cellular processes. The complex consists of three enzymes and a number of cofactors. Pyruvate dehydrogenase E1 is a thiamine diphosphate- and Mg^{2+}-dependent enzyme catalyzing the first step of the multistep process associated with PDHc [35].

The Institute of Pesticide and Organic Chemistry of Central China Normal University recently reported novel cyclic methylphosphonates (Figure 8) that target pyruvate dehydrogenase complex (PDHc) using molecular docking and three-dimensional quantitative structure-activity relationship studies [36]. Early acetylphosphinates and acetylphosphonates analogs had relatively low herbicidal activity, but these structures served as the basis for structural optimization to generate 1-(substituted phenoxyacetoxy)alkylphosphonate derivatives with notably higher herbicidal activities (Figure 8) [36]. Herbicidal activity is proportional to inhibition of PDHc E1.

Recent development reported that these PDHc inhibitors were most effective against broadleaf weeds and active at rates ranging from 50 and 300 ai g/ha, whereas they had no effect on maize and rice even at 900–1200 ai g/ha. Some of these compounds also had activity on sedge weeds when applied at 225–375 ai g/ha [37].

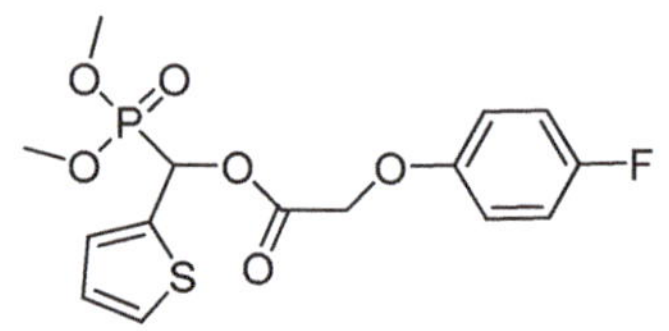

Figure 8. Structure of a potent 1-(substituted phenoxyacetoxy)alkylphosphonate that targets pyruvate dehydrogenase complex (PDHc).

2.5. *Imadazoleglycerol Phosphate Dehydratase (IGPD)*

Imadazoleglycerol phosphate dehydratase (IGPD) catalyzes an important step in histidine biosynthesis in plants and microorganisms. It has been studied for many years as a potential target for herbicides, since this enzyme does not exist in animals. A class of phloem-mobile herbicides (the triazole-phosphonates) act as potent inhibitors of IGPD [38]. Syngenta has been working on this target site for many years. The triazole phosphonate inhibitor 2-hydroxy-3-(1,2,4-triazol-1-yl) propylphosphonate (Figure 9) is structurally similar to the proposed diazafulvene intermediate in IGPD catalysis [39]. Several triazole phosphonate inhibitors have activities similar to glyphosate [40]

Structurally, triazole phosphonate inhibitors consists of three parts, the triazole head, an hydroxylated linker and a phosphate mimick. The position of the hydroxy group alters the binding of the molecules to the catalytic domain of IGPD forming either 5- or 6-membered ring chelates with one of the Mn atom [41].

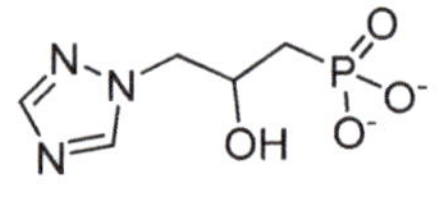

Figure 9. Structure of 2-hydroxy-3-(1,2,4-triazol-1-yl) propylphosphonate, an herbicidal inhibitor of imadazoleglycerol phosphate dehydratase (IGPD).

2.6. *Dihydroorotate Dehydrogenase (DHODH)*

De novo pyrimidine nucleotide biosynthesis (also known as the orotate pathway) consists of six enzymatic steps leading to the formation of uridine monophosphate from carbamoylphosphate, aspartate, and 5-phosphoribosyl-1-pyrophosphate. Because of the central role of nucleotides, inhibition of this pathway is lethal to most organisms. The fourth step is catalyzed by dihydroorotate dehydrogenase (DHODH), which carries out the ubiquinone-mediated oxidation of dihydroorotate to orotate [42].

All plant DHODHs are flavoproteins located on the outer surface of the inner mitochondrial membrane. Plant DHODHs have different substrate specificity and inhibition from the animal form of this enzyme [43]. FMC Agricultural Solutions recently announced a new herbicide chemical class (aryl pyrrolidinone anilide) targeting DHODH. The common chemical name of the flagship molecule currently being developed was provisionally approved as tetflupyrolimet (Figure 10). This potent herbicide is selective for grass control in rice. Sensitive plants treated with tetflupyrolimet have no chlorosis but develop a unique stunting phenotype suggesting that they are lacking a key molecule for growth (pyrimidine). The target site was discovered using a combination of forward genetic screens and metabolomics approaches and confirmed by determining intrinsic affinities of specific analogs using biochemical methods [17]. Structure-activity studies determined that the 3*S*-4*R* enantiomer is the active form of this aryl pyrrolidinone anilide, and the 3*R*-4*S* enantiomer had no herbicidal activity. Additionally, the presence of the electron withdrawing groups (fluorine) on the two benzyl rings and the alkylation (methyl group) of the γ-lactam heterocycle are required for herbicidal activity. Tetflupyrolimet competes for the quinone binding site on DHODH. The activity of tetflupyrolimet was about 10-fold greater on the foxtail DHODH enzyme (I_{50} = 3 nM) compared to rice (I_{50} = 33 nM). However, selectivity for rice is much greater than 10-fold, suggesting that differential metabolism may also contribute to tolerance in rice. Additional work demonstrated that tetflupyrolimet was much less active on animal DHODH. Commercialization of this product is projected to be in 2024.

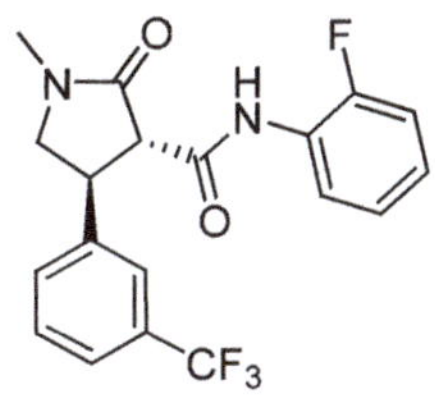

Figure 10. Structure of tetflupyrolimet, an aryl pyrrolidinone anilide targeting Dihydroorotate dehydrogenase, a key enzyme in pyrimidine biosynthesis.

2.7. Peptide Deformylase

In higher plants, synthesis of plastid encoded proteins is initiated with *N*-formylmethionine. Removal of the N-formyl group by a peptide deformylase and the methionine by methionine amino peptidase is necessary to produce the mature protein. The initiator methionine is sometimes retained [44]. Peptide deformylase is the target of actinonin, an hydroxamic acid microbial metabolite produced by soil actinomycetes (Figure 11) [45]. This unique MOA has received a lot of interest and the herbicidal activity of actinonin has been patented, but no commercial product has been developed. Plants treated with actinonin are stunted with bleached foliage which ultimately develop necrotic lesions. It has proved effective on many important weed species [46,47].

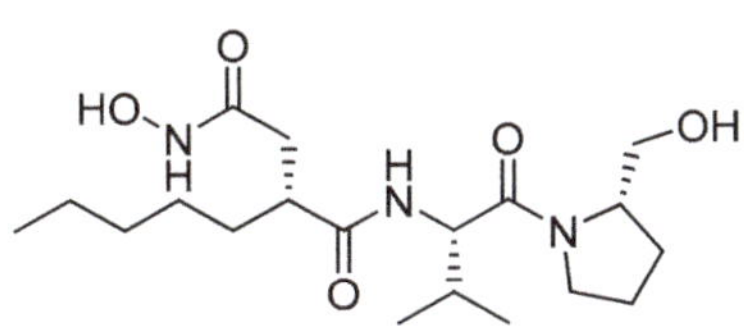

Figure 11. Structure of actinonin, a microbial metabolite targeting chloroplastic peptide deformylase.

2.8. DNA Gyrase

DNA gyrases are prokaryotic Type II topoisomerases that were thought to be absent from most eukaryotes. However, ancestral forms of DNA gyrases may be present in certain organelles of plants and apicomplexans, although their exact functions in replication are not well understood. A research group from the University of Western Australia investigated DNA gyrase as a potential herbicide target site by testing the activity of a number of compounds likely to interact with this enzyme. Several molecules, including the antimicrobial ciprofloxacin (Figure 12), were herbicidal by inhibiting the function of gyrase in higher plants [48]. Three genes (*ATGYRA*, *ATGYRB1*, *ATGYRB2*) encoding for plant gyrases were identified in *Arabidopsis thaliana*. Forward genetic approaches led to the discovery of a point mutation in *ATGYRA* that confers resistance to ciprofloxacin, thus confirming that this gene encodes a functional organelle-localized DNA gyrase that is the target of quinolone antimalarial drugs [49].

Subsequent work exploring the activity of ciprofloxacin analogs on DNA gyrase led to the characterization of the pharmacophore scaffold required for activity and the discovery of structures with improved herbicidal efficacy and diminished antibacterial activity, relative to ciprofloxacin. The optimized experimental analog 44 (Figure 12) had an ethyl side chain and a piperidine ring instead of a cyclopropyl side chain and a piperazine ring attached to the fluoroquinolone scaffold. This molecule was slightly less herbicidal than ciprofloxacin, but its specificity for plant DNA gyrase was superior, leading to a 600-fold increase in selectivity for plants relative to other organisms [50].

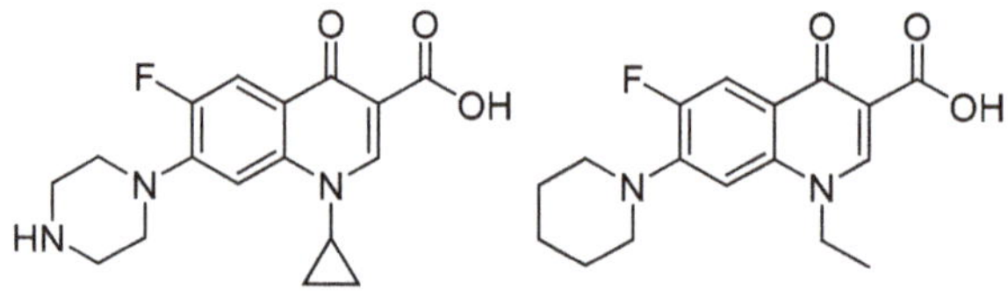

Figure 12. Structure of the fluoroquinolone ciprofloxacin and structure-optimized analog 44 with increased specificity against plants DNA gyrase and better herbicidal profile.

2.9. *Dihydrofolate Reductase (DHFR)*

The biosynthesis of folate has been the target for pharmaceutical and agrochemical discovery (Figure 13). Folate is an important metabolite required for the synthesis of numerous compounds necessary for plant growth and development. To date, asulam (Figure 13) is only one commercial herbicide to inhibit this pathway by targeting 7,8-dihydropteroate synthetase [51]. This carbamate herbicide is an analogue of 4-aminobenzoate, one of the substrates of 7,8-dihydropteroate synthase, and its selectivity is based on differential metabolic degradation.

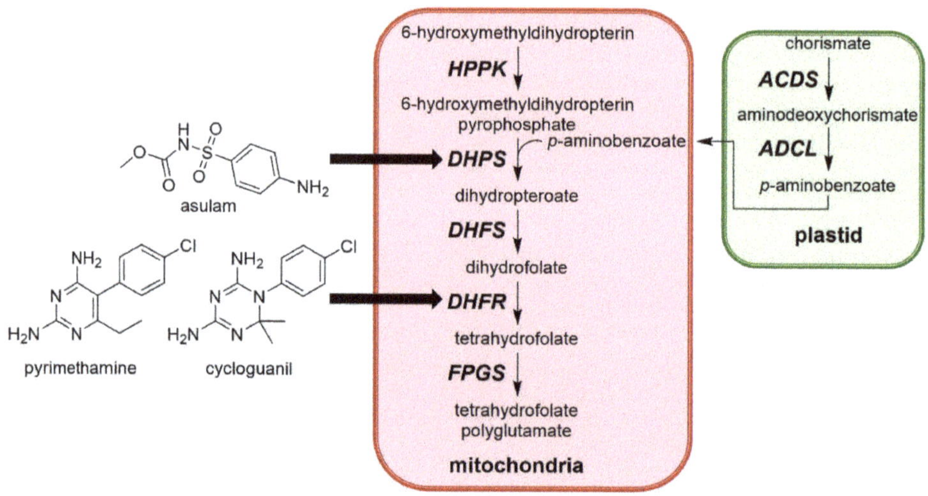

Figure 13. Simplified biosynthesis of folate in higher plants and structures of herbicidal compounds targeting this pathway. HPPK: 6-hydroxymethyldihydropterin pyrophosphokinase; DHPS: dihydropteroate synthase; DHFS: dihydrofolate synthetase; DHFR: dihydrofolate reductase; FPGS: folylpolyglutamate synthetase; ADCS: aminodeoxychorismate synthase; ADCL: aminodeoxychorismate lyase.

Another enzyme in this pathway, dihydrofolate reductase (DHFR) (Figure 12) has already received some interest as a target for drug development due to its essential role in the synthesis of DNA precursors and some amino acids. A recent study by a group at the University of Western Australia identified DHFR as a potential new target for herbicides based on the herbicidal activity of antimalarial compounds such as pyrimethamine and cycloguanil (Figure 13) [52]. The requirement of two of the three isoforms of DHFR for seed development was identified by knockout mutant analysis. Validation of this enzyme as a new target site was confirmed by screening mutated *Arabidopsis thaliana* seeds for resistance to these antimalarial compounds. A G137D mutation in the isoform 1 of DHFR and a A71V mutation in isoform 2 of DHFR imparted resistance, confirming that the herbicidal activity associated with the antimalarial molecules were due to inhibition of DHFR. This discovery sets the stage for high throughput screening of chemical libraries to identify molecules with better herbicidal profile.

3. New Insight on Known Mechanisms of Action

3.1. New Insight on Glufosinate Mechanism of Action

The MOA of glufosinate has been studied extensively. While the inhibition of glutamine synthetase and subsequent accumulation of ammonia, disruption of amino acid balance, and reduction of both the light and dark reaction of photosynthesis are well documented, these did not account for the rapid desiccation of the foliage induced by glufosinate. New insight on the factors contributing to the contact activity of glufosinate has been reported by the Weed Research Laboratory at Colorado State University. Glufosinate triggers a rapid and massive production of reactive oxygen species (ROS) driving the catastrophic lipid peroxidation of the cell membranes and rapid cell death (Figure 14A) [53]. The effect was proportional to absorption of the herbicide. Interestingly, young leaves were less sensitive to glufosinate. While older leaves absorbed more glufosinate than younger tissues (Figure 14B), similar levels of glutamine synthetase inhibition and ammonia accumulation were observed (Figure 14C,D), indicating that ammonia accumulation was not responsible for the toxicity of this herbicide. In contrast, glufosinate induced a rapid and massive accumulation of ROS in older tissue and almost no ROS in younger tissue (Figure 14E), which correlated directly with the level of injury.

Figure 14. New insight into the mechanisms of action (MOA) of glufosinate. (**A**) Older leaves are more sensitive to glufosinate than meristematic tissue and younger leaves. (**B**) Absorption of glufosinate. (**C**) Inhibition of glutamine synthetase (GS). (**D**) Ammonia accumulation. (**E**) Reactive oxygen species (ROS) accumulation. Reproduced from Takano et al. 2019 with permission.

3.2. New Insight on Slow-Binding Properties of HPPD Inhibitors

As mentioned in Section 2.2, inhibition of plastoquinone biosynthesis has been recognized as an excellent target for new herbicide research since the discovery that triketone herbicides inhibit HPPD. These herbicides bind slowly but very tightly (nearly irreversibly) to the catalytic site by coordinating with Fe atom involved in catalysis [54]. A research group based at Central China Normal University provided new molecular insights into the mechanism of 4-hydroxyphenylpyruvate dioxygenase (HPPD) inhibition using enzyme kinetics, X-ray crystallography of *Arabidopsis thaliana* HPPD complexed with herbicides, and computational simulations approaches. This work dissected the interaction between ligand and receptor to discover a novel quinazoline-2.4-dione herbicide benquitrione (Y13161) (Figure 15) [55]. Their analysis suggests that the slow binding properties of HPPD inhibitors may be related to steric hindrance requiring a conformational change on the enzyme upon herbicide binding. Benquitrione has excellent herbicidal activity that compares favorably with that of mesotrione. This molecule also demonstrated selectivity on corn and sorghum, whereas the mesotrione caused injury to sorghum. Finally, the structural features of benquitrione can also serve as a template to develop the next generation of high performance HPPD-inhibiting herbicides.

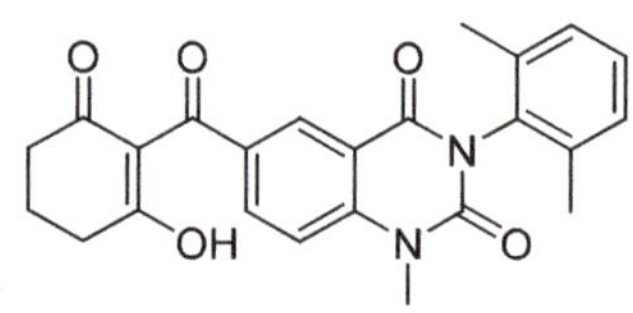

Figure 15. Structure of novel HPPD inhibitor Y13161 (benquitrione).

4. Promising New Chemistry

4.1. Isoxazolopyridine Herbicides

BASF recently reported a new class of herbicides based on an isoxazolopyridine (OXP) backbone (Figure 16). Some of the compounds have selectivity on monocotyledonous crops while providing excellent post-emergence control of dicot weeds and good activity on some grasses. Sensitive plants treated with OXP herbicides develop necrosis on the foliage and the compounds appear to have systemic activity, with phloem-mobility in dicots but limited translocation in monocots.

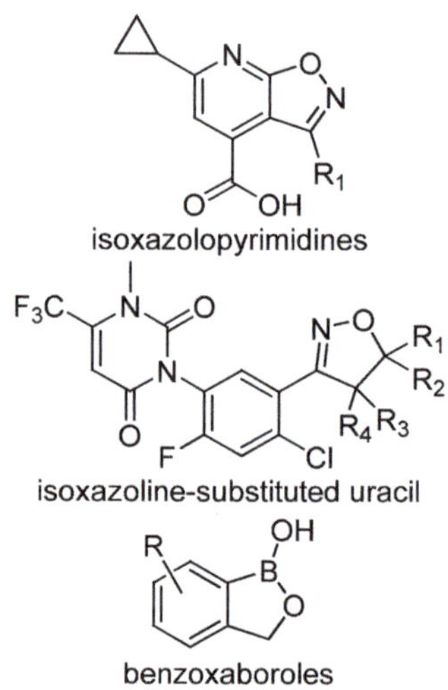

Figure 16. Structural features of the chemical classes mentioned in Section 4.

Biochemical/physiological studies excluded known MOA. Therefore, isoxazolopyridines may have a novel mechanism of action on plant photosystems that involves carbohydrate metabolism within the chloroplast. Cellular thermal shift assays suggest that light harvesting protein may be the potential target. Interestingly, a L218V mutation on the D1 protein of photosystem II from metamitron tolerant *Chenopodium album* (Common lambsquarters) provided partial resistance to OXPs, whereas the well-known S264G mutation did not protect plants against the activity of OXPs.

4.2. Isoxazoline-Substituted Uracil Herbicides

Sinochem Agrochemicals R&D Co. Ltd. developed novel uracil herbicides containing an isoxazolopyrimidine ring (Figure 16). These compounds effectively control a number of economically important monocot and dicots weeds and appears to be safe to wheat, corn and rice. One of the most promising molecules (SYP-1604) can be used alone or in combination with other herbicides. SYP-1604 spectrum of activity outperforms saflufenacil when applied post and compared favorably to flumioxazin when applied pre. Little is known about its MOA, but it appears to act as an inhibitor of protoporphyrinogen oxidase.

4.3. Benzoxaboroles Herbicides

Benzoxaboroles (Figure 16) are derivatives of boronic acids that were first described over 50 years ago. Interest in these compounds renewed in 2006, after reports of the sugar-binding properties of certain benzoxaboroles were made public. Consequently, most of benzoxaboroles have been described over the last decade [56]. Interest in these structures widened due to their range of biological activities, including the commercialization of two pharmaceutical products. Benzoxaboroles have recently been considered as starting points for new herbicides. Scientists at Corteva explored the chemical space occupied by benzoxaboroles and uncovered a promising area for new herbicide discovery. A limiting factor is the relatively high pka of boronic acid (8.8), but structures with constricted rings, such as the benzoxaboroles, can be designed with lower pka values A number of benzoxaboroles were identified herbicidal hits in initial screens, and plants treated with this type of chemistry developed unique symptoms. However, rates required for activity were high (1–4 kg ha^{-1}). Symptoms of plants treated with this class of chemistry varied from stunting to bleaching and necrosis on the leaf margin. While no target site has been identified or reported to date, structure-activity work demonstrated that adding a methyl group enhanced the activity. Replacing this methyl group with cyano group was even more potent. Interestingly, increasing the steric bulk by adding a phenyl ring was advantageous, but decoration with electron withdrawing groups (i.e., chlorophenyl) has a negative impact. Finally, adding alkyl spacer to the phenyl ring positively modulated activity. While no herbicides have arisen from this chemical class, isoxaboroles may lead to new scaffolds to develop new phloem-mobile molecules.

5. Conclusions

The time for innovative MOA and chemistry targeting these is overdue. The current crisis experienced by farmers facing difficulties managing weeds that have evolved resistance to many of the existing MOA must be addressed, and new weed management tools are necessary. While the current renewed interest in research and development programs observed in the agrichemical industry, as well as academic and governmental institutions is a positive development, there is no silver bullet chemistry ready to enter the marketplace. Many of the MOA and associated chemistries described in this review are at least 5 years away from commercialization or will fail to reach commercialization due to the many hurdles facing such a process. The aggregation of the agrichemical industry is certain to continue, further limiting diversity in creativity and discovery. One may hope that the few startup companies using truly innovative approaches to herbicide discovery will provide platforms to explore new chemical spaces and biochemical processes. It may also be time to incorporate non-chemical means of weed control, such as mechanical seed destruction (e.g., Harrington seed destructor), robotic weed management, and precision farming.

Funding: This work was funded by the USDA National Institute of Food and Agriculture, Hatch Project 1016591, COL00785.

Acknowledgments: The author thanks presenters at the Herbicide Mechanisms of Action and Resistance session at the 2019 IUPAC congress in Ghent, Belgium and the New Herbicides and Their Modes of Action symposium at the 2019 ACS national meeting in San Diego for sharing some of their most recent discoveries.

Conflicts of Interest: The author declares no conflict of interest.

References

1. Sterling, T.M.; Hall, J.C. Mechanism of action of natural auxins and the auxinic herbicides. In *Herbicide Activity: Toxicology, Biochemistry and Molecular Biology*; Roe, R.M., Burton, J.D., Kuhr, R.J., Eds.; IOS Press: Amsterdam, The Netherlands, 1997; pp. 111–141.
2. Pannell, D.J.; Tillie, P.; Rodríguez-Cerezo, E.; Ervin, D.; Frisvold, G.B. Herbicide resistance: Economic and environmental challenges. *AgBioForum* **2017**, *19*, 136–155.
3. Shaw, D.R. The "Wicked" nature of the herbicide resistance problem. *Weed Sci.* **2016**, *64*, 552–558. [CrossRef]
4. Ervin, D.E.; Breshears, E.H.; Frisvold, G.B.; Hurley, T.; Dentzman, K.E.; Gunsolus, J.L.; Jussaume, R.A.; Owen, M.D.K.; Norsworthy, J.K.; Al Mamun, M.M.; et al. Farmer attitudes toward cooperative approaches to herbicide resistance management: A common pool ecosystem service challenge. *Ecol. Econ.* **2019**, *157*, 237–245. [CrossRef]
5. Davis, A.S.; Frisvold, G.B. Are herbicides a once in a century method of weed control? *Pest Manag. Sci.* **2017**, *73*, 2209–2220. [CrossRef] [PubMed]
6. Duke, S.O. Why have no new herbicide modes of action appeared in recent years? *Pest Manag. Sci.* **2012**, *68*, 505–512. [CrossRef] [PubMed]
7. Dayan, F.E. Is there a natural route to the next generation of herbicides? *Outlooks Pest Manag.* **2018**, *29*, 54–57. [CrossRef]
8. Duke, S.O. The history and current status of glyphosate. *Pest Manag. Sci.* **2018**, *74*, 1027–1034. [CrossRef]
9. McDougall, P. *Agrochemical Research and Development: The Cost of New Product Discovery, Development and Registratoin*; Pathhead: Midlothian, UK, 2016; p. 43.
10. Peters, B.; Strek, H.J. Herbicide discovery in light of rapidly spreading resistance and ever-increasing regulatory hurdles. *Pest Manag. Sci.* **2018**, *74*, 2211–2215. [CrossRef]
11. Copping, L.G. The evolution of crop protection companies. *Outlooks Pest Manag.* **2018**, *29*, 25–37. [CrossRef]
12. Gerwick, B.C. Thirty years of herbicide discovery: surveying the past and contemplating the future. In *Chapters VII–IX in Agrow Report*; Informa: London, UK, 2010; pp. VII–IX.
13. Gandy, M.N.; Corral, M.G.; Mylne, J.S.; Stubbs, K.A. An interactive database to explore herbicide physicochemical properties. *Org. Biomol. Chem.* **2015**, *13*, 5586–5590. [CrossRef]
14. Dayan, F.E.; Barker, A.; Bough, R.; Ortiz, M.; Takano, H.; Duke, S.O. Herbicide mechanisms of action and resistance. In *Comprehensive Biotechnology*, 3rd ed.; Grodzinski, B., Ed.; Elsevier: Amsterdam, The Netherlands, 2019; Volume 4, in press.
15. Shaner, D.L. *Herbicide Handbook*, 10th ed.; Weed Science Society of America: Lawrence, KS, USA, 2014; p. 513.
16. Beckie, H.J.; Harker, K.N. Our top 10 herbicide-resistant weed management practices. *Pest Manag. Sci.* **2017**, *73*, 1045–1052. [CrossRef] [PubMed]
17. Dayan, F.E.; Haesaert, G.; Van Leeuwen, T.; Holden-Dye, L.; Crossthwaite, A.; Nauen, R. Pesticides modes of action and resistance: A perspective from the 2019 IUPAC congress. *Outlooks Pest Manag.* **2019**, *30*, 157–163. [CrossRef]
18. Campe, R.; Hollenbach, E.; Kämmerer, L.; Hendriks, J.; Höffken, H.W.; Kraus, H.; Lerchl, J.; Mietzner, T.; Tresch, S.; Witschel, M.; et al. A new herbicidal site of action: Cinmethylin binds to acyl-ACP thioesterase and inhibits plant fatty acid biosynthesis. *Pestic. Biochem. Physiol.* **2018**, *148*, 116–125. [CrossRef] [PubMed]
19. Dayan, F.E.; Romagni, J.G.; Duke, S.O. Herbicides: Cinmethylin. In *Encyclopedia of Agrochemicals*; Plimmer, J.R., Gammon, D.W., Ragsdale, N.N., Eds.; John Wiley & Sons: New York, NY, USA, 2003; Volume 2, pp. 754–757.
20. Norris, S.R.; Barrette, T.R.; DellaPenna, D. Genetic dissection of carotenoid synthesis in *Arabidopsis* defines plastoquinone as an essential component of phytoene desaturation. *Plant Cell* **1995**, *7*, 2139–2149. [PubMed]
21. Liu, M.; Lu, S. Plastoquinone and ubiquinone in plants: Biosynthesis, physiological function and metabolic engineering. *Front. Plant Sci.* **2016**, *7*, 1898. [CrossRef] [PubMed]

22. Ohara, K.; Sasaki, K.; Yazaki, K. Two solanesyl diphosphate synthases with different subcellular localizations and their respective physiological roles in *Oryza sativa*. *J. Experiment. Bot.* **2010**, *61*, 2683–2692. [CrossRef] [PubMed]
23. Sadre, R.; Frentzen, M.; Saeed, M.; Hawkes, T. Catalytic reactions of the homogentisate prenyl transferase involved in plastoquinone-9 biosynthesis. *J. Biol. Chem.* **2010**, *285*, 18191–18198. [CrossRef] [PubMed]
24. Shino, M.; Hamada, T.; Shigematsu, Y.; Hirase, K.; Banba, S. Action mechanism of bleaching herbicide cyclopyrimorate, a novel homogentisate solanesyltransferase inhibitor. *J. Pestic. Sci.* **2018**, *43*, 233–239. [CrossRef]
25. Yan, Y.; Liu, Q.; Zang, X.; Yuan, S.; Bat-Erdene, U.; Nguyen, C.; Gan, J.; Zhou, J.; Jacobsen, S.E.; Tang, Y. Resistance-gene-directed discovery of a natural-product herbicide with a new mode of action. *Nature* **2018**, *559*, 415–418. [CrossRef] [PubMed]
26. Duke, S.O.; Stidham, M.A.; Dayan, F.E. A novel genomic approach to herbicide and herbicide mode of action discovery. *Pest Manag. Sci.* **2019**, *75*, 314–317. [CrossRef] [PubMed]
27. Tohge, T.; Watanabe, M.; Hoefgen, R.; Fernie, A.R. Shikimate and phenylalanine biosynthesis in the green lineage. *Front. Plant Sci.* **2013**, *4*, 62. [CrossRef] [PubMed]
28. Brilisauer, K.; Rapp, J.; Rath, P.; Schöllhorn, A.; Bleul, L.; Weiß, E.; Stahl, M.; Grond, S.; Forchhammer, K. Cyanobacterial antimetabolite 7-deoxy-sedoheptulose blocks the shikimate pathway to inhibit the growth of prototrophic organisms. *Nat. Commun.* **2019**, *10*, 545. [CrossRef] [PubMed]
29. Dayan, F.E.; Owens, D.K.; Corniani, N.; Silva, F.M.L.; Watson, S.B.; Howell, J.L.; Shaner, D.L. Biochemical markers and enzyme assays for herbicide mode of action and resistance studies. *Weed Sci.* **2015**, *63*, 23–63. [CrossRef]
30. Johnson, L.N. The regulation of protein phosphorylation. *Biochem. Soc. Trans.* **2009**, *37*, 627–641. [CrossRef] [PubMed]
31. DeLong, A. Switching the flip: Protein phosphatase roles in signaling pathways. *Curr. Opin. Plant Biol.* **2006**, *9*, 470–477. [CrossRef]
32. Uhrig, R.G.; Labandera, A.M.; Moorhead, G.B. *Arabidopsis* PPP family of serine/threonine protein phosphatases: Many targets but few engines. *Trends Plant Sci.* **2013**, *18*, 505–513. [CrossRef]
33. Bajsa, J.; Pan, Z.; Dayan, F.E.; Owens, D.K.; Duke, S.O. Validation of serine-threonine protein phosphatase as the herbicide target site of endothall. *Pestic. Biochem. Physiol.* **2012**, *102*, 38–44. [CrossRef]
34. Ortiz, M.F.; Nissen, S.J.; Gray, C.J. Endothall behavior in *Myriophyllum spicatum* and *Hydrilla verticillata*. *Pest Manag. Sci.* **2019**, in press. [CrossRef]
35. Jordan, F.; Nemeria, N.; Guo, F.; Baburina, I.; Gao, Y.; Kahyaoglu, A.; Li, H.; Wang, J.; Yi, J.; Guest, J.R.; et al. Regulation of thiamin diphosphate-dependent 2-oxo acid decarboxylases by substrate and thiamin diphosphate.Mg(II) – evidence for tertiary and quaternary interactions. *Biochim. Biophys. Acta - Prot. Struct. Mol. Enzymol.* **1998**, *1385*, 287–306. [CrossRef]
36. Peng, H.; Wang, T.; Xie, P.; Chen, T.; He, H.W.; Wan, J. Molecular docking and three-dimensional quantitative structure–activity relationship studies on the binding modes of herbicidal 1-(substituted phenoxyacetoxy)alkylphosphonates to the E1 component of pyruvate dehydrogenase. *J. Agric. Food Chem.* **2007**, *55*, 1871–1880. [CrossRef]
37. He, H.-W.; Peng, H.; Wang, T.; Wang, C.; Yuan, J.L.; Chen, T.; He, J.; Tan, X. α-(Substituted-phenoxyacetoxy)-α-heterocyclylmethylphosphonates: Synthesis, herbicidal activity, inhibition on pyruvate dehydrogenase complex (PDHc), and application as postemergent herbicide against broadleaf weeds. *J. Agric. Food Chem.* **2013**, *61*, 2479–2488. [CrossRef] [PubMed]
38. Gohda, K.; Kimura, Y.; Mori, I.; Ohta, D.; Kikuchi, T. Theoretical evidence of the existence of a diazafulvene intermediate in the reaction pathway of imidazoleglycerol phosphate dehydratase: Design of a novel and potent heterocycle structure for the inhibitor on the basis of the electronic structure-activity relationship study. *Biochim. Biophys. Acta — Prot. Struct. Mol. Enzymol.* **1998**, *1385*, 107–114. [CrossRef]
39. Glynn, S.E.; Baker, P.J.; Sedelnikova, S.E.; Davies, C.L.; Eadsforth, T.C.; Levy, C.W.; Rodgers, H.F.; Blackburn, G.M.; Hawkes, T.R.; Viner, R.; et al. Structure and mechanism of imidazoleglycerol-phosphate dehydratase. *Structure* **2005**, *13*, 1809–1817. [CrossRef] [PubMed]
40. Cox, J.M. Substituted propyl phosphonic acid derivatives and their use as herbicides. EP Patent 78613, 11 May 1983.

41. Bisson, C.; Britton, K.L.; Sedelnikova, S.E.; Rodgers, H.F.; Eadsforth, T.C.; Viner, R.C.; Hawkes, T.R.; Baker, P.J.; Rice, D.W. Crystal structures reveal that the reaction mechanism of imidazoleglycerol-phosphate dehydratase is controlled by switching Mn(II) coordination. *Structure* **2015**, *23*, 1236–1245. [CrossRef] [PubMed]
42. Zrenner, R.; Stitt, M.; Sonnewald, U.; Boldt, R. Pyrimidine and purine biosynthesis and degradation in plants. *Annu. Rev. Plant Biol.* **2006**, *57*, 805–836. [CrossRef] [PubMed]
43. Ullrich, A.; Knecht, W.; Piskur, J.; Löffler, M. Plant dihydroorotate dehydrogenase differs significantly in substrate specificity and inhibition from the animal enzymes. *FEBS Lett.* **2002**, *529*, 346–350. [CrossRef]
44. Chen, D.Z.; Patel, D.V.; Hackbarth, C.J.; Wang, W.; Dreyer, G.; Young, D.C.; Margolis, P.S.; Wu, C.; Ni, Z.J.; Trias, J.; et al. Actinonin, a naturally occurring antibacterial agent, is a potent deformylase inhibitor. *Biochemistry* **2000**, *39*, 1256–1262. [CrossRef] [PubMed]
45. Dayan, F.E.; Duke, S.O. Natural compounds as next generation herbicides. *Plant Physiol.* **2014**, *166*, 1090–1105. [CrossRef] [PubMed]
46. Fernández-San Millán, A.; Obregón, P.; Veramendi, J. Over-expression of peptide deformylase in chloroplasts confers actinonin resistance, but is not a suitable selective marker system for plastid transformation. *Transgenic Res.* **2011**, *20*, 613–624. [CrossRef]
47. Hou, C.X.; Dirk, L.M.A.; Goodman, J.P.; Williams, M.A. Metabolism of the peptide deformylase inhibitor actinonin in tobacco. *Weed Sci.* **2006**, *54*, 246–254. [CrossRef]
48. Wall, M.K.; Mitchenall, L.A.; Maxwell, A. *Arabidopsis thaliana* DNA gyrase is targeted to chloroplasts and mitochondria. *Proc. Natl. Acad. Sci. USA* **2004**, *101*, 7821–7826. [CrossRef] [PubMed]
49. Evans-Roberts, K.M.; Mitchenall, L.A.; Wall, M.K.; Leroux, J.; Mylne, J.S.; Maxwell, A. DNA gyrase is the target for the quinolone drug ciprofloxacin in *Arabidopsis thaliana*. *J. Biol. Chem.* **2016**, *291*, 3136–3144. [CrossRef] [PubMed]
50. Wallace, M.D.; Waraich, N.F.; Debowski, A.W.; Corral, M.G.; Maxwell, A.; Mylne, J.S.; Stubbs, K.A. Developing ciprofloxacin analogues against plant DNA gyrase: A novel herbicide mode of action. *Chem. Commun.* **2018**, *54*, 1869–1872. [CrossRef] [PubMed]
51. Veerasekaran, P.; Kirkwood, R.C.; Parnell, E.W. Studies of the mechanism of action of asulam in plants. Part II: Effect of asulam on the biosynthesis of folic acid. *Pestic. Sci.* **1981**, *12*, 330–338. [CrossRef]
52. Corral, M.G.; Haywood, J.; Stehl, L.H.; Stubbs, K.A.; Murcha, M.W.; Mylne, J.S. Targeting plant DIHYDROFOLATE REDUCTASE with antifolates and mechanisms for genetic resistance. *Plant J.* **2018**, *95*, 727–742. [CrossRef] [PubMed]
53. Takano, H.K.; Beffa, R.; Preston, C.; Westra, P.; Dayan, F.E. Reactive oxygen species trigger the fast action of glufosinate. *Planta* **2019**, *249*, 1837–1849. [CrossRef] [PubMed]
54. Ellis, M.K.; Whitfield, A.C.; Gowans, L.A.; Auton, T.R.; Provan, W.M.; Lock, E.A.; Smith, L.L. Inhibition of 4-hydroxyphenylpyruvate dioxygenase by 2-(2-nitro-4-trifluoromethylbenzoyl)-cyclohexane-1,3-dione and 2-(2-chloro-4-methanesulfonylbenzoyl)-cyclohexane-1,3-dione. *Toxicol. Appl. Pharmacol.* **1995**, *133*, 12–19. [CrossRef] [PubMed]
55. Wang, D.W.; Lin, H.Y.; Cao, R.J.; Ming, Z.Z.; Chen, T.; Hao, G.F.; Yang, W.C.; Yang, G.F. Design, synthesis and herbicidal activity of novel quinazoline-2,4-diones as 4-hydroxyphenylpyruvate dioxygenase inhibitors. *Pest Manag. Sci.* **2015**, *71*, 1122–1132. [CrossRef]
56. Adamczyk-Woźniak, A.; Borys, K.M.; Sporzyński, A. Recent developments in the chemistry and biological applications of benzoxaboroles. *Chem. Rev.* **2015**, *115*, 5224–5247. [CrossRef]

plants

MDPI

Review

Herbicide Resistance Traits in Maize and Soybean: Current Status and Future Outlook

Vijay K. Nandula

Crop Production Systems Research Unit, Agricultural Research Service, United States Department of Agriculture, Stoneville, MS 38776, USA; vijay.nandula@usda.gov

Received: 16 August 2019; Accepted: 6 September 2019; Published: 9 September 2019

Abstract: This article reviews, focusing on maize and soybean, previous efforts to develop nontransgenic herbicide-resistant crops (HRCs), currently available transgenic HRC traits and technologies, as well as future chemical weed management options over the horizon. Since the mid twentieth century, herbicides rapidly replaced all other means of weed management. Overreliance on 'herbicide-only' weed control strategies hastened evolution of HR weed species. Glyphosate-resistant (GR) crop technology revolutionized weed management in agronomic crops, but GR weeds, led by Palmer amaranth, severely reduced returns from various cropping systems and affected the bottom line of growers across the world. An additional problem was the lack of commercialization of a new herbicide mode of action since the 1990s. Auxinic HRCs offer a short-term alternative for management of GR Palmer amaranth and other weed species. New HRCs stacked with multiple herbicide resistance traits and at least two new herbicide modes of action expected to be available in the mid-2020s provide new chemical options for weed management in row crops in the next decade.

Keywords: corn; herbicide resistance trait; maize; soybean

1. Introduction

Weeds cause extensive losses amounting to billions of US$ [1] through increased production costs, decreased quality and quantity of produce, reduced aesthetic value of landscapes that they thrive in, health effects on humans and pets, and other undesirable effects such as fuel for forest fires, etc. Over the past several centuries, weeds have been controlled with mechanical, biological, and cultural tools. Chemical weed control with inorganic compounds was extensively practiced in the late-nineteenth to mid-twentieth century, with earliest evidence even pointing back to the Roman era [2]. The real 'Chemical Era' of weed control started in the 1940s with the discovery of 2,4-dichorophenoxyacetic acid (2,4-D) during World War II chemical warfare research [2]. Since then, several herbicides belonging to different chemical classes and possessing diverse modes of action have been synthesized and commercialized around the world. Herbicides rapidly replaced all other means of weed management due to their superior efficacy, relatively low cost, selectivity, and targeted weed control. There has been at least one herbicide labeled for every cropping system imagined. Herbicides provided advantages such as increased productivity, improved quality of produce, reduced drudgery of hand weeding, and reduced soil erosion and top soil loss due to reduced cultivation and tillage (enhanced by less fossil fuel use). Overreliance on herbicides alone pushed weed species toward evolving resistance to herbicides. The astronomical cost of commercializing a new herbicide active ingredient (cost of discovery, development, and regulatory approval of a new synthetic pesticide was estimated to be $280 million in 2016 [3] coupled with the paucity of new herbicide modes of action [3] steered the agrochemical industry toward engineering/development of crops resistant to 'currently' available (subject to change) herbicides. This review covers maize (*Zea mays* L.) and soybean (*Glycine max* (L.) Merr.) only with discussion of earlier efforts to develop herbicide-resistant crops (HRCs), currently available HRC technologies, and future developments in the HRC arena.

2. Early Efforts

HRCs can be classified as nontransgenic (traditional genetic methods of selection of resistance traits) and transgenic (genetically engineered). Nontransgenic HRCs were developed using conventional breeding techniques (Table 1) such as seed mutagenesis (soybean resistant to sulfonylurea herbicides, disclosed in 1987), pollen mutagenesis (maize resistant to imidazolinone herbicides, released in 1992), and mutations in tissue culture (maize resistant to imidazolinone herbicides, revealed in 1991; to sethoxydim, an acetyl-CoA carboxylase (ACCase) inhibitor, released in 1992; soybean resistant to metribuzin, a triazine herbicide, disclosed in 1996) [4,5]. Agronomic performance of nontransgenic HRCs met with modest acceptance in the marketplace and often did not reach the expectations of growers and commodity groups. Scientists began to look at alternative ways to develop HRCs as weed management tools, to manage a broad spectrum of weeds, with superior agronomic characteristics.

Table 1. Nontransgenic herbicide-resistant maize and soybean.

Selection Method	Herbicide Family	Crop	Year of Disclosure
Seed mutagenesis	Sulfonylurea	Soybean	1987
Pollen mutagenesis	Imidazolinone	Maize	1992
Tissue culture	ACCase inhibitor	Maize	1992
	Imidazolinone	Maize	1991
	Triazine	Soybean	1996

Adapted from [4,5]. ACCase, acetyl-CoA carboxylase.

3. Current Transgenic HRCs

Currently commercialized transgenic HRCs (some with associated herbicide formulations registered by the US Environmental Protection Agency (EPA)) are summarized in Table 2. The following sections describe these HRCs separated by crop and herbicide mode of action. The context and issues pertaining to each resistant trait are limited to the geographical region where they have been commercialized. For example, glyphosate-resistant (GR) traits and related aspects are common across North America (Canada and US) and South America (Brazil and Argentina), whereas dicamba-resistant crop technologies are restricted to the US.

Table 2. Current transgenic herbicide-resistant maize and soybean and associated trait genes.

Crop	Resistance Trait	Trait Gene	Trait Designation	First Sales	Trade Name
Maize	Glyphosate	Three modified maize *epsps*	GA21	1998	Roundup Ready®
		Two *cp4 epsps*	NK603	2001	Roundup Ready® 2
	Glufosinate	*pat*	T14, T25	1996	LibertyLink System®
	2,4-D	*tfdA*	DHT1	2019	Enlist™ Weed Control System
	AOPP	*aad*	DHT1	2019	Enlist™ Weed Control System
Soybean	Glyphosate	*cp4 epsps*	GTS 40-3-2	1996	Roundup Ready®
		cp4 epsps	MON89788	2009	Roundup Ready® 2 Yield
	Glufosinate	*pat*	A2704-12	2009	LibertyLink System®
	Dicamba	*dmo*	MON87708	2017	Roundup Ready 2 Xtend®
	2,4-D	*tfdA*	DHT2	2019	Enlist™ Weed Control System

Partly adapted from [5,6]. AOPP: Aryloxyphenoxypropionate; 2,4-D: 2,4-Dichorophenoxyacetic acid.

4. Maize

The real breakthrough occurred in the 1990s with the commercial release of glyphosate-resistant (GR) crops. These crops allowed the application of glyphosate multiple times in the growing season without the risk of crop injury. Glyphosate was, hitherto, used nonselectively for weed control in vineyards, orchards, rights-of-way, industrial areas, and railroads. It has been deemed as "a once-in-a-century herbicide" [7] for its broad weed spectrum, reasonable cost, favorable environmental properties, and association with the widely popular GR crops. In susceptible plants, glyphosate inhibits 5-enolpyruvylshikimate-3-phosphate synthase (EPSPS), a key enzyme in the shikimate pathway responsible for the biosynthesis of aromatic amino acids and several secondary metabolites in the phenylpropanoid pathway.

GR maize was introduced in 1998 [5,6]. Transformation of maize plants with *CP4* (an *Agrobacterium* species strain) *EPSPS* and e35S promoter produced plants with vegetative resistance to glyphosate but reduced male fertility [8]. Therefore, the first-generation GR maize, trademarked as Roundup Ready® (RR) trait, GA21 utilized the rice actin 1 promoter driving the gene for a GR form of maize EPSPS (*TIPS-EPSPS*) [8], (*ZM-EPSPS*) [5,6]. A new event NK603, with two copies of a slightly modified *EPSPS CP4* gene, was developed to improve maize tolerance to glyphosate at both vegetative and reproductive stages, and was commercially released in 2001 in a breeding stack with glufosinate and four insect resistance traits [5].

Glufosinate inhibits the enzyme glutamine synthetase (GS), which catalyzes assimilation of ammonia with glutamate to form glutamine [8]. Glufosinate resistance is due to metabolic inactivation by an acetyltransferase enzyme that catalyzes the acetylation of glufosinate [5]. Two glufosinate resistance genes, *bar* and *pat*, encode homologous enzymes [9]. Both genes were isolated from soil microorganisms, *pat* from *Streptomyces viridochromogenes* and *bar* from *Streptomyces hygroscopicus* [6]. Glufosinate-resistant maize was commercialized for the first time in 1996 stacked with *Bt* insect resistance, as a stand-alone trait in 1997 [6], and was combined with GR maize as a 'double stacked trait' in the mid-2000s.

2,4-D is an auxin herbicide with phytotoxic action limited to broadleaf weed species. However, 2,4-D-resistant maize was developed in tandem with 2,4-D-resistant soybean, trademarked as Enlist™ Weed Control System by Corteva Agriscience (process described in a later section) and deregulated by the US Department of Agriculture (USDA) and an associated low volatility 2,4-D choline formulation registered by EPA for use only in 2,4-D-resistant crops, both in 2014, but not commercially launched in the US until 2018. It is to be noted that 2,4-D had been labeled for use in maize over the past several decades, both as preemergence and postemergence applications. A type of aryloxyalkanoate dioxygenase (AAD) enzyme was identified that provided resistance to 2,4-D as well as a class of ACCase inhibiting herbicides, popularly known as 'fops' belonging to the aryloxyphenoxypropionate (AOPP) chemical family, for example, quizalofop [10,11]. 2,4-D and the 'fop' herbicides possess an identical bond that facilitates their metabolism by a common enzyme. Due to concerns that GR grass weeds would run amuck before commercialization of 2,4-D-resistant crops, 2,4-D-resistant maize was promoted to control GR grass. Another class of ACCase inhibiting herbicides, the 'dims' belonging to the cyclohexanedione chemical family, lack the above bond and can be used to manage volunteer 2,4-D-resistant maize. Invariably, most transgenic maize HRCs on the market also carry insect-resistance traits (*Bt* trait), which will not be discussed here.

5. Soybean

In 1996, GR soybean was the first GR crop to be commercialized. The first generation of GR soybean, event 40-3-2, were the most successful outcome of over-expressing the glyphosate-insensitive *CP4 EPSPS* in all tissues using strong, constitutive viral promoters such as e35S or FMV from cauliflower or figwort mosaic viruses, respectively [8]. The first-generation GR soybean went off patent in 2015, which means individuals can grow them and save seed for re-use as long as the seed has no other trait or varietal patents [12]. Although the first-generation GR soybean has been phased out of the seed stock (of formerly Monsanto Co., now Bayer Crop Science), some institutions in Missouri and Arkansas have done breeding with this older trait and developed cultivars exhibiting the trait.

The second generation of GR soybean was commercialized in 2009 with a broader launch in 2010 as Roundup Ready 2 Yield® (RR2Y) by Monsanto Co. [8]. Several seed companies still sell RR2Y cultivars and they were available in 2019 [12]. The RR2Y event, MON89788 contained the same *CP4*

EPSPS as GTS 40-3-2, but with the gene inserted at a different site in an elite variety "A3244" with a different promoter and regulatory elements to enhance expression in the sensitive tissues [5].

Glufosinate-resistant soybean, with the *pat* gene and the CaMV 35S promoter, was publicly released for sale in 2009 as a promising tool to combat GR weeds, especially tall water hemp (*Amaranthus tuberculatus* Moq. Sauer) and Palmer amaranth (*Amaranthus palmeri* S. Watson) [5]. The glufosinate-resistant trait in soybean has been a good candidate for stacking in other herbicide-resistant soybean cultivars (discussed in a later section).

From the 2019 seed sales season, the glufosinate resistance technology now rests in the hands of BASF Crop Protection, who purchased it from Bayer Crop Science as part of anti-trust remediation [12]. Glufosinate-resistant crops including soybean have been steadily gaining market share as GR weeds spread across the southern and midwestern US, approaching 20% of the soybean market share. Low seed prices coupled with availability of generic glufosinate herbicides make this technology, labeled as LibertyLink System®, a viable option for soybean growers and is available on BASF's Credenz® soybean platform and other independent seed companies totaling 78 licensees [12].

Dicamba-resistant soybean, Roundup Ready 2 Xtend® (RR2Xtend) from formerly Monsanto Co., now Bayer Crop Science, was deregulated by the Animal and Plant Health Inspection Service (APHIS) of the USDA in 2015. Dicamba monooxygenase (DMO), from the soil bacterium *Pseudomonas maltophilia* (strain DI-6), encodes for Rieske nonheme monooxygenase that converts dicamba to 3-6-dichlorosalicylic acid (DCSA) [13]. The genetically engineered version of the *DMO* gene for expression in higher plants used the FLt36 promoter from peanut chlorotic streak virus, a translational enhancer from the tobacco etch virus (TEV), a chloroplast transit peptide–coding region from the pea Rubisco small subunit gene for chloroplast localization of DMO, and a terminator region from the pea Rubisco small subunit gene (*rbcS3'*) [13].

Formulations of dicamba specifically labeled for use in RR2Xtend were not registered until fall 2016. Three dicamba formulations, XtendiMax® and FeXapan®, both containing the diglycolamine (DGA) salt of dicamba, and Engenia® comprising the BAPMA (N, N-Bis-(aminopropyl) methylamine) salt of dicamba were registered for use in the US by EPA in 2016 until 2018; in 2018 registration was extended until December 2020. In 2017, additional restrictions were implemented toward application of the above formulations which were labeled for use only in dicamba-resistant crops. In 2019, Tavium® containing dicamba DGA salt plus *S*-metolachlor was registered by EPA. In 2016, illegal/off-target/off-label applications of dicamba via formulations other than XtendiMax®, FeXapan®, and Engenia® were made on dicamba-resistant soybean and cotton (*Gossypium hirsutum* L.) in AR, MO, TN, MS, and several other states, resulting in injury to non-dicamba-resistant crops and sensitive flora across the landscape from dicamba drift (volatile/vapor drift and/or physical drift due to droplet movement owing to temperature inversion and other factors). The issue of injury to non-dicamba-resistant crops from dicamba drift was compounded multifold in 2017 when registered dicamba applications in dicamba-resistant soybean were made over large swaths of the cropping area. Dicamba had been labeled for use in maize over the past several decades, both as preemergence and postemergence applications. However, dicamba has not been applied in the middle of the growing season, when temperatures are usually higher than during preplant or early crop growing conditions, or when several sensitive plant species are present, prior to commercialization of dicamba-resistant crops. In the 2017 growing season, a total of 1.44 million ha of dicamba-injured soybean were estimated from 2708 official dicamba-related injury investigations as reported by individual state departments of agriculture and state extension weed scientists in the US (Figure 1) [14]. It was believed, by several row-crop production practitioners, that the soybean hectarage reported above is a gross underestimation. In 2018, there were fewer complaints of soybean injury compared to 2017, probably, due to more growers planting dicamba-resistant soybean as an insurance against injury, neighboring growers settling disputes off the record, a marked improvement in efficiency of applications, or a combination of more than one of the above reasons. Records of injury from dicamba drift in 2018 and 2019 are available elsewhere in the literature.

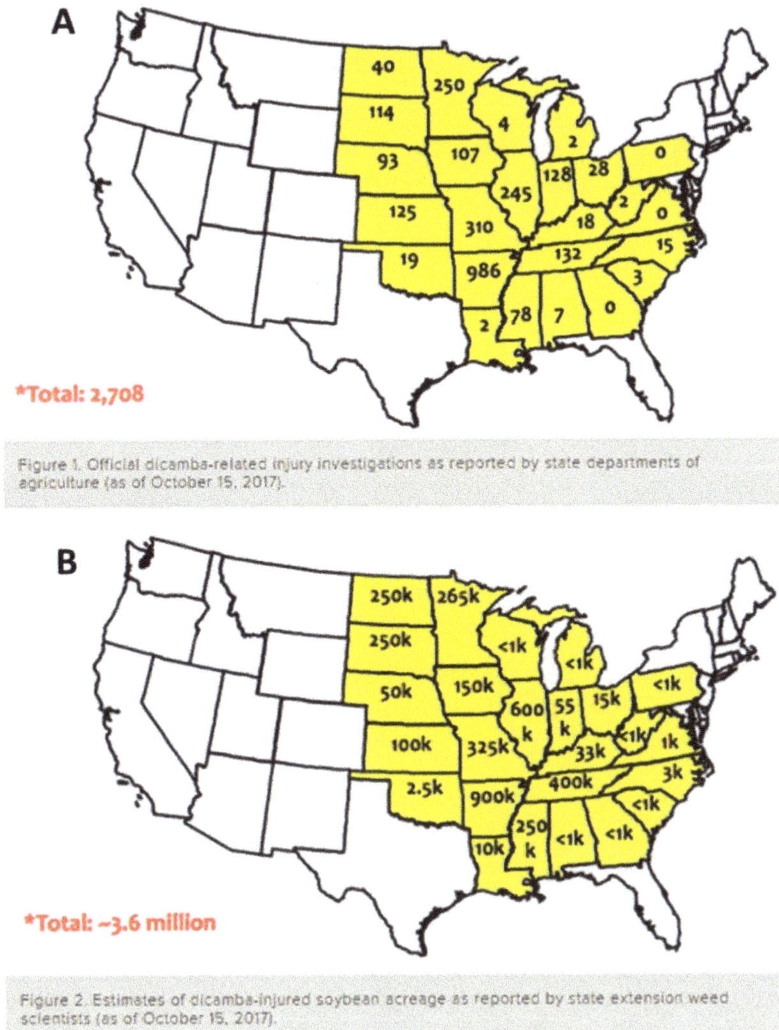

Figure 1. Official dicamba-related injury investigations as reported by state departments of agriculture (**A**) and estimates of dicamba-injured soybean acreage as reported by state extension weed scientists (**B**) in the 2017 growing season in the US.

The RR2Xtend crop hectarage estimates ranged from 16.2 to 20.2 million ha in 2018 [12]. The RR2Xtend soybean trait was available from a variety of seed companies via licensing agreements in 2019. It allows growers to spray dicamba and glyphosate postemergence to the crop.

Another group of auxin HRCs are the 2,4-D- resistant crop technologies, developed by formerly Dow AgroSciences and MS Technologies, now managed by Corteva Agriscience. Several species of bacteria possess families of *tfdA* genes that are known to produce 2,4-D-metabolizing enzymes [15,16]. A *tfdA* transgene, originally isolated from the bacterium *Alcaligenes eutrophus* [17], conferred resistance to 2,4-D when expressed in cotton. This gene, independently discovered in bacteria isolated from soil exposed to 2,4-D [18], increased tolerance in grapes (*Vitis vinifera* L.) to 2,4-D by 100-fold.

The *aad-1* gene isolated from a gram-negative soil bacteria, *Sphingobium herbicidovorans*, codes for a Fe(II) and 2-ketoglutarate-dependent dioxygenase that degrades the alkanoate side chains of both 2,4-D and members of the AOPP class of ACCase inhibitors to a hydroxyl [10]. Another gene sequence called *aad-12*, isolated from *Delftia acidovorans*, codes for a 2-ketoglutarate-dependent dioxygenase that inactivates phenoxyacetate auxins such as 2,4-D and pyridinyloxyacetate auxins such as triclopyr or fluroxypyr, but not commercial AOPPs [11]. The 2,4-D resistance traits were coded DHT1 for maize and DHT2 for soybean [19].

2,4-D-resistant soybean, trademarked as Enlist E3™, contain a single molecular stack providing resistance to glufosinate and glyphosate were commercialized in 2019. The original Enlist soybean with 2,4-D and glyphosate resistance was deregulated in 2014 but was not planted due to international import limitations. Another type of Enlist soybean, Enlist Soy + RR2Y, also confers tolerance to these herbicides, but Corteva Agriscience and MS Technologies focused their 2019 commercial offerings on Enlist E3 soybean in the US, Canada, and Brazil after international import restrictions were lifted by China and Philippines in early 2019 [12,20]. In 2018, some farmers in Indiana, Illinois, and Ohio planted Enlist E3 soybean within a stewarded closed-loop system based on an agreement between then Dow AgroSciences and Archer Daniels Midland (ADM) Company, as well as seed production acres. Corteva Agriscience is working with more than 100 independent seed companies to broadly license the Enlist E3 soybean trait, which will enable wide availability for 2020 and beyond [20].

2,4-D formulations labeled for use with the Enlist Weed Control System include Enlist™ Duo (2,4-D choline salt + dimethylammonium salt of glyphosate) and Enlist™ One (2,4-D choline salt) with COLEX-D™ Technology. 2,4-D products that do not contain COLEX-D™ Technology are not authorized for use in conjunction with Enlist crops.

Herbicides that inhibit the enzyme 4-hydroxyphenylpyruvate dioxygenase (HPPD) represent the last group of commercialized chemical weed control products with a unique mode of action. A new herbicide resistance technology, designated as GT27™ Soybean Performance System, originally developed by Bayer Crop Science and associated entities, confers resistance to glyphosate and a new HPPD-inhibiting herbicide isoxaflutole (ALITE 27, previously Balance Bean) applied preemergence only, pending EPA registration as of July 2019. The trait was commercially available in 2019 as a standalone product offered by some independent seed companies. However, BASF, which now owns the LibertyLink® technology, has made this technology available in 2019 as LibertyLink® GT27™ stack through its Credenz® soybean platform (maturity groups from 0 to 4.5) providing the option to use glyphosate, ALITE 27, and glufosinate. The major risk is that growers will be drawn to applying cheaper generic HPPD inhibitors on the LL GT27 soybean, inviting crop injury.

MGI soybean, carrying resistance to mesotrione, glufosinate, and isoxaflutole, were codeveloped by BASF and Syngenta [21]. The MGI soybean trait, that received approval from China, is scheduled for commercial launch in 2020, contingent on all regulatory approvals.

6. Future HRCs and Related Technologies

Although the USDA deregulated the dicamba resistance trait in 2016, dicamba-resistant maize, trademarked as XtendFlex® maize by Bayer has not been commercialized. Bayer has applied to the EPA regarding expanding uses of its XtendiMax dicamba formulation to dicamba resistant maize. Dicamba-resistant maize, in a three-way stack with glyphosate and glufosinate resistance, could be in the market in the next few years and is projected to be planted on 32 million ha or 89% of the US maize area [22]. Similarly, dicamba-resistant soybean and cotton exceeded expectations with a combined area of 24 million ha in 2019 as against initial estimates of 16 to 16.8 million ha [22].

A choline formulation of dicamba, to be applied only in RR2Xtend crops, is under development, but has not been registered with the EPA for commercialization yet.

Escalating costs of discovery and development of new herbicide active ingredients with new modes of action have resulted in a drought for these much-needed tools required to manage both wild-type and herbicide-resistant weed populations around the world. A January 2019 announcement by the FMC Corporation was welcome news [23]. The company will introduce two new herbicide modes of action over the next decade. An unnamed active molecule from a chemical group has been targeted for rice (*Oryza sativa* L.) and could be commercialized in five years. Other molecules from this group will be developed for maize and soybean. Another molecule, yet un-named, with another mode of action and targeting Palmer amaranth in maize and soybean will be launched, approximately, in 2026.

It is hoped and anticipated that additional herbicide modes of action will be announced by research and discovery groups in major agrochemical companies in the next few years. For example, Bayer Crop Science announced the potential for a new maize and soybean herbicide mode of action in the late 2020s [24].

7. Worldwide Use of Transgenic Crops

The data discussed in this section have been gleaned from the website of the International Service for the Acquisition of Agri-Biotech Applications [25]. Transgenic crop area in 2017 attained new record-high adoption at 189.8 million ha worldwide—an increase of 4.7 million ha (11.6 million A) or 3% from 185.1 million ha in 2016. The average transgenic crop adoption rate reached ≥93% or close to saturation in the US, Canada, Brazil, Argentina, and India. The global area of transgenic crops has increased ~112-fold from 1.7 million ha in 1996 to 189.8 million ha in 2017 making transgenic crops the fastest adopted crop technology. An accumulated 2.3 billion ha was achieved in 22 years (1996–2017) of transgenic crop commercialization. A total of 67 countries adopted transgenic crops, with 24 countries planting and 43 additional countries importing transgenic produce.

Transgenic soybean covered 50% of the global transgenic crop area, occupying 94.1 million ha, and corn was second with 59.7 million ha. The area planted with transgenic crops with stacked traits (insect resistance and herbicide resistance) increased by 3% and occupied 41% of the global transgenic crop area and only herbicide resistance traits were planted to 47% of the global area.

It is most likely that the approvals for transgenic crops in the countries that are members of the European Union (EU) are not for growing in most of the EU countries but for importation of the harvested crops for animal feed. The number of approvals of HRC events is summarized in Table 3.

Table 3. Global approvals of herbicide-resistant maize and soybean transgenic events [25].

Event	Crop	# of approvals
NK603	Maize	55 approvals in 26 non-EU countries plus 28 EU countries
GTS 40-3-2	Soybean	54 approvals in 27 non-EU countries plus 28 EU countries
GA21	Maize	50 approvals in 24 non-EU countries plus 28 EU countries
A2704-12	Soybean	43 approvals in 23 non-EU countries plus 28 EU countries
T25	Maize	41 approvals in 20 non-EU countries plus 28 EU countries

8. Conclusions

Chemical weed control will continue to remain the dominant method of weed management in the short term, despite the evolution of GR and other HR weeds, including 2,4-D and dicamba-resistant Palmer amaranth and tall water hemp [26,27]. New maize, soybean, and other crop technologies, providing resistance to multiple herbicides and improved formulations, and new herbicide modes of action (expected in the next five to ten years) offer additional tools to growers toward effective and sustainable weed management. Public and private land managers must implement nonchemical control strategies such as cultural, biological, and mechanical means wherever and whenever possible to sustain the rapidly depleting herbicide portfolio as well as to prepare for new patterns of weed emergence, growth, and resistance evolution in response to prevailing weather conditions. Pending an unforeseen revolutionary technology, chemical weed control will be necessary to meet current and future challenges of weed management.

Author Contributions: V.K.N. solely contributed to the writing and editing of this review article. A few colleagues edited earlier versions of this manuscript.

Funding: This research received no external funding.

Conflicts of Interest: The author declares no conflict of interest.

References

1. Crop Loss. Available online: http://wssa.net/wssa/weed/croploss-2/ (accessed on 9 August 2019).
2. Timmons, F.L. A history of weed control in the United States and Canada. *Weed Sci.* **2005**, *53*, 748–761. [CrossRef]
3. Marrone, P.G. Pesticidal natural products—Status and future potential. *Pest Manag. Sci.* **2019**. [CrossRef] [PubMed]
4. Duke, S.O. Taking stock of herbicide-resistant crops ten years after introduction. *Pest Manag. Sci.* **2005**, *61*, 211–218. [CrossRef] [PubMed]
5. Green, J.M.; Castle, L.A. Transitioning from single to multiple herbicide-resistant crops. In *Glyphosate Resistance in Crops and Weeds: History, Development, and Management*; Nandula, V.K., Ed.; John Wiley & Sons, Inc.: Hoboken, NJ, USA, 2010; pp. 67–92.
6. Green, J.M. Evolution of glyphosate-resistant crop technology. *Weed Sci.* **2009**, *57*, 108–117. [CrossRef]
7. Duke, S.O.; Powles, S.B. Glyphosate: A once-in-a-century herbicide. *Pest Manag. Sci.* **2008**, *64*, 319–325. [CrossRef] [PubMed]
8. Feng, P.C.C.; CaJacob, C.A.; Martino-Catt, S.J.; Cerny, R.E.; Elmore, G.A.; Heck, G.R.; Huang, J.; Kruger, W.M.; Malven, M.; Miklos, J.A.; et al. Glyphosate-resistant crops: Developing the next generation products. In *Glyphosate Resistance in Crops and Weeds: History, Development, and Management*; Nandula, V.K., Ed.; John Wiley & Sons, Inc.: Hoboken, NJ, USA, 2010; pp. 45–66.
9. Herouet, C.; Esdaile, D.J.; Mallyon, B.A.; Debruyne, E.; Schulz, A.; Currier, T.; Hendicks, K.; van der Klis, R.J.; Rouan, D. Safety evaluation of the phosphinothricin acetyltransferase proteins encoded by the pat and bar sequences that confer tolerance to glufosinate-ammonium herbicide in transgenic plants. *Regul. Toxicol. Pharmacol.* **2005**, *41*, 134–149. [CrossRef] [PubMed]
10. Wright, T.R.; Lira, J.M.; Merlo, D.J.; Hopkins, N. Novel Herbicide Resistance Genes. World Intellectual Property Organization Patent WO2005107437A2, 17 November 2005.
11. Wright, T.R.; Lira, J.M.; Walsh, T.A.; Merlo, D.J.; Jayakumar, P.S.; Lin, G. Novel Herbicide Resistance Genes. World Intellectual Property Organization Patent WO/2007/053482, 5 October 2007.
12. Picking Beans: A Look at the Many 2019 Herbicide-Tolerant Soybean Options. Available online: https://www.dtnpf.com/agriculture/web/ag/crops/article/2018/10/02/look-many-2019-herbicide-tolerant (accessed on 9 August 2019).
13. Behrens, M.R.; Mutlu, N.; Chakraborty, S.; Dumitru, R.; Jiang, W.Z.; Lavallee, B.J.; Herman, P.L.; Clemente, T.E.; Weeks, D.P. Dicamba resistance: Enlarging and preserving biotechnology-based weed management strategies. *Science* **2007**, *316*, 1185–1188. [CrossRef] [PubMed]
14. A Final Report on Dicamba-Injured Soybean Acres. Available online: https://ipm.missouri.edu/IPCM/2017/10/final_report_dicamba_injured_soybean/ (accessed on 9 August 2019).
15. Lyon, B.R.; Cousins, Y.L.; Llewellyn, D.J.; Dennis, E.S. Cotton plants transformed with a bacterial degradation gene are protected from accidental spray drift damage by the herbicide 2,4-dichlorophenoxyacetic acid. *Transgenic Res.* **1993**, *3*, 162–169. [CrossRef]
16. Streber, W.R.; Willmitzer, L. Transgenic tobacco expressing a bacterial etoxifying enzyme are resistant to 2,4-D. *Bio/technology* **1989**, *8*, 811–816.
17. Laurent, F.; Debrauwer, L.; Rathahao, E.; Scalla, R. 2,4-Dichlorophenoxyacetic acid metabolism in transgenic tolerant cotton (*Gossypium hirsutum*). *J. Agric. Food Chem.* **2000**, *48*, 5307–5311. [CrossRef] [PubMed]
18. Skirvin, R.M.; Norton, M.A.; Farrand, S.K.; Mulwa, R.M.S. Grape plant named "Improved Chancellor". U.S. Patent Application USPP20428P3, 20 October 2009.
19. Simpson, D.M.; Wright, T.R.; Chambers, R.S.; Peterson, M.A.; Cui, C.; Robinson, A.E.; Richburg, J.S.; Ruen, D.C.; Ferguson, S.; Maddy, B.E. Introduction to Dow AgroSciences herbicide tolerance traits. In Proceedings of the 2008 Annual Meeting of the Weed Science Society of America, Chicago, IL, USA, 4–7 February 2008; Volume 48, p. 115.

20. Enlist Weed Control System. Available online: https://www.enlist.com/en/news.html (accessed on 9 August 2019).
21. Corn, Soybean Trait Options for 2019 and Beyond. Available online: https://farmweeknow.com/blogs-corn-soybean-trait-options-2019-beyond-19929 (accessed on 9 August 2019).
22. More Dicamba to Come? Dicamba-Stacked Corn on the Horizon. Available online: https://www.dtnpf.com/agriculture/web/ag/crops/article/2019/03/21/dicamba-stacked-corn-horizon (accessed on 5 September 2019).
23. FMC to Introduce Two New Herbicide Modes of Action in Next Decade. Available online: https://www.agprofessional.com/article/fmc-introduce-two-new-herbicide-modes-action-next-decade (accessed on 12 August 2019).
24. New Corn and Soybean Herbicide Site of Action May be Coming in Late 2020s. Available online: https://www.agriculture.com/crops/new-corn-and-soybean-herbicide-site-of-action-may-be-coming-in-late-2020s (accessed on 12 August 2019).
25. Biotech Crop Adoption Surges as Economic Benefits Accumulate in 22 Years. Available online: http://www.isaaa.org/resources/publications/briefs/53/executivesummary/default.asp (accessed on 12 August 2019).
26. Palmer Pigweed Resistance to 2,4-D and Dicamba Confirmed in Kansas. Available online: https://web.archive.org/web/20190905153145/https://agfaxweedsolutions.com/2019/03/05/palmer-pigweed-resistant-to-24-d-and-dicamba-confirmed-in-kansas/ (accessed on 5 September 2019).
27. International Survey of Herbicide Resistant Weeds. Available online: https://web.archive.org/web/20190905154516/http://weedscience.com/Summary/MOA.aspx?MOAID=24 (accessed on 5 September 2019).

plants

MDPI

Review

Herbicide Resistance Management: Recent Developments and Trends

Hugh J. Beckie *, Michael B. Ashworth and Ken C. Flower

Australian Herbicide Resistance Initiative (AHRI), School of Agriculture and Environment, The University of Western Australia, Crawley, WA 6009, Australia; mike.ashworth@uwa.edu.au (M.B.A.); ken.flower@uwa.edu.au (K.C.F.)

* Correspondence: hugh.beckie@uwa.edu.au; Tel.: +61-8-6488-4615

Received: 3 May 2019; Accepted: 5 June 2019; Published: 8 June 2019

Abstract: This review covers recent developments and trends in herbicide-resistant (HR) weed management in agronomic field crops. In countries where input-intensive agriculture is practiced, these developments and trends over the past decade include renewed efforts by the agrichemical industry in herbicide discovery, cultivation of crops with combined (stacked) HR traits, increasing reliance on preemergence vs. postemergence herbicides, breeding for weed-competitive crop cultivars, expansion of harvest weed seed control practices, and advances in site-specific or precision weed management. The unifying framework or strategy underlying these developments and trends is mitigation of viable weed seeds into the soil seed bank and maintaining low weed seed banks to minimize population proliferation, evolution of resistance to additional herbicidal sites of action, and spread. A key question going forward is: how much weed control is enough to consistently achieve the goal of low weed seed banks? The vision for future HR weed management programs must be sustained crop production and profitability with reduced herbicide (particularly glyphosate) dependency.

Keywords: best management practices; crop competition; herbicide resistance; integrated weed management; precision weed management; site-specific weed management

1. Introduction

Over the past decade, the most significant change in strategy in herbicide-resistant (HR) weed management globally has been the increased focus on reducing the weed seed bank and maintaining low seed bank levels by whatever means possible. During the 1960s to 1990s, much research was devoted to determining the effect of weed density and time of emergence relative to the crop on crop revenue (yield by price) loss. These results were used to develop models or decision-support systems for herbicide application. Those initial weed economic threshold models with a single crop season focus became more sophisticated by considering longer time periods and weed seed bank implications, i.e., economic optimum thresholds or injury levels [1–3]. Today, the threshold concept, while still used in entomology and pathology, is no longer recommended by weed scientists or practitioners. The reality for most growers today is that the weed seed banks in their fields are likely resistant to one or more herbicide site(s) of action (SOA). A zero tolerance policy ('take no prisoners') is now being advocated [4,5]. However, it may not be agronomically and economically feasible or necessary to achieve this goal for all weed species and in all global cropping systems.

As detailed in the following sections, recent developments and trends related to HR weed management include renewed efforts in herbicide discovery by the agrichemical industry following a prolonged period from the late 1980s to present day when no new herbicide SOA was commercialized. In this intervening period, industry strategy has been to use existing chemistry for new, expanded uses by introducing single or combined (stacked) HR traits into our major agronomic crops, most notably soybean [*Glycine max* (L.) Merr.], maize (*Zea mays* L.), and cotton (*Gossypium hirsutum* L.) [6]. Although

HR-trait stacking offers growers increased flexibility to manage HR weeds, the consensus of weed science academics is that this solution is not sustainable in the long-term with current practices and will inevitably lead to increased incidence of multiple-HR populations.

Another major trend related to herbicide usage is increasing prominence of preemergence (PRE) herbicides with soil-residual activity to fill the void left by postemergence (POST) products, such as acetyl-CoA carboxylase (ACCase) or acetolactate synthase (ALS) inhibitors, due to loss in efficacy because of widespread evolved resistance. Therefore, the pendulum is now swinging in the other direction. In the 1970s and 1980s, PRE herbicide application requiring soil incorporation dropped off markedly in favour of POST herbicide application more suited to low soil disturbance no-tillage cropping systems [7].

Additionally, we review two non-herbicidal HR weed management developments—one already making a significant impact (harvest weed seed control, HWSC) and one potentially posed to do so in the near future (weed-competitive crop cultivars). Although research and development into site-specific weed management (SSWM) (or precision weed management) have been ongoing for over 20 years, the adoption of this technology in agronomic field crops is still very low despite the demonstrated economic benefits [8]. Nevertheless, the significant potential economic and environmental benefits of SSWM cannot be ignored. We outline a relatively simple initial approach to SSWM that may help to spur grower adoption in the next decade.

2. Herbicide Discovery: Renewed Efforts

Herbicides remain an essential component of integrated weed management (IWM) systems in conventional agriculture. During the 1950s to 1970s, a new herbicide SOA was frequently introduced into the marketplace (Table 1 [9]). This pace of introduction slowed considerably in the 1980s and came to a standstill by the end of the decade. Presently, a herbicide with a new SOA has not been commercialized in more than 30 years.

A key reason was the widespread adoption of glyphosate-resistant (GR) crops in the mid-1990s and concomitant increased use of glyphosate herbicide (Roundup™). Patent expiration in 2000 led to cheaper generic products and a marked increase in glyphosate usage. Glyphosate usage now surpasses that of any other herbicide SOA [10,11]. In addition, consolidation of the agrichemical industry into a few key players still supporting herbicide discovery programs and increasingly restrictive regulations have contributed to the current situation [12]. Today, we have a perfect storm of increasingly widespread global occurrence of multiple-HR weed populations in the face of increasingly limited herbicide SOAs due to weed resistance, uncertain economic returns from investment in herbicide discovery and development, and herbicide regulatory requirements or restrictions.

Discovering, developing, registering, and commercializing a new herbicide is a risky and costly venture, approaching USD $300 million [13]. On average, over 160,000 candidate compounds are screened to successfully introduce a new herbicide, with a development time from discovery to registration exceeding 10 years [14]. Nevertheless, industry has increased its efforts in recent years to find new molecules that may be promising future herbicides [15,16].

One promising direction is the discovery and development of novel herbicides based on products that are by-products of microorganisms or extracts of plants [17]. Research with natural phytotoxins has demonstrated several potential novel herbicide target sites; these compounds may lead to discovery of new SOAs [18]. Additionally, herbicidal properties of many antimalarial drugs suggest a possible source of compounds having novel SOAs [19]. Agrichemical companies have also adapted their discovery process to increase the probability of finding a new herbicide SOA, including molecular and enzyme structural modelling and SOA studies combined with a highly automated screening process [13]. Today's molecular and other tools allow much more targeted and informed screening through genomics and metabolomics [12]. Collectively, these new or adapted technologies and processes used in herbicide discovery offer some optimism for new herbicide SOA by 2025 [20]. In the meantime, reliance on old herbicides registered for new and expanded uses (described in the following section) will have to bridge the gap.

Table 1. Global introduction of herbicide site of action groups (adapted from Heap [9]; WSSA: Weed Science Society of America system used in the USA and Canada; HRAC: Herbicide Resistance Action Committee system used in all other countries except Australia, which has its own system).

Decade	Site of Action *	Example	WSSA	HRAC	Australia
1930s	Uncouplers (membrane disruption)	Dinoterb	24	M	Z
1940s	Synthetic auxins	2,4-D	4	O	I
	Auxin transport inhibitors	Diflufenzopyr	19	P	P
	Mitosis inhibitors	Propham	23	K2	E
1950s	Microtubule assembly inhibitors	Trifluralin	3	K1	D
	PS-II inhibitors	Atrazine	5	C1	C
	PS-II inhibitors (ureas and amides)	Chlorotoluron	7	C2	C
	Lipid inhibitors	Triallate	8	N	J
	Carotenoid biosynthesis inhibitors	Amitrole	11	F3	Q
	Nucleic acid inhibitors	MSMA	17	Z	Z
	PS-I electron diverters	Paraquat	22	D	L
1960s	PS-II inhibitors (nitriles)	Bromoxynil	6	C3	C
	PPO inhibitors	Oxyfluorfen	14	E	G
	VLCFA inhibitors	Metolachlor	15	K3	K
	Lipid inhibitors	Ethofumesate	16	N	J
	DHP synthase inhibitors	Asulam	18	I	R
	Cellulose inhibitors	Dichlobenil	20	L	I,O,Z
1970s	ACCase inhibitors	Diclofop	1	A	A
	ALS inhibitors	Chlorsulfuron	2	B	B
	Cell elongtion inhibitors	Difenzoquat	8	Z	Z
	EPSPS inhibitors	Glyphosate	9	G	M
	Glutamine synthase inhibitors	Glufosinate	10	H	N
	Carotenoid biosynthesis inhibitors (PDS)	Diflufenican	12	F1	F
	Antimicrotubule mitotic disrupters	Flamprop	25	Z	Z
1980s	DOXP inhibitors	Clomazone	13	F4	Q
	Cellulose inhibitors	Dichlobenil	21,26	L	I,O,Z
	HPPD inhibitors	Isoxaflutole	27	F2	H

* Abbreviations: ACCase: acetyl-CoA carboxylase; ALS: acetolactate synthase; DHP: dihydropteroate; DOXP: 4-deoxy-D-xylulose-5-phosphate synthase; EPSPS: 5-enolpyruvlshikimate-3-phosphate synthase; HPPD: hydroxyphenyl-pyruvate-dioxygenase; PDS: phytoene desaturase; PPO: protoporphyrinogen oxidase; PS: photosystem; VLCFA: very long chain fatty acid.

3. Herbicide-Resistant (HR) Crops: The Changing State of Trait Adoption

Soybean in the United States (U.S) perhaps best illustrates the changing adoption of cultivars with different HR traits (Table 2). The planted area of soybean in the U.S. in 2018 was 36.3 million ha [21]. Since 2014, the level of adoption of HR soybean in the U.S. has remained constant at 94% [22]. However, that statistic masks the changes occurring in adoption of different HR traits [23].

The dominant trait is still GR. The original Roundup Ready (RR) soybean went off patent in 2015. In 2009, RR2 soybean were introduced into the market. Although Bayer is transitioning to the RR2 Xtend (dicamba resistance) platform, many seed companies still sell RR2 cultivars. Currently, cultivars with the sole RR trait account for 30% of the U.S. market. The GR cultivars with the Xtend trait are expected to reach close to 50% of market share in 2019. This high adoption rate of the Xtend trait in just three years reflects its importance to growers in managing GR and other HR weeds. However, the dicamba herbicides (e.g., XtendiMax™) registered for use in these soybean cultivars are classified as a 'restricted use pesticide' by the U.S. Environmental Protection Agency. Therefore, they can only be sold to, and used by certified applicators in an effort to reduce potential injury to sensitive cultivars or crops due to off-target movement.

Table 2. Herbicide resistance (HR) traits in cultivars of major agronomic crops.

HR Trait	Soybean	Maize	Cotton	Rice	Canola	Wheat
ACCase inhibitor		X		X		X
ALS inhibitor		X		X	X	X
Triazine *					X	
Glyphosate	X	X	X		X	
Glufosinate	X	X	X		X	
Glyphosate+glufosinate	X	X	X			
Glyphosate+triazine*					X	
Glyphosate+dicamba	X		X			
Glyphosate+2,4-D+APP (ACCase)		X				
Glyphosate+isoxaflutole	X					
Glyphosate+glufosinate+dicamba	X		X			
Glyphosate+glufosinate+2,4-D	X		X			
Glyphosate+isoxaflutole+glufosinate	X					

* Australia only. Abbreviations: ACCase: acetyl-CoA carboxylase; ALS: acetolactate synthase; APP: aryloxyphen oxypropionate.

The LibertyLink™ (LL) soybean system, introduced in 2009, confers resistance to glufosinate (Liberty™) herbicide. In the merger with Monsanto, the technology was divested by Bayer to BASF (Credenz™ soybean platform). As with the RR2 Xtend soybean system, the LL system has been gaining market share over the past five years (currently about 20%) due to the need to manage the increasing incidence of GR weeds such as *Amaranthus* spp. This increased market share is also the result of lower seed prices in recent years and the availability of generic glufosinate herbicide products. Soybean cultivars with the combined RR+LL and RR+LL+dicamba (Xtend flex) traits are also available.

Going forward, GT27 soybean, developed by MS Technologies, Bayer and Mertec LLC, confer tolerance to glyphosate and isoxaflutole, an hydroxyphenyl-pyruvate-dioxygenase (HPPD)-inhibiting herbicide [23]. BASF introduced LL GT27, which will offer tolerance to three SOAs: glyphosate, glufosinate, and PRE-applied isoxaflutole. Another three-way stacked soybean system to be released is Enlist E3 developed by Dow AgroSciences and MS Technologies (now Corteva Agriscience), tolerant to glyphosate, glufosinate, and 2,4-D.

A company's decision to develop and introduce an HR trait for a particular crop depends primarily on the prospect for regulatory approval, market size (i.e., potential revenue), future market access, and grower demand (e.g., weed control needs). For example, frequent outcrossing between rice and related weed, red rice (*Oryza sativa* L.) was a consideration impacting the introduction of transgenic cultivars. Crops such as canola that are grown on a relatively small area globally are of lesser priority for investment in new HR-trait cultivars. Monsanto abandoned plans to commercialize GR wheat (*Triticum aestivum* L.) in 2004 largely because of lack of market acceptance of transgenic cultivars. Therefore, soybean, maize and cotton will continue to be the tier-1 level for investment, development, and introduction of stacked-HR trait cultivars.

With the rapid transition towards soybean and other crop cultivars with stacked (two- and three-way) HR traits, it will become more challenging for growers to choose the right system or systems to control HR and non-HR weeds on their farm. Comprehensive training for growers in proper stewardship practices of these technologies will be critical to optimizing and prolonging their benefits and minimizing risks as they are repeatedly and widely deployed across millions of hectares of cropland annually. Seed retailers and agronomists also require professional development training in this area as they often advise and influence a grower's decision regarding herbicide options best tailored to their specific weed problems.

4. Increasing Prominence of Preemergence Herbicides

Over the past decade, there has been increasing reliance on preplant or PRE herbicides with short to long soil-residual activity to manage weed populations resistant to glyphosate and many POST herbicides such as ACCase or ALS inhibitors. For example, the use of preplant and PRE herbicides in U.S. soybean from 2000 to 2015 increased from 25 to 70% of crop area, largely in response to GR weeds [24].

The PRE herbicide SOAs commonly used in field crops include microtubule assembly inhibitors (dinitroanilines such as trifluralin), photosystem-II inhibitors (e.g., metribuzin), protoporphyrinogen oxidase (PPO) inhibitors (e.g., saflufenacil, flumioxazin, sulfentrazone), fat synthesis inhibitors (e.g., prosulfocarb, triallate), and very long chain fatty acid (VLCFA) inhibitors (e.g., pyroxasulfone [25]).

Ideally, PRE herbicides will control weed cohorts very early in the growing season, thereby lessening potential crop yield loss due to weed competition and the selection pressure for resistance evolution from any in-crop (vegetative stage) or pre-harvest herbicide treatments that may still be required for acceptable season-long weed control. A PRE herbicide application is especially important for wide-row crops (e.g., maize, soybean, cotton) [26] or pulse crops (e.g., chickpea, *Cicer arientum* L.; lentil, *Lens culinaris* Medik.) with slow early-season growth and development or slow canopy closure. To proactively or reactively manage HR weeds in a growing season, the recommended herbicide treatment program should ideally comprise multiple SOAs applied as needed during one or more application windows (pre-seeding, in-crop, post-harvest) in mixtures or in sequence as required [27,28].

One case study that exemplifies this trend towards PRE herbicide usage is annual ryegrass (*Lolium rigidum* Gaud.) control in wheat in Australia. Traditionally, it was advised that growers should delay sowing in fields with high weed densities to maximize weed control before sowing and reduce the seed bank by the use of glyphosate or paraquat prior to sowing [29]. Today, however, wheat is often dry-seeded in a no-till system in April or early May to optimize yield potential. Common PRE herbicide treatments (soil incorporated by the seeding operation) in wheat include trifluralin, prosulfocarb + *S*-metolachlor, triallate, or pyroxasulfone. Growers are advised to integrate PRE herbicide application with multiple, effective cultural practices that optimize herbicide efficacy and crop suppression of weeds. With traditional delayed seeding to facilitate preplant weed control, weeds that establish in these late-sown winter crops can be more competitive due to the reduced crop growth rate associated with reduced soil and air temperatures. With the availability of highly-effective PRE herbicides that control initial weed cohorts, early seeding may be the best strategy to reduce weed seed production, as later germinating cohorts are suppressed by a larger crop canopy [30]. The effective control of summer annual weed species plus machinery innovations in no-tillage seeding now allow growers to efficiently establish crops earlier and make better use of available water resulting in increased yields [31,32].

To optimize the performance of these soil-applied herbicides, growers need to be aware of the multiple factors influencing their efficacy, including soil temperature; soil moisture; crop residue type, abundance and distribution; and degree of soil disturbance by tillage (ranging from minimal by disc openers to complete inversion by moldboard plow) and its impact on the distribution of soil organic carbon (organic matter) and texture (e.g., clay fraction) in the soil profile. A basic knowledge of the physical and chemical properties of each herbicide is needed to understand how it will likely behave in the soil environment of a grower's field. Because these herbicides may be applied a few days before seeding or at the time of seeding (soil incorporation by sowing), the type of seeding equipment and operational parameters can affect PRE herbicide spatial (horizontal and vertical) distribution, which, in turn, is influenced by water distribution and abundance in the soil profile. When PRE herbicide application occurs in dry soil, research is needed to better understand the interaction between time of PRE herbicide application and time of rainfall (or time of seeding) in terms of crop injury and weed control efficacy. Additionally, more research is needed on agronomic practices that optimize the effectiveness and utility of PRE herbicides in conservation-tillage systems, such as crop seeding rate, row spacing, and fertilizer placement and timing.

5. Plant Breeders Consider Weed Competitiveness

Crop competitiveness against weeds is an important pillar of IWM and therefore HR weed management [33,34]. Weed scientists have long lamented that crop breeders have neglected to consider weed competitiveness in their breeding programs. Crop variety guides typically include yield, maturity, quality, and disease tolerance ratings, but provide no information on the degree of weed competitiveness. In the weed science community, breeding and selecting weed-suppressive crop genotypes have been a high priority for many years [35,36].

With the continual increase in incidence and complexity of herbicide resistance in weed populations and declining availability of effective herbicide tools, plant breeders are now researching and developing germplasm with enhanced weed competitiveness (Figure 1; [37]). Research has identified both above- and below-ground traits that confer enhanced weed competitiveness. These traits include increased plant height, greater early vigour, and resource-competitive root systems. Early-season vigour is enabled by alternative dwarfing genes that do not reduce coleoptile length, which is a metric of weed competitiveness. Early ground cover (canopy closure) has been shown to be strongly correlated with the level of weed suppression. Germplasm that expresses weed-competitiveness traits are being developed in high-yielding genetic backgrounds. Weed-suppressive cereal (e.g., rice, wheat, barley (*Hordeum vulgare* L.)) cultivars will become increasingly available to growers over the next decade [37,38].

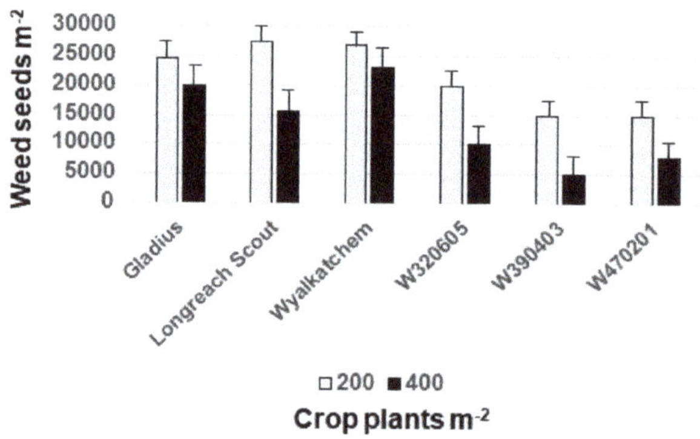

Figure 1. Weed seed production in field plots of wheat lines W320605, W390403 and W470201 with enhanced weed competitiveness compared with three commercial wheat cultivars (adapted from Rebetzke [37]).

In future crop variety guides, we envision the inclusion of weed-competiveness ratings, with data collected annually from variety trial programs. Field phenomics/phenotyping uses non-destructive assessments (e.g., normalized difference vegetative index, NDVI) of crop growth vigour, such as percentage ground cover, canopy closure, or leaf area index. A first step, however, is validation of such assessments through controlled environment experiments on relative growth rates as well as small-plot experiments (weedy and weed-free conditions) measuring key parameters correlated with crop competitiveness [39]. Furthermore, evaluation of the weed-competitiveness rating system in the field should be performed using agronomic practices frequently used by growers. In the future, integrating agronomic practices such as fertilizer timing and placement, crop seeding rate and crop row spacing with weed-competitive cultivars will enhance the ability of the crop to suppress weeds and the opportunity for the grower to reduce their dependency on herbicide inputs.

6. Harvest Weed Seed Control (HWSC) Gaining Momentum Globally

HWSC is now an established, widely adopted weed management tool used by Australian grain growers following extensive research, development, and extension efforts. Various HWSC practices can greatly reduce the viability of weed seeds in the chaff fraction emitted by a combine harvester. These HWSC practices include the following (reviewed in Walsh et al. [40]): (1) narrow-windrow burning; (2) chaff cart towed behind the combine harvester; (3) bale-direct system (baler towed behind the combine harvester); (4) chaff-lining (chaff funneled into a narrow band behind the combine harvester); (5) chaff tram-lining (chaff directed onto the combine harvester wheel tracks; and (6) weed seed destruction (e.g., integrated HSD [41]; Figure 2). The adoption of narrow-windrow burning is significantly greater than that of the other HWSC practices [42]. Use of chaff carts, or to a lesser extent balers towed behind the combine harvester, is favoured by growers with livestock. However, the low cost of chaff-lining as well as the falling cost and improved engineering of weed seed destruction systems over time will likely increase grower adoption of these systems globally in the future.

Figure 2. Mechanical seed destruction unit (iHSD) mounted in a commercial combine harvester.

Weed species most amenable to HWSC practices are those that retain their seeds in sufficient quantities until harvest (i.e., limited seed shatter) at a plant height (ca. 15 cm or higher) suitable for collection by the combine harvester. In Australia, two of the most problematic weeds, annual ryegrass and wild radish (*Raphanus raphanistrum* L.) meet both criteria well. In the last five years, substantial research and development efforts have been conducted in Canada and the U.S. to evaluate the potential of HWSC for key troublesome weed species, such as wild oat (*Avena fatua* L.) and cleavers (*Galium* spp.) in the Great Plains [43–45], and Palmer amaranth (*Amaranthus palmeri* S. Watson) and barnyardgrass (*Echinochloa* spp.) in the Midwest or southern U.S. [46]. Overall, results to date are encouraging. For example, only three of 20 weed species evaluated rated low potential for HWSC based on the criteria listed above (summarized in Walsh et al. [40]). In Europe, research is examining the potential for HWSC in management of HR blackgrass (*Alopecurus myosuroides* Huds.), despite substantial seed shatter before harvest, and other economically important weed species.

Ongoing field evaluation of different HWSC practices in different cropping systems across North America and elsewhere will facilitate grower awareness and future adoption of this important non-herbicidal weed management tool. An equally important consideration in these evaluations is the monitoring of weed species shifts over time as an inevitable consequence of the varying efficacy of HWSC practices on different weed species. In addition, weed populations may quickly adapt to recurrent HWSC practices in a field. Such adaptation may include earlier flowering and seed maturity, reduced seed retention on the plant, or plants that reproduce below harvest cutting height [47]. As with any recurrent weed management tool, whether herbicidal or not, we recommend close monitoring of changes in the phenology of weed populations as an adaptive response to reduce or evade exposure to this potential strong selection force.

Similar to herbicides, HWSC is not a stand-alone tool for weed management. Even on weed species well suited to seed capture or destruction at harvest such as annual ryegrass, weed control efficacy averages about 60% [48], not $\geq$ 80% efficacy as typical for most herbicides. Therefore, HWSC is just one component of an IWM system, which ideally comprises combined (stacked) effective non-herbicidal practices that provide synergistic suppression of weed growth and fecundity.

7. Is Site-Specific Weed Management (SSWM) in Agronomic Field Crops Set to Take Off?

There is much hype surrounding 'digital' or 'precision' agriculture in relation to yield mapping, fertilizer application, etc. Currently, SSWM is mostly used in high-value, irrigated, traditionally labour-intensive horticultural crops [49]. To date, the rate of grower adoption of SSWM in larger-scale dryland agronomic field crops has been very slow despite the demonstrated economic benefits [8,50]. Are recent major corporate investments, such as John Deere acquisition of Blue River Technology or Monsanto (now Bayer) acquisition of The Climate Corporation (Climate FieldView™ software), a prelude to the take-off in development and adoption of SSWM in agronomic crops?

Today, real-time weed control in fallow fields ('green on brown', e.g., WEED-It™ or WeedSeeker™ sprayers) is becoming more widespread in semi-arid to arid cropping regions (Figure 3). Herbicide savings of up to 90% have been reported [51]. Real-time weed detection/recognition and control in agronomic field crops (i.e., 'green on green') requires seamless integration and high performance of sensors, data processing, and actuation systems. Continuing technological advances in computer vision, robotics, machine learning, etc. are advancing this objective despite the many challenges that range from sensing weed vs. crop plants accurately (e.g., grass weeds in a cereal crop) while moving at speeds of up to 25 km h^{-1} to efficiently process and analyze large amounts of generated data [50,52]. Realizing the full potential utility of 'big' data for weed control is still conditional upon establishment of organizational, ethical and legal arrangements of data sharing [53].

Figure 3. Autonomous site-specific weed control with a real-time weed detection and application sprayer in a fallow field.

Weeds often are not distributed evenly within fields, but tend to occur in patches of varying size, shape and density [54]. Weed patches are a visible manifestation of the underlying seed bank, which needs to be reduced over the period that seeds remain dormant or viable. Anecdotal observations of growers and agronomists suggest that weed populations are becoming less abundant and more patchy where HWSC has been frequently used in a field. Increasing occurrence of weed populations with a more patchy distribution as a result of recurrent HWSC is supported by modelling simulations. Recurrent HWSC in fields with multiple-HR weed populations was predicted to result in greater weed patchiness (spatial heterogeneity) across a field [55]. Under these circumstances, the cost-effectiveness of spraying the entire cropped field with a herbicide is questionable. In this case, a relatively simple initial approach to SSWM may be prudent (Figure 4).

Figure 4. Growing season opportunities (semiarid southern hemisphere example) to map weed patches in wheat at pre-flower development stage using drone with mounted camera or near crop maturity using light detection and ranging (LiDAR) unit that could be mounted on a combine harvester; additionally, controlling weeds in fallow phase using a real-time weed detection sprayer.

A first step in an elementary SSWM approach could be the creation of a weed map at crop maturity or time of harvest, using a global positioning system (GPS) unit mounted on a tractor or combine harvester. Such a map can be created manually (i.e, visually marked by the operator) or via mounted sensors with accompanying datalogger. Such sensors could include cameras, such as RGB (red, green, blue) or multispectral/color-infrared (3–7 bands around 100-nm width in the visible and infrared region); structure from motion; or light detection and ranging (LiDAR) system [50]. Such a monitoring system would be best suited to weed species with low propagule dispersal favouring patch stability or plants protruding above the crop canopy. Remote sensing in weed detection using unmanned aerial vehicles (UAVs) or drones can map weed patches in crops at early to late phenological stages and provide high spatial resolution images in a timely manner with relatively low operational costs [50,56,57]. Efficient data processing and geographic information system (GIS) mapping are required, with demonstrated accuracy (i.e., ground-truthing) of mapped weed patches. Until more technological advances are made in real-time weed control in agronomic crop fields, sequencing the operations of weed patch mapping and management is a less risky interim strategy.

Seeding and spraying farm equipment are becoming increasingly capable of the variable rate application of inputs. Therefore, these weed patches could be targeted with greater crop seeding rates or highly effective herbicide treatments. Further research is needed to expand the use pattern of effective herbicides to enable label registration. As weed distributions would be mapped annually, growers would be able to continually adapt to changing weed distributions or newly evolving HR weed patches. With supporting research and development, this simple initial approach to SSWM may bridge the grower implementation disconnect that has characterized SSWM. The challenge for researchers is to work with growers and their agronomists and advisers to demonstrate (1) how existing technology can be adapted and used to easily and reliably map and manage weed patches in fields; and (2) sufficient return on investment. Grower experience and confidence gained using a relatively simple SSWM system may be an invaluable prerequisite for future adoption of more technologically-advanced real-time SSWM systems.

8. Conclusions and Future Outlook

The history of HR weed management globally shows that growers deal with the problem after it occurs, not before [58]. Furthermore, the extent of changes to their farming system depend upon the magnitude (distribution and abundance) and complexity (cross- and multiple-resistance patterns) of their HR weed populations in their fields. An increasing number of growers are now facing the prospect of changing crops or crop rotations to manage their HR weeds with remaining effective herbicides. For example, the percentage of farms in the United Kingdom adopting spring cropping as a means to manage HR grass weeds, such as blackgrass, increased from 32 to 81% from 2000 to 2016 [59]. Annual legume (pulse) crops are usually the weak link in the rotation because of their poor weed competitiveness combined with few registered herbicides or dependency on a few herbicide SOAs such as ALS inhibitors. These crop types are at greatest risk of declining planting area and production. The experience in Australia over the past decade is that effective PRE herbicides combined with agronomic practices to promote crop competition and minimize weed seed set or seed bank replenishment have generally resulted in sustained low weed seed bank levels of problematic weeds and profitable grain crop production.

The vision for the future of HR weed management globally should center on reduced herbicide dependency, especially glyphosate. Non-herbicidal alternatives are often adopted to compensate for reduced herbicide efficacy due to increasing incidence of resistance, rather than as a partial replacement for herbicides [59]. The metric for successful IWM must be low weed seed banks concurrent with reduced herbicide use. What we are already witnessing is the merging of conventional and organic weed management strategies and tactics because of increasing incidence and impact of HR weed populations, increasing societal pressure on reducing pesticide use in food and feed production systems from perceived health and environmental perspectives, and increasing regulatory costs, requirements or restrictions surrounding pesticide registration and usage. Will herbicides be a 'once in a century' method of weed control due to widespread multiple-HR populations in many major cropping systems globally [60]? A key research question going forward is how much weed control is enough to consistently achieve the goal of low weed seed banks? This question implies reduced herbicide use, as measured by both treated area times the number of applications and herbicide loading (kg ha^{-1}). Standardized methodology can be utilized to better assess the economic feasibility, impact, and consequences of reduced pesticide use [61]. To adequately address this question requires multi-site, medium-term (4–8 years) large-plot or landscape/field-level farming systems projects. Comparing key economic, agronomic, and environmental indices among farming systems, each with their unique crop sequence and combinations of weed management practices, can help identify more ecologically sustainable weed management systems. In addition, weed surveys and associated grower management questionnaires can help identify best management practices for maintaining low weed seed banks in reduced-herbicide farming systems.

Author Contributions: All author conributed equally with respect to conceptualization, original draft preparation, review and editing.

Funding: This research received no external funding.

Conflicts of Interest: The authors declare no conflict of interest.

References

1. Jones, R.E.; Medd, R.W. Economic thresholds and the case for longer term approaches to population management of weeds. *Weed Technol.* **2000**, *14*, 337–350. [CrossRef]
2. Lacoste, M.; Powles, S. RIM: Anatomy of a weed management decision support system for adaptation and wider application. *Weed Sci.* **2015**, *63*, 676–689. [CrossRef]
3. Somerville, G.J.; Powles, S.B.; Walsh, M.J.; Renton, M. How do spatial heterogeneity and dispersal in weed population models affect predictions of herbicide resistance evolution? *Ecol. Model.* **2017**, *362*, 37–53. [CrossRef]
4. Barber, L.T.; Smith, K.L.; Scott, R.C.; Norsworthy, J.K.; Vangilder, A.M. *Zero Tolerance: A Community-Based Program for Glyphosate-Resistant Palmer Amaranth Management*; University of Arkansas Cooperative Extension Service Bulletin FSA2177: Fayetteville, AR, USA, 2015.

5. Bayer CropScience. How a Zero Tolerance Approach can Mimimize Weed Resistance. 2019. Available online: https://www.cropscience.bayer.us/learning-center/articles/how-a-zero-tolerance-approach-can-minimize-weed-resistance (accessed on 1 May 2019).
6. Green, J.M.; Owen, M.D.K. Herbicide-resistant crops: Utilities and limitations for herbicide-resistant weed management. *J. Agric. Food Chem.* **2011**, *59*, 5819–5829. [CrossRef] [PubMed]
7. Fernandez-Cornejo, J.; Hallahan, C.; Nehring, R.; Wechsler, S. Conservation tillage, herbicide use, and genetically engineered crops in the United States: The case of soybeans. *AgBioForum* **2012**, *15*, 231–241.
8. Katalin, T.-G.; Eniko, L.; István, T. Economic benefits of precision weed control and why its uptake is so slow. *Stud. Agric. Econ.* **2013**, *115*, 40–46.
9. Heap, I.M. International Survey of Herbicide Resistant Weeds. 2019. Available online: http://www.weedscience.org (accessed on 1 May 2019).
10. Duke, S.O. The history and current status of glyphosate. *Pest Manag. Sci.* **2018**, *74*, 1027–1034. [CrossRef]
11. Benbrook, C.M. Trends in glyphosate herbicide use in the United States and globally. *Environ. Sci. Eur.* **2016**, *28*, 3. [CrossRef]
12. Duke, S.O.; Stidham, M.A.; Dayan, F.E. A novel genomic approach to herbicide and herbicide mode of action discovery. *Pest Manag. Sci.* **2019**, *75*, 314–317. [CrossRef]
13. Peters, B.; Strek, H.J. Herbicide discovery in light of rapidly spreading resistance and ever-increasing regulatory hurdles. *Pest Manag. Sci.* **2018**, *74*, 2211–2215. [CrossRef]
14. Phillips McDougall. *The Cost of New Agrochemical Product Discovery, Development and Registration in 1995, 2000, 2005–8 and 2010 to 2014—R&D Expenditure in 2014 and Expectations for 2019*; R&D Report; Phillips McDougall: Pathhead, UK, 2016; 41p.
15. Dickmann, R. *Herbicide Discovery a Global Expedition*; GroundCover Issue 135; Grains Research and Development Corporation: Canberra, Australia, 2018.
16. Begemann, S. *FMC to Introduce Two New Herbicide Modes of Action in Next Decade*; AgProfessional Farm Journal: Lenexa, KS, USA, 2019.
17. Westwood, J.H.; Charudattan, R.; Duke, S.O.; Fennimore, S.A.; Marrone, P.; Slaughter, D.C.; Swanton, C.; Zollinger, R. Weed management in 2050: Perspectives on the future of weed science. *Weed Sci.* **2017**, *66*, 275–285. [CrossRef]
18. Dayan, F.E.; Duke, S.O. Natural compounds as next-generation herbicides. *Plant Physiol.* **2014**, *166*, 1090–1105. [CrossRef]
19. Corral, M.G.; Leroux, J.; Stubbs, K.A.; Mylne, J.S. Herbicidal properties of antimalarial drugs. *Sci. Rep.* **2017**, *7*, e45871. [CrossRef]
20. Kraehmer, H.; van Almsick, A.; Beffa, R.; Dietrich, H.; Eckes, P.; Hacker, E.; Hain, R.; Strek, H.J.; Stuebler, H.; Willms, L. Herbicides as weed control agents: State of the art: II. recent achievements. *Plant Physiol.* **2014**, *166*, 1132–1148. [CrossRef]
21. [USDA-NASS] United States Department of Agriculture–National Agricultural Statistics Service. *USDA Reports Soybean, Corn Acreage Down*; 2018. Available online: https://www.nass.usda.gov/Newsroom/2018/06-29-2018.php (accessed on 1 May 2019).
22. Wechsler, S.J. Trends in the Adoption of Genetically Engineered Corn, Cotton, and Soybeans. 2018. Available online: https://www.ers.usda.gov/amber-waves/2018/december/trends-in-the-adoption-of-genetically-engineered-corn-cotton-and-soybeans (accessed on 1 May 2019).
23. Unglesbee, E. Soybean Herbicide Tolerant-Trait Systems for 2019—Breaking It Down. 2018. Available online: https://agfax.com/2018/10/03/soybean-seed-choices-for-2019-breaking-it-down-dtn (accessed on 1 May 2019).
24. Peterson, M.A.; Collavo, A.; Ovejero, R.; Shivrain, V.; Walsh, M.J. The challenge of herbicide resistance around the world: A current summary. *Pest Manag. Sci.* **2018**, *74*, 2246–2259. [CrossRef] [PubMed]
25. Walsh, M.J.; Fowler, T.M.; Crowe, B.; Ambe, T.; Powles, S.B. The potential for pyroxasulfone to selectively control resistant and susceptible rigid ryegrass (*Lolium rigidum*) biotypes in Australian grain crop production systems. *Weed Technol.* **2011**, *25*, 30–37. [CrossRef]
26. Webster, T.M.; Sosnoskie, L.M. Loss of glyphosate efficacy: A changing weed spectrum in Georgia cotton. *Weed Sci.* **2010**, *58*, 73–79. [CrossRef]
27. Beckie, H.J.; Reboud, X. Selecting for weed resistance: Herbicide rotation and mixture. *Weed Technol.* **2009**, *23*, 363–370. [CrossRef]

28. Owen, M.D. Diverse approaches to herbicide-resistant weed management. *Weed Sci.* **2016**, *64*, 570–584. [CrossRef]
29. Gill, G.S.; Holmes, J.E. Efficacy of cultural control methods for combating herbicide-resistant *Lolium rigidum*. *Pestic. Sci.* **1997**, *51*, 352–358. [CrossRef]
30. Gomez-Macpherson, H.; Richards, R.A. Effect of sowing time on yield and agronomic characteristics of wheat in south-eastern Australia. *Aust. J. Agric. Res.* **1995**, *46*, 1381–1399. [CrossRef]
31. Flower, K.; Crabtree, B.; Butler, G. No-till cropping systems in Australia. In *No-Till Farming Systems*; Goddard, T., Zoebisch, M., Gan, Y., Ellis, W., Watson, A., Sombatpanit, S., Eds.; Special Publication No. 3; World Association of Soil and Water Conservation: Bankok, Thailand, 2008; pp. 457–467.
32. Ashworth, M.; Desbiolles, J.; Tola, E. *Disc Seeding in Zero-Till Farming Systems—A Review of Technology and Paddock Issues*; Western Australian No-Tillage Farmers Association: Northam, Australia, 2010; 226p.
33. Beckie, H.J.; Harker, K.N. Our top 10 herbicide-resistant weed management practices. *Pest Manag. Sci.* **2017**, *73*, 1045–1052. [CrossRef] [PubMed]
34. WeedSmart. The Big 6. 2019. Available online: http://www.weedsmart.org.au (accessed on 1 May 2019).
35. Liebman, M.; Baraibar, B.; Buckley, Y.; Childs, D.; Christensen, S.; Cousens, R.; Eizenberg, H.; Heijting, S.; Loddo, D.; Morotto, A., Jr.; et al. Ecologically sustainable weed management: How do we get from proof-of-concept to adoption. *Ecol. Appl.* **2016**, *26*, 1352–1369. [CrossRef] [PubMed]
36. Andrew, I.K.S.; Storkey, J.; Sparkes, D.L. A review of the potential for competitive cereal cultivars as a tool in integrated weed management. *Weed Res.* **2015**, *55*, 239–248. [CrossRef] [PubMed]
37. Rebetzke, G. Weed-Competitive Plants on the Way. GroundCover Supplement Issue 127, Grains Research and Development Corporation, Canberra, Australia. 2017. Available online: https://grdc.com.au/resources-and-publications/groundcover/ground-cover-supplements/groundcover-issue-127-wheat-prebreeding/weed-competitive-plants-on-the-way (accessed on 1 May 2019).
38. Worthington, M.; Reberg-Horton, C. Breeding cereal crops for enhanced weed suppression: Optimizing allelopathy and competitive ability. *J. Chem. Ecol.* **2013**, *39*, 213–231. [CrossRef] [PubMed]
39. Widderick, M.; Lemerle, D.; Bell, K.; Storrie, A.; Hashem, A.; Taylor, C.; Osten, V.; Johansen, C.; Cook, T. *Crop Competition Management Options for Herbicide-resistant Weeds: Literature Review, Meta-analysis, Gap Analysis and Trial Design Recommendations*. GRDC Project DAQ00197. 2015; 150p. Available online: https://grdc.com.au/research/reports/report?id=6253 (accessed on 1 May 2019).
40. Walsh, M.J.; Broster, J.C.; Schwartz-Lazaro, L.M.; Norsworthy, J.K.; Davis, A.S.; Tidemann, B.D.; Beckie, H.J.; Lyon, D.J.; Soni, N.; Neve, P.; et al. Opportunities and challenges for harvest weed seed control in global cropping systems. *Pest Manag. Sci.* **2018**, *74*, 2235–2245. [CrossRef] [PubMed]
41. Walsh, M.J.; Harrington, R.B.; Powles, S.B. Harrington Seed Destructor: A new nonchemical weed control tool for global grain crops. *Crop Sci.* **2012**, *52*, 1343–1347. [CrossRef]
42. Walsh, M.J.; Ouzman, J.; Newman, P.; Powles, S.; Llewellyn, R. High levels of adoption indicate that harvest weed seed control is now an established weed control practice in Australian cropping. *Weed Technol.* **2017**, *31*, 1–7. [CrossRef]
43. Tidemann, B.D.; Hall, L.M.; Harker, K.N.; Beckie, H.J. Factors affecting weed seed devitalization with the Harrington Seed Destructor. *Weed Sci.* **2017**, *65*, 650–658. [CrossRef]
44. Tidemann, B.D.; Hall, L.M.; Harker, K.N.; Beckie, H.J.; Johnson, E.N.; Stevenson, F.C. Suitability of wild oat (*Avena fatua*), false cleavers (*Galium spurium*), and volunteer canola (*Brassica napus*) for harvest weed seed control in western Canada. *Weed Sci.* **2017**, *65*, 769–777. [CrossRef]
45. Beckie, H.J.; Blackshaw, R.E.; Harker, K.N.; Tidemann, B.D. Weed seed shatter in spring wheat in Alberta. *Can. J. Plant Sci.* **2018**, *98*, 107–114. [CrossRef]
46. Schwartz-Lazaro, L.M.; Green, J.K.; Norsworthy, J.K. Seed retention of Palmer amaranth (*Amaranthus palmeri*) and barnyardgrass (*Echinochloa crus-galli*) in soybean. *Weed Technol.* **2017**, *31*, 617–622. [CrossRef]
47. Ashworth, M.B.; Walsh, M.J.; Flower, K.C.; Vila-Aiub, M.M.; Powles, S.B. Directional selection for flowering time leads to adaptive evolution in *Raphanus raphanistrum* (wild radish). *Evol. Appl.* **2016**, *9*, 619–629. [CrossRef] [PubMed]
48. Walsh, M.J.; Aves, C.; Powles, S.B. Harvest weed seed control systems are similarly effective on rigid ryegrass. *Weed Technol.* **2017**, *31*, 178–183. [CrossRef]
49. Young, S.L.; Pierce, F.J. *Automation: The Future of Weed Control in Cropping Systems*; Springer: New York, NY, USA, 2014; 265p.

50. Fernández-Quintanilla, C.; Peña, J.M.; Andújar, D.; Dorado, J.; Ribeiro, A.; López-Granados, F. Is the current state of the art of weed monitoring suitable for site-specific weed management in arable crops? *Weed Res.* **2018**, *58*, 259–272. [CrossRef]
51. Cameron, J.; Storrie, A. *Summer Fallow Weed Management: A Reference Manual for Grain Growers and Advisers in the Southern and Western Grains Regions of Australia; GRDC Publ. ICN00012*. 2014; 71p. Available online: https://grdc.com.au/__data/assets/pdf_file/0028/98632/summer-fallow-weedmanagement-manual.pdf.pdf (accessed on 1 May 2019).
52. Partel, V.; Kakarla, S.C.; Ampatzidis, Y. Development and evaluation of a low-cost and smart technology for precision weed management utilizing artificial intelligence. *Comput. Electron. Agric.* **2019**, *157*, 339–350. [CrossRef]
53. Van Evert, F.K.; Fountas, S.; Jakovetic, D.; Crnojevic, V.; Travlos, I.; Kempenaar, C. Big data for weed control and crop protection. *Weed Res.* **2017**, *57*, 218–233. [CrossRef]
54. Rew, L.J.; Cousens, R.D. Spatial distributions of weeds in arable crops: Are current sampling and analytical methods appropriate? *Weed Res.* **2001**, *41*, 1–18. [CrossRef]
55. Somerville, G.J.; Powles, S.B.; Walsh, M.J.; Renton, M. Modeling the impact of harvest weed seed control on herbicide-resistance evolution. *Weed Sci.* **2018**, *66*, 395–403. [CrossRef]
56. Castillejo-González, I.L.; Peña-Barragán, J.M.; Jurado-Expósito, M.; Mesas-Carrascosa, F.J.; López-Granados, F. Evaluation of pixel- and object-based approaches for mapping wild oat (*Avena sterilis*) weed patches in wheat fields using QuickBird imagery for site-specific management. *Eur. J. Agron.* **2014**, *59*, 57–66. [CrossRef]
57. Huang, Y.; Reddy, K.N.; Fletcher, R.S.; Pennington, D. UAV low-altitude remote sensing for precision weed management. *Weed Technol.* **2017**, *32*, 2–6. [CrossRef]
58. Beckie, H.J. Herbicide-resistant weeds: Management tactics and practices. *Weed Technol.* **2006**, *20*, 793–814. [CrossRef]
59. Moss, S. Integrated weed management (IWM): Why are farmers reluctant to adopt non-chemical alternatives to herbicides? *Pest. Manag. Sci.* **2019**, *75*, 1205–1211. [CrossRef] [PubMed]
60. Davis, A.S.; Frisvold, G.B. Are herbicides a once in a century method of weed control? *Pest Manag. Sci.* **2017**, *73*, 2209–2220. [CrossRef]
61. Frisvold, G.B. How low can you go? Estimating impacts of reduced pesticide use. *Pest Manag. Sci.* **2019**, *75*, 1223–1233. [CrossRef] [PubMed]

plants

MDPI

Review

Fitness of Herbicide-Resistant Weeds: Current Knowledge and Implications for Management

Martin M. Vila-Aiub [1,2]

[1] IFEVA, CONICET, Department of Ecology, Faculty of Agronomy, University of Buenos Aires (UBA), Buenos Aires 1417, Argentina; vila@ifeva.edu.ar

[2] School of Agriculture & Environment, University of Western Australia (UWA), Crawley, WA 6009, Australia

Received: 3 September 2019; Accepted: 29 October 2019; Published: 1 November 2019

Abstract: Herbicide resistance is the ultimate evidence of the extraordinary capacity of weeds to evolve under stressful conditions. Despite the extraordinary plant fitness advantage endowed by herbicide resistance mutations in agroecosystems under herbicide selection, resistance mutations are predicted to exhibit an adaptation cost (i.e., fitness cost), relative to the susceptible wild-type, in herbicide untreated conditions. Fitness costs associated with herbicide resistance mutations are not universal and their expression depends on the particular mutation, genetic background, dominance of the fitness cost, and environmental conditions. The detrimental effects of herbicide resistance mutations on plant fitness may arise as a direct impact on fitness-related traits and/or coevolution with changes in other life history traits that ultimately may lead to fitness costs under particular ecological conditions. This brings the idea that a "lower adaptive value" of herbicide resistance mutations represents an opportunity for the design of resistance management practices that could minimize the evolution of herbicide resistance. It is evident that the challenge for weed management practices aiming to control, minimize, or even reverse the frequency of resistance mutations in the agricultural landscape is to "create" those agroecological conditions that could expose, exploit, and exacerbate those life history and/or fitness traits affecting the evolution of herbicide resistance mutations. Ideally, resistance management should implement a wide range of cultural practices leading to environmentally mediated fitness costs associated with herbicide resistance mutations.

Keywords: resistance mutation; fitness benefit; fitness cost; resistance management

1. Weeds in Agroecosystems

Agroecosystems are particular environments characterized by frequent, extensive, and intense disturbances and stress imposed by humans. As such, agroecosystems constitute an environment that selects for traits that maximise reproductive capacity, also called *r*- adaptive strategy (*r*, *per capita* rate of increase), in which high dispersion and growth rates, high resource allocation to reproduction, and a short life cycle are key to maximize plant fitness [1,2]. Adapted plants (i.e., weeds) of this particular disturbed and stressful environment embody the optimal phenotype and represent a major constraint to the quality and yield of grain crops [3] and agriculture sustainability [4].

Synthetic herbicides were developed and introduced into agroecosystems 70 years ago and continue today as the main agricultural tool to reduce weed densities securing global food production [5–7]. Both the global extension of agriculture frontiers and the substantial increase in herbicide reliance over the last decades combine to exert the strongest selection pressure ever experienced by weeds [8–10]. And this has inexorably led to herbicide resistance evolution in an ever-increasing list of weed species [11,12].

Herbicide resistance is the ultimate evidence of the extraordinary capacity of weeds to evolve under stressful conditions [9,11,13]. Herbicide resistance genes originate from random DNA mutations that endow a remarkable advantage to survive and reproduce, and therefore, are rapidly selected for and enriched in weed populations under herbicide treatment [14]. In particular, glyphosate resistance evolution has showed an alarming rate of increase among weeds in recent years [12].

2. Theoretical Considerations on Fitness Costs

Herbicide resistance is an adaptive evolutionary process in response to new environmental conditions (i.e., weed chemical control) in the agroecosystem. Herbicide resistance alleles are beneficial mutations that rapidly spread in weed populations under recurrent herbicide exposure [15,16]. These resistance mutations establish diverse defence mechanisms that protect plants from herbicide damage in different ways [11,17]. Some resistance mutations lead to amino acid substitutions in the herbicide target enzyme that change its configuration and geometry, altering distances to ligand H+ atoms and C- and N-terminal tails and water molecules [18–20]. These structural changes significantly reduce herbicide binding into the target enzyme, and thus confer resistance at the whole plant level (i.e., target site resistance mechanism). Alternatively, over-expression of the particular gene-encoding herbicide target enzyme increases its synthesis, which makes the herbicide insufficient to disrupt the normal plant metabolism [21]. Other mutations are responsible for regulating resistance mechanisms that minimize the amount of herbicide reaching the herbicide target site (i.e., non-target site resistance mechanism) [11,17]. For instance, enhanced herbicide metabolism (cytochrome P450 monooxygenases (CYP-450s), glutathione S-transferases (GSTs), or aldo-keto reductases (AKR)), reduced herbicide leaf uptake and translocation, and herbicide vacuolar sequestration are among the non-target site resistance mechanisms usually found in herbicide-resistant weeds [11,17,22]. However, whereas the biochemical basis associated with these non-target site resistance mechanisms has been elucidated, the molecular basis remains unknown.

Herbicide resistance mutations may pre-exist or arise spontaneously (de novo) within weed populations, and the rate at which they occur is very low [16,23,24], but see [25]. Despite the extraordinary plant fitness advantage endowed by herbicide resistance mutations in agroecosystems under herbicide selection, resistance mutations remain very rare traits in herbicide unselected weed populations [24–26].

A possible explanation for the low frequency of herbicide resistance alleles in unselected weed populations is their selective disadvantage imposed by associated fitness trade-offs. The prediction of a "lower adaptive value" or "deleterious effect" of resistance mutations in the original agroecosystem environment under no herbicide treatment would represent a cost of adaptation (i.e., fitness cost) which would limit the evolution of herbicide resistance by natural selection [27]. A fitness cost is the reduced success in contributing individuals to the next generation due to both or either impaired fecundity or survival [9,28]. After all, a fitness cost is the ultimate outcome of all genetic, biochemical, and physiological changes driven by a particular herbicide resistance mutation interacting within a particular genetic and ecological background [29].

Predicted fitness costs associated with resistance mutations have been a central paradigm in evolutionary ecology of herbicide resistance [15,28,30–33]. The fundamental evolutionary principle behind fitness costs is based on the resource-based allocation theory that predicts that plants divert resources into different functions to maximize their ecological success under the selection imposed by the environment [34–37]. As environmental resources are limited, any diversion of resources to one plant function would imply a decrease in allocation into other functions [36]. This theory underlies the trade-off usually found in plants between reproduction and defence functions [1,38,39]. It has been within this evolutionary context that herbicide resistance mutations, encoding for sophisticated defence mechanisms against herbicides, have been sought to divert resources and thus attract fitness costs [40].

Certainly, potential for allocation-based fitness costs in herbicide-resistant weeds correspond to resistance mutations responsible for herbicide metabolism via increased activity of endogenous detoxifying enzymes (e.g., cytochrome P450 monooxygenases) [41], reduced herbicide translocation within plants via vacuolar sequestration, or increased over-expression/duplication of herbicide resistance genes [11,21]. Provided that these herbicide resistance defence mechanisms require a diversion of resources to operate, it would be predictable the expression of associated fitness costs in plants carrying these resistance mechanisms.

On the other hand, it would be less predictable the expression of allocation-based fitness costs associated with target site resistance mutations leading to changes in the structure and geometry of the herbicide target site enzyme due to changes in the amino acid sequence [29,32]. However, changes in catalytic activity, natural substrate affinity, and/or feedback inhibition of the mutated herbicide target site enzyme may alter normal plant metabolism, resulting in a whole plant fitness cost [29].

3. Fitness Costs Associated with Herbicide Resistance Mutations Are Not Universal

Numerous studies reviewing the existence of herbicide resistance fitness costs and their biochemical, molecular, physiological, and ecological mechanisms have been published elsewhere [28–32,42–44]. Despite a sound theoretical background, these studies have concluded that there is no universality in the expression of fitness costs associated with herbicide resistance mutations. Rather, these studies have determined that fitness cost expression in herbicide-resistant weeds depend on the particular herbicide resistance mutation [29,45–48], dominance of the fitness cost [49], genetic background [50], and environmental conditions [51,52].

An example of the complex biochemical, genetic, and environmental dependence of fitness costs is given by target site EPSPS mutations endowing resistance to glyphosate. A common DNA point mutation (*EPSPS* CCA to TCA) endowing moderate glyphosate resistance in several weed species leads the change of Pro to Ser (Pro-106-Ser) in the EPSPS enzyme [53]. This single amino acid substitution has been shown to not alter EPSPS kinetics and metabolism in *Eleusine indica*, rendering glyphosate resistant plants as fit as the glyphosate susceptible ones [54,55]. On the contrary, when another single EPSPS substitution (*EPSPS* ACT to ATT) replaces Thr for Ile (Thr-102-Ile) and combines in addition to the Pro-106-Ser EPSPS mutation (i.e., double EPSPS TIPS resistance mutation), two contrasting effects on *E. indica* resistant plants arise. Whereas the TIPS mutation shows a clear beneficial effect under glyphosate selection as the level of glyphosate resistance increases notoriously compared to the single Pro-106-Ser mutation, it shows a very high deleterious effect on plant fitness in environments under no glyphosate selection [54,55]. The fitness cost associated with the resistance TIPS mutation is only observed in homozygous resistant (RR) but not heterozygous resistant (RS) plants, and the magnitude of the cost increases significantly under interspecific plant resource competition [55]. The high fitness cost observed in plants with the homozygous TIPS mutations is likely due to the reduced EPSPS catalytic efficiency (Vmax), accumulation in excess of carbon-rich shikimate and quinate acids, and unbalanced polar metabolites from glycolysis and starch and sucrose metabolism [55].

4. Fitness Costs May Arise as Direct Effects of the Herbicide Resistance Mutations vs. Pleiotropic Effects on Other Plant Traits

The anticipated detrimental effects of herbicide resistance mutations on plant fitness may arise as a direct impact on fitness-related traits (e.g., reduced pollen viability) and/or co-evolution of loci interactions (e.g., resistance and non-resistance alleles) contributing to changes in other plant traits (e.g., seed dormancy) that ultimately may, in particular ecological environments, lead to fitness costs [15,44].

An example of a direct effect on fitness is the point mutation in the chloroplastic *psbA* gene, resulting in the amino acid substitution of serine to glycine (Ser-264-Gly) in the catalytic site of D1 protein. This mutation endows resistance to triazine herbicides, but also reduces the substrate affinity reducing the electron transfer rate in the photosystem II (PSII) (reviewed in [33,56,57]). This physiological change has been shown to decrease photosynthesis rate with direct negative effects on vegetative and reproductive growth rates in triazine resistant plants [33].

In other cases, herbicide resistance mutations have been shown to alter morphological, developmental, or phenological traits in weeds without necessarily a direct impact on plant fitness per se [58–66]. The changes in these traits are best thought of as changes in life history characters due to either subtle pleiotropic effects of resistance mutations or their coevolution with non-resistance life history traits in response to the wide range of selective factors operating in agroecosystems [9,67].

For instance, in some weedy grasses, particular ACCase resistance mutations have been shown to coevolve with higher levels of seed dormancy, absence of germination in dark conditions, and/or delayed seed germination (Table 1) [64–66]. Similarly, herbicide resistance mutations and mechanisms endowing resistance to different herbicide classes have been shown to covary with changes in plant size, root anatomy, leaf appearance rate, plant height, number of tillers, outcrossing vs. selfing mating rates, susceptibility to herbivory and diseases, and flowering time (Table 1) [58,60–62,68,69]. The coevolution of herbicide resistance and changes in life history traits is an adaptive response to the agroecosystem to maximize fitness of herbicide-resistant weed populations (i.e., population size, genetic diversity), and thus, the spread of herbicide resistance mutations [44].

Can we anticipate direct vs. indirect effects of herbicide resistance mutations on fitness? A number of traits, further from fitness (i.e., survival and fecundity) [28], related to development, phenology, metabolism, physiology, and morphology have been associated with a number of dissimilar herbicide resistance traits [59,60,66,70–75]. Inferences from these and other studies on the causal relationship between these life history and resistance traits are difficult to make. Are these modified life history traits a direct consequence of pleiotropic effects of herbicide resistance mutations, or the result of confounded effects driven by the presence of multiple resistance traits within populations, or local adaptation to particular environments which leads non-resistance loci to co-segregate with the resistance trait?

On the contrary, when herbicide resistance mutations drive changes in the architecture and structure of herbicide target enzymes, altering central kinetic parameters, direct detrimental effects on plant fitness are more likely to express and be predicted (reviewed in [29,32,45]). Changes in activity, substrate affinity, reaction speed, and/or feedback inhibition in herbicide target enzymes are strong predictors of the expression of fitness costs in herbicide-resistant plants. Similarly, energy constraints driven by herbicide resistance traits (e.g., gene amplification, enhanced metabolism) that are theoretically associated with higher cell energy budgets would also lead to the expression of direct fitness costs. However, current evidence suggests that estimation of cell energy budgets associated with these resistance mutations is necessary before any generalization.

5. Effects of Fitness Costs on the Equilibrium Frequency of Herbicide Resistance Mutations

In the agricultural landscape, there are dynamic, fluctuating, and diverse agroecological conditions imposed by the matrix of herbicide-treated and untreated areas. It is within these contrasting environments where the fitness of plants carrying herbicide mutations is defined and shaped by a suite of selection forces.

The beneficial effect of herbicide resistance mutations is realized in herbicide-treated areas due to the extraordinary survival advantage they confer, relative to the susceptible wild-type. Inevitably, herbicide resistance mutations will spread over time in continuously herbicide-treated environments. Fitness costs associated with a particular resistance mutation under particular ecological conditions, however, will disclose a deleterious effect as an adaptive disadvantage, relative to the susceptible wild-type, in the herbicide untreated area [15,42]. Thus, a resistance mutation expressing a fitness cost in a particular herbicide untreated ecological environment will exhibit limits to evolve by natural selection. Overall, the contrasting beneficial and deleterious effects of herbicide resistance mutations are the mechanisms that maintain resistance polymorphisms at the agricultural landscape scale with high and low frequencies of resistance mutations in herbicide-treated and untreated areas, respectively. As a result, fitness benefit and cost play a fundamental role in predicting the spread of herbicide resistance mutations and determining their equilibrium frequencies at the agricultural landscape level.

The recent identification of a glyphosate resistance double mutation in *E. indica* can illustrate the impact of contrasting resistance benefits and costs associated with herbicide resistance mutations on their final equilibrium frequencies. Within a single *E. indica* population, it has been observed that individuals with the glyphosate resistance EPSPS Pro-106-Ser mutation coexist with plants which exhibit, in addition to the Pro-106-Ser mutation, a second EPSPS mutation, Thr-102-Ile—this double mutation is known as TIPS [54]. Under glyphosate selection, the single EPSPS Pro-106-Ser and TIPS mutations have been shown to endow, respectively, a moderate and high level of glyphosate resistance at both EPSPS and plant levels [54,76]. Remarkably, the high level of glyphosate resistance conferred by the EPSPS TIPS mutation is shared in both homozygous and heterozygous TIPS plants [76]. However, under no glyphosate selection, *E. indica* plants homozygous for the TIPS mutation pay an extremely high fitness cost which, interestingly, is not observed in plants heterozygous for the TIPS mutations, nor in individuals with the single EPSPS Pro-106-Ser mutation [55]. Thus, it may be predicted that from a very low allele frequency of 1×10^{-10} and after 50 generations under recurrent selection with recommended field glyphosate doses (1080 g ha^{-1}), the frequency of the EPSPS TIPS allele will enrich and be nearly fixed (final freq. = 0.9) in the treated population at the expense of the wild-type (WT) (final freq. = 2.4×10^{-9}) and Pro-106-Ser (final freq. = 0.09) alleles, which will become nearly extinct (Figure 1A).

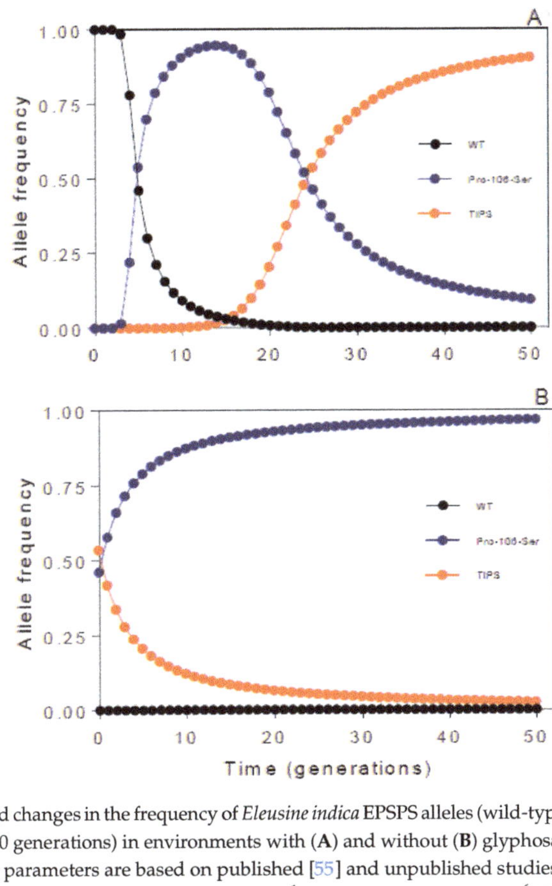

Figure 1. Predicted changes in the frequency of *Eleusine indica* EPSPS alleles (wild-type (WT), Pro-106-Ser, TIPS) over time (50 generations) in environments with (**A**) and without (**B**) glyphosate selection (1080 g ha^{-1}). Simulation parameters are based on published [55] and unpublished studies. Input parameters in (**A**): Initial allele frequency (WT = 9.99999×10^{-1}, Pro-106-Ser = 1.00×10^{-6}, TIPS = 1.00×10^{-10}); genotype fitness (WT/WT = 0.02, WT/Pro-106-Ser = 0.5, WT/TIPS = 0.6, Pro-106-Ser/Pro-106-Ser = 0.6, Pro-106-Ser/TIPS = 0.99, TIPS/TIPS = 0.99). Input parameters in (**B**): Initial allele frequency (WT = 9.26×10^{-4}, Pro-106-Ser = 0.463, TIPS = 0.536); genotype fitness (WT/WT = 0.99, WT/Pro-106-Ser = 0.99, WT/TIPS = 0.99, Pro-106-Ser/Pro-106-Ser = 0.99, Pro-106-Ser/TIPS = 0.99, TIPS/TIPS = 0.30). Simulations were run for 50 generations using Populus software [82], assuming no further mutational events, genetic drift, and allele migration events.

Table 1. Examples of herbicide-resistant weeds where resistance mutations have been associated with decreased fitness and/or altered life history traits.

Resistance Mutation/Trait	Weed Species	Fitness/Life History Trait	Environment	Biochemical/Physiological Change	Reference
ACCase/ALS CYP-450 metabolism	*Lolium rigidum*	Reduced RGR *, fecundity	Crop competition		[58,77]
ACCase/ALS target site resistance and CYP-450 metabolism	*L. rigidum*	Higher seed dormancy	Controlled conditions		[66]
EPSPS over-expression	*L. perenne*	Reduced height, leaf area, fecundity	Intra-specific competition in rain fed conditions		[78]
EPSPS TIPS mutation	*Eleusine indica*	Reduced RGR, fecundity	Crop competition	Reduced EPSPS Vmax Altered C-rich metabolite levels	[55]
ACCase Ile-1781-Leu	*L. rigidum*	Light requirement for seed germination	Controlled conditions	Changes in sensitivity of phytochrome B	[64] (Vila-Aiub et al. unpublished)
ACCase 2078	*Alopecurus myosuroides*	Lower germination rate	Wheat competition	Reduced ACCase activity	[48]
psbA Ser-264-Gly	Many broadleaf species	Reduced RGR, fecundity	Controlled and field conditions	Reduced QB affinity, inefficient PSII electron transport, lower photosynthesis	Reviewed in [33]
psbA Ser-264-Gly	*Amaranthus powelii*	Higher susceptibility to herbivory	Field conditions	Higher leaf N concentration	[68,79]
ALS Trp-574-Leu	*A. powelli*	Smaller roots, reduced leaf area and RGR	Intra-specific competition	Likely impaired ALS function	[62]
Glyphosate resistance	*Ipomoea purpurea*	Higher selfing rate	Controlled and field conditions	Lower anther–stigma distance	[61]
EPSPS amplification	*Kochia scoparia*	Delayed flowering	Controlled conditions		[71]
AUX/IAA KsIAA16 Gly-73-Asn	*K. scoparia*	Reduced RGR, leaf area, height, fecundity	Controlled conditions		[80,81]

* RGR: Relative growth rate.

If the equilibrium allelic frequencies attained after 25 years of glyphosate use (Figure 1A) are considered as the starting point (WT = 9.26×10^{-4}, Pro-106-Ser = 0.46, TIPS = 0.53) for an environment where glyphosate is discontinued for 50 generations, a basic simulation exercise predicts that, whereas the frequency of the WT allele will show a negligible increase (final freq. = 0.002), the frequency of the single Pro-106-Ser mutation will be nearly fixed (final freq. = 0.97) at the expense of the TIPS mutation (final freq. = 0.028) (Figure 1B).

6. Implications of Fitness Costs to Resistance Management

The idea of a "lower adaptive value" of herbicide resistance mutations is often seen as an opportunity for the design of resistance management practices that could minimize the evolution of herbicide resistance [28,83,84]. However, a number of realizations need to be made to understand whether weed management practices can realistically exploit the predicted cost of adaptation of herbicide resistance mutations.

Firstly, excluding a very few exceptions (see [29,48,55,56,62]), fitness costs, regardless of whether they arise as a direct effect of the resistance mutation vs. their coevolution with changes in life history traits, are environmentally dependent, meaning that they will solely express under certain ecological conditions. They may not always express, as either compensatory molecular evolution of costs is possible [85] or the "right" ecological conditions to bring the resistance mutation at disadvantage may not be present.

Secondly, the detrimental effect of fitness costs on plants carrying the resistance mutation, relative to the susceptible wild-type, is immediately masked under conditions of continuous herbicide treatment. It is straightforward then that for a management practice to exploit a fitness cost of a resistance mutation, no herbicide treatments need to be considered during successive growing seasons (i.e., herbicide "off" years), single entire season (i.e., herbicide "off" and "on" years) or, at least, a limited time window within a growing season (e.g., non-chemical fallow). The final impact of removing the herbicide selective benefit on the frequency of the resistance mutations at the landscape level will be a function of the magnitude of the fitness cost (negligible, moderate, high), period of time (single vs. several generations/growing seasons) under no herbicide use, and area covered by the resistant population (single field vs. farm vs. regional area).

It is evident that the challenge for weed management practices aiming to control, minimize, or even reverse the frequency of resistance mutations in the agricultural landscape is to "create" those agroecological conditions that could expose, exploit, and exacerbate those life history and/or fitness traits affecting the evolution of herbicide resistance mutations [9,67]. Ideally, resistance management should implement a wide range of practices leading to environmentally mediated fitness costs associated with herbicide resistance mutations.

Generation of dynamic spatial and temporal diverse agroecological conditions are possible through the implementation of management practices, such as the use of grazed and ungrazed pasture phases [86], cover crops [87], choice of competitive cultivars [88], changes in cultural practices such as seeding and harvest time, crop row spacing, density and orientation [89], implementation of soil tillage [83], and management of fence lines and field margins [90], just to mention a few. These diverse agroecological conditions imply environmental changes in fluctuating temperatures and light intensity and quality (red/far red ratio) at soil levels and under crop canopies and demand of plant resources together, which in turn may bring changes in soil chemical and physical properties. More diversified agricultural landscapes will likely select against herbicide resistance mutations through exploitation of fitness costs compared to agroecosystems with simplified weed management practices.

Funding: This research received no external funding.

Acknowledgments: The author thanks the useful comments provided by the referees.

Conflicts of Interest: The author declares no conflicts of interest.

References

1. Grime, J.P. Evidence for the existence of three primary strategies in plants and its relevance to ecological and evolutionary theory. *Am. Nat.* **1977**, *111*, 1169–1194. [CrossRef]
2. MacArthur, R.H.; Wilson, E.O. *The Theory of Island Biogeography*; Princeton University Press: Princeton, NJ, USA, 1967.
3. Cousens, R.; Mortimer, M. *Dynamics of Weed Populations*; Cambridge University Press: Cambridge, UK, 1995.
4. Oerke, E.C. Crop losses to pests. *J. Agric. Sci.* **2006**, *144*, 31–43. [CrossRef]
5. Powles, S. Global Herbicide Resistance Challenge. *Pest Manag. Sci.* **2014**, *70*, 1305. [CrossRef] [PubMed]
6. Powles, S.B. Evolved glyphosate-resistant weeds around the world: Lessons to be learnt. *Pest Manag. Sci.* **2008**, *64*, 360–365. [CrossRef]
7. National Research Council. *The Future Role of Pesticides in US Agriculture*, 1st ed.; National Academic Press: Washington, DC, USA, 2000.
8. Palumbi, S.R. Evolution-Humans as the world's greatest evolutionary force. *Science* **2001**, *293*, 1786–1790. [CrossRef]
9. Neve, P.; Vila-Aiub, M.; Roux, F. Evolutionary-thinking in agricultural weed management. *New Phytol.* **2009**, *184*, 783–793. [CrossRef]
10. James, C. *Executive Summary of Global Status of Commercialized Biotech/GM Crops*; ISAAA International Service for the Acquisition of the Agri-Biotech Applications: Ithaca, NY, USA, 2016; pp. 1–15.
11. Powles, S.B.; Yu, Q. Evolution in action: Plants resistant to herbicides. *Ann. Rev. Plant Biol.* **2010**, *61*, 317–347. [CrossRef]
12. Heap, I. The International Survey of Herbicide Resistant Weeds. Available online: www.weedscience.com (accessed on 22 May 2019).
13. Neve, P.; Busi, R.; Renton, M.; Vila-Aiub, M.M. Expanding the eco-evolutionary context of herbicide resistance research. *Pest Manag. Sci.* **2014**, *70*, 1385–1393. [CrossRef]
14. Maxwell, B.D.; Mortimer, A.M. Selection for Herbicide Resistance. In *Herbicide Resistance in Plants: Biology and Biochemistry*; Powles, S.B., Holtum, J.A.M., Eds.; Lewis: Boca Raton, FL, USA, 1994; pp. 1–25.
15. Lenormand, T.; Harmand, N.; Gallet, R. Cost of resistance: An unreasonably expensive concept. *Rethink. Ecol.* **2018**, *3*, 51–70. [CrossRef]
16. Gressel, J.; Levy, A.A. Agriculture: The selector of improbable mutations. *Proc. Natl. Acad. Sci. USA* **2006**, *103*, 12215–12216. [CrossRef]
17. Délye, C.; Jasieniuk, M.; Le Corre, V. Deciphering the evolution of herbicide resistance in weeds. *Trends Genet.* **2013**, *29*, 649–658. [CrossRef] [PubMed]
18. McCourt, J.A.; Pang, S.S.; King-Scott, J.; Guddat, L.W.; Duggleby, R.G. Herbicide-binding sites revealed in the structure of plant acetohydroxyacid synthase. *Proc. Natl. Acad. Sci. USA* **2006**, *103*, 569–573. [CrossRef]
19. Schönbrunn, E.; Eschenburg, S.; Shuttleworth, W.A.; Schloss, J.V.; Amrhein, N.; Evans, J.N.; Kabsch, W. Interaction of the herbicide glyphosate with its target enzyme 5-enolpyruvylshikimate 3-phosphate synthase in atomic detail. *Proc. Natl. Acad. Sci. USA* **2001**, *98*, 1376–1380. [CrossRef] [PubMed]
20. Zhang, H.; Tweel, B.; Tong, L. Molecular basis for the inhibition of the carboxyltransferase domain of acetyl-coenzyme-A carboxylase by haloxyfop and diclofop. *Proc. Natl. Acad. Sci. USA* **2004**, *101*, 5910–5915. [CrossRef] [PubMed]
21. Gaines, T.A.; Patterson, E.L.; Neve, P. Molecular mechanisms of adaptive evolution revealed by global selection for glyphosate resistance. *New Phytol.* **2019**, *223*, 1770–1775. [CrossRef]
22. Pan, L.; Yu, Q.; Han, H.; Mao, L.; Nyporko, A.; Fan, L.; Powles, S.B. Aldo-keto reductase metabolizes glyphosate and confers glyphosate resistance in Echinochloa colona. *Plant Physiol.* **2019**. [CrossRef]
23. Casale, F.A.; Giacomini, D.A.; Tranel, P.J. Empirical investigation of mutation rate for herbicide resistance. *Weed Sci.* **2019**, *67*, 1–8. [CrossRef]
24. Busi, R.; Gaines, T.A.; Walsh, M.J.; Powles, S.B. Understanding the potential for resistance evolution to the new herbicide pyroxasulfone: Field selection at high doses versus recurrent selection at low doses. *Weed Res.* **2012**, *52*, 489–499. [CrossRef]
25. Neve, P.; Powles, S. High survival frequencies at low herbicide use rates in populations of Lolium rigidum result in rapid evolution of herbicide resistance. *Heredity* **2005**, *95*, 485–492. [CrossRef]

26. Preston, C.; Powles, S.B. Evolution of herbicide resistance in weeds: Initial frequency of target site-based resistance to acetolactate synthase-inhibiting herbicides in Lolium rigidum. *Heredity* **2002**, *88*, 8–13. [CrossRef]
27. Fisher, R.A. *The Genetical Theory of Natural Selection*; Dover Publications: New York, NY, USA, 1958; p. 291.
28. Cousens, R.D.; Fournier-Level, A. Herbicide resistance costs: What are we actually measuring and why? *Pest Manag. Sci.* **2018**, *74*, 1539–1546. [CrossRef] [PubMed]
29. Vila-Aiub, M.M.; Yu, Q.; Powles, S.B. Do plants pay a fitness cost to be resistant to glyphosate? *New Phytol.* **2019**, *223*, 532–547. [CrossRef] [PubMed]
30. Bergelson, J.; Purrington, C.B. Surveying patterns in the cost of resistance in plants. *Am. Nat.* **1996**, *148*, 536–558. [CrossRef]
31. Vila-Aiub, M.M.; Neve, P.; Roux, F. A unified approach to the estimation and interpretation of resistance costs in plants. *Heredity* **2011**, *107*, 386–394. [CrossRef] [PubMed]
32. Vila-Aiub, M.M.; Neve, P.; Powles, S.B. Fitness costs associated with evolved herbicide resistance alleles in plants. *New Phytol.* **2009**, *184*, 751–767. [CrossRef]
33. Holt, J.S.; Thill, D.C. Growth and Productivity of Resistant Plants. In *Herbicide Resistance in Plants. Biology and Biochemistry*; Powles, S.B., Holtum, J.A.M., Eds.; Lewis Publishers: Boca Raton, FL, USA, 1994; pp. 299–316.
34. Harper, J. *Population Biology of Plants*; Academic Press: London, UK, 1977.
35. Solbrig, O.T. Plant Traits and Adaptive Strategies: Their Role in Ecosystem Function. In *Biodiversity and Ecosystem Function*; Schulze, E.D., Mooney, H.A., Eds.; Springer: Berlin, Germany, 1994; pp. 97–116.
36. Lerdau, M.; Gershenzon, J. Allocation Theory and Chemical Defense. In *Plant Resource Allocation*; Bazzaz, F., Grace, J., Eds.; Academic Press: London, UK, 1997; pp. 265–277.
37. Strauss, S.Y.; Rudgers, J.A.; Lau, J.A.; Irwin, R.E. Direct and ecological costs of resistance to herbivory. *Trends Ecol. Evol.* **2002**, *17*, 278–285. [CrossRef]
38. Coley, P.D.; Bryant, J.P.; Chapin, F.S. Resource availability and plant antiherbivore defense. *Science* **1985**, *230*, 895–899. [CrossRef]
39. Chapin, F.S., III; Autumn, K.; Pugnaire, F. Evolution of suites of traits in response to environmental-stress. *Am. Nat.* **1993**, *142*, S78–S92. [CrossRef]
40. Herms, D.A.; Mattson, W.J. The dilemma of plants—To grow or defend. *Q. Rev. Biol.* **1992**, *67*, 283–335. [CrossRef]
41. Yu, Q.; Powles, S. Metabolism-based herbicide resistance and cross-resistance in crop weeds: A threat to herbicide sustainability and global crop production. *Plant Physiol.* **2014**, *166*, 1106–1118. [CrossRef]
42. Vila-Aiub, M.M.; Gundel, P.E.; Preston, C. Experimental methods for estimation of plant fitness costs associated with herbicide-resistance genes. *Weed Sci.* **2015**, *63* (Suppl. 1), 203–216. [CrossRef]
43. Keshtkar, E.; Abdolshahi, R.; Sasanfar, H.; Zand, E.; Beffa, R.; Dayan, F.E.; Kudsk, P. Assessing Fitness Costs from a Herbicide-Resistance Management Perspective: A Review and Insight. *Weed Sci.* **2019**, *67*, 1–12. [CrossRef]
44. Baucom, R.G. Evolutionary and ecological insights from herbicide resistant weeds: What have we learned about plant adaptation, and what is left to uncover? *New Phytol.* **2019**, *223*, 68–82. [CrossRef] [PubMed]
45. Yu, Q.; Han, H.; Vila-Aiub, M.M.; Powles, S.B. AHAS herbicide resistance endowing mutations: Effect on AHAS functionality and plant growth. *J. Exp. Bot.* **2010**, *61*, 3925–3934. [CrossRef]
46. Ashigh, J.; Tardif, F. An Ala_{205}Val substitution in acetohydroxyacid synthase of Eastern black nightshade (Solanum ptychanthum) reduces sensitivity to herbicides and feedback inhibition. *Weed Sci.* **2007**, *55*, 558–565. [CrossRef]
47. Purrington, C.B.; Bergelson, J. Exploring the physiological basis of costs of herbicide resistance in Arabidopsis thaliana. *Am. Nat.* **1999**, *154*, S82–S91. [CrossRef]
48. Menchari, Y.; Chauvel, B.; Darmency, H.; Délye, C. Fitness costs associated with three mutant acetyl-coenzyme A carboxylase alleles endowing herbicide resistance in black-grass Alopecurus myosuroides. *J. Appl. Ecol.* **2008**, *45*, 939–947. [CrossRef]
49. Roux, F.; Gasquez, J.; Reboud, X. The dominance of the herbicide resistance cost in several Arabidopsis thaliana mutant lines. *Genetics* **2004**, *166*, 449–460. [CrossRef]
50. Paris, M.; Roux, F.; Berard, A.; Reboud, X. The effects of the genetic background on herbicide resistance fitness cost and its associated dominance in Arabidopsis thaliana. *Heredity* **2008**, *101*, 499–506. [CrossRef]
51. Frenkel, E.; Matzrafi, M.; Rubin, B.; Peleg, Z. Effects of environmental conditions on the fitness penalty in herbicide resistant brachypodium hybridum. *Front. Plant Sci.* **2017**, *8*, 1–10. [CrossRef]

52. Williams, M.M.I.; Jordan, N.; Yerkes, C. The fitness cost of triazine resistance in jimsonweed (Datura stramonium L.). *Am. Midl. Nat.* **1995**, *133*, 131–137.
53. Sammons, R.D.; Gaines, T.A. Glyphosate resistance: State of knowledge. *Pest Manag. Sci.* **2014**, *70*, 1367–1377. [CrossRef] [PubMed]
54. Yu, Q.; Jalaludin, A.; Han, H.; Chen, M.; Sammons, R.D.; Powles, S.B. Evolution of a double amino acid substitution in the EPSP synthase in Eleusine indica conferring high level glyphosate resistance. *Plant Physiol.* **2015**, *167*, 1440–1447. [CrossRef] [PubMed]
55. Han, H.; Vila-Aiub, M.M.; Jalaludin, A.; Yu, Q.; Powles, S.B. A double EPSPS gene mutation endowing glyphosate resistance shows a remarkably high resistance cost. *Plant Cell Environ.* **2017**, *40*, 3031–3042. [CrossRef]
56. Gronwald, J.W. Resistance to Photosystem II Inhibiting Herbicides. In *Herbicide Resistance in Plants. Biology and Biochemistry*; Powles, S.B., Holtum, J.A.M., Eds.; CRC Press: Boca Raton, FL, USA, 1994; pp. 27–60.
57. Devine, M.D.; Shukla, A. Altered target sites as a mechanism of herbicide resistance. *Crop Prot.* **2000**, *19*, 881–889. [CrossRef]
58. Vila-Aiub, M.M.; Neve, P.; Powles, S.B. Resistance cost of a cytochrome P450 herbicide metabolism mechanism but not an ACCase target site mutation in a multiple resistant Lolium rigidum population. *New Phytol.* **2005**, *167*, 787–796. [CrossRef]
59. Bravo, W.; Leon, R.G.; Ferrell, J.A.; Mulvaney, M.J.; Wood, C.W. Differentiation of life-history traits among Palmer amaranth populations (Amaranthus palmeri) and its relation to cropping systems and glyphosate sensitivity. *Weed Sci.* **2017**, *65*, 339–349. [CrossRef]
60. Van Etten, M.L.; Kuester, A.; Chang, S.M.; Baucom, R.S. Fitness costs of herbicide resistance across natural populations of the common morning glory, Ipomoea purpurea. *Evolution* **2016**, *70*, 2199–2210. [CrossRef]
61. Kuester, A.; Fall, E.; Chang, S.M.; Baucom, R.S. Shifts in outcrossing rates and changes to floral traits are associated with the evolution of herbicide resistance in the common morning glory. *Ecol. Lett.* **2017**, *20*, 41–49. [CrossRef]
62. Tardif, F.J.; Rajcan, I.; Costea, M. A mutation in the herbicide target site acetohydroxyacid synthase produces morphological and structural alterations and reduces fitness in Amaranthus powellii. *New Phytol.* **2006**, *169*, 251–264. [CrossRef]
63. Comont, D.; Knight, C.; Crook, L.; Hull, R.; Beffa, R.; Neve, P. Alterations in Life-History Associated With Non-target-site Herbicide Resistance in Alopecurus myosuroides. *Front. Plant Sci.* **2019**, *223*, 1584–1594. [CrossRef] [PubMed]
64. Vila-Aiub, M.M.; Neve, P.; Steadman, K.J.; Powles, S.B. Ecological fitness of a multiple herbicide-resistant Lolium rigidum population: Dynamics of seed germination and seedling emergence of resistant and susceptible phenotypes. *J. Appl. Ecol.* **2005**, *42*, 288–298. [CrossRef]
65. Délye, C.; Menchari, Y.; Michel, S.; Cadet, É.; Le Corre, V. A new insight into arable weed adaptive evolution: Mutations endowing herbicide resistance also affect germination dynamics and seedling emergence. *Ann. Bot.* **2013**, *111*, 681–691. [CrossRef] [PubMed]
66. Owen, M.J.; Michael, P.J.; Renton, M.; Steadman, K.J.; Powles, S.B. Towards large-scale prediction of Lolium rigidum emergence. II. Correlation between dormancy and herbicide resistance levels suggests an impact of cropping systems. *Weed Res.* **2011**, *51*, 133–141. [CrossRef]
67. Bagavathiannan, M.V.; Davis, A.S. An ecological perspective on managing weeds during the great selection for herbicide resistance. *Pest Manag. Sci.* **2018**, *74*, 2277–2286. [CrossRef]
68. Gassmann, A.J. Resistance to herbicide and susceptibility to herbivores: Environmental variation in the magnitude of an ecological trade-off. *Oecologia* **2005**, *145*, 575–585. [CrossRef]
69. Salzmann, D.; Handley, R.J.; Mueller-Scharer, H. Functional significance of triazine-herbicide resistance in defence of Senecio vulgaris against a rust fungus. *Basic Appl. Ecol.* **2008**, *9*, 577–587. [CrossRef]
70. Mithila, J.; McLean, M.D.; Chen, S.; Christopher Hall, J. Development of near-isogenic lines and identification of markers linked to auxinic herbicide resistance in wild mustard (Sinapis arvensis L.). *Pest Manag. Sci.* **2012**, *68*, 548–556. [CrossRef]
71. Martin, S.L.; Benedict, L.; Sauder, C.A.; Wei, W.; da Costa, L.O.; Hall, L.M.; Beckie, H.J. Glyphosate resistance reduces kochia fitness: Comparison of segregating resistant and susceptible F2 populations. *Plant Sci.* **2017**, *261*, 69–79. [CrossRef]

72. Kumar, V.; Jha, P.; Lim, C.A.; Stahlman, P.W. Differential Germination Characteristics of Dicamba-Resistant Kochia (Bassia scoparia) Populations in Response to Temperature. *Weed Sci.* **2018**, *66*, 1–8. [CrossRef]
73. Osipitan, O.A.; Dille, J.A. Fitness Outcomes Related to Glyphosate Resistance in Kochia (Kochia scoparia): What Life History Stage to Examine? *Front. Plant Sci.* **2017**, *8*, 1–13. [CrossRef] [PubMed]
74. Beckie, H.J.; Blackshaw, R.E.; Leeson, J.Y.; Stahlman, P.W.; Gaines, T.A.; Johnson, E.N. Seedbank persistence, germination and early growth of glyphosate-resistant Kochia scoparia. *Weed Res.* **2018**, *58*, 177–187. [CrossRef]
75. Baucom, R.S.; Mauricio, R. Fitness costs and benefits of novel herbicide tolerance in a noxious weed. *Proc. Natl. Acad. Sci. USA* **2004**, *101*, 13386–13390. [CrossRef] [PubMed]
76. Vila-Aiub, M.; García, F.; Han, H.; Jalaludin, A.; Yu, Q.; Powles, S.B. Resistance Benefit Endowed by a Double EPSPS Glyphosate Resistance Mutation (TIPS) in Eleusine Indica. In *Resistance 2019*; Rothamsted Research: Herts, UK, 2019.
77. Vila-Aiub, M.M.; Neve, P.; Powles, S.B. Evidence for an ecological cost of enhanced herbicide metabolism in Lolium rigidum. *J. Ecol.* **2009**, *97*, 772–780. [CrossRef]
78. Yanniccari, M.; Vila-Aiub, M.; Istilart, C.; Acciaresi, H.; Castro, A.M. Glyphosate resistance in perennial ryegrass (Lolium perenne L.) is associated with a fitness penalty. *Weed Sci.* **2016**, *64*, 71–79. [CrossRef]
79. Gassmann, A.J.; Futuyma, D.J. Consequence of herbivory for the fitness cost of herbicide resistance: Photosynthetic variation in the context of plant-herbivore interactions. *J. Evol. Biol.* **2005**, *18*, 447–454. [CrossRef]
80. Murphy, B.P.; Tranel, P.J. Target-Site Mutations Conferring Herbicide Resistance. *Plants* **2019**, *8*, 382. [CrossRef]
81. LeClere, S.; Wu, C.; Westra, P.; Sammons, R.D. Cross-resistance to dicamba, 2, 4-D, and fluroxypyr in Kochia scoparia is endowed by a mutation in an AUX/IAA gene. *Proc. Natl. Acad. Sci. USA* **2018**, *115*, 2911–2920. [CrossRef]
82. Alstad, D. *Basic Populus Models of Ecology*; Prentice Hall: Upper Saddle River, NJ, USA, 2001.
83. Colbach, N.; Chauvel, B.; Darmency, H.; Délye, C.; Le Corre, V. Choosing the best cropping systems to target pleiotropic effects when managing single-gene herbicide resistance in grass weeds. A blackgrass simulation study. *Pest Manag. Sci.* **2016**, *72*, 1910–1925. [CrossRef]
84. Jordan, N.; Kelrick, M.; Brooks, J.; Kinerk, W. Biorational management tactics to select against triazine-resistant Amaranthus hybridus: A field trial. *J. Appl. Ecol.* **1999**, *36*, 123–132. [CrossRef]
85. Uyenoyama, M. Pleiotropy and the Evolution of Genetic Systems Conferring Resistance to Pesticides. In *Pesticide Resistance. Strategies and Tactics for Management*; Glass, E., Ed.; National Academy of Sciences: Washington, DC, USA, 1986; pp. 207–221.
86. Doole, G.J. Optimal management of annual ryegrass (Lolium rigidum Gaud.) in phase rotations in the Western Australian Wheatbelt. *Aust. J. Agric. Resour. Econ.* **2008**, *52*, 339–362. [CrossRef]
87. Osipitan, O.A.; Dille, J.A.; Assefa, Y.; Radicetti, E.; Ayeni, A.; Knezevic, S.Z. Impact of Cover Crop Management on Level of Weed Suppression: A Meta-Analysis. *Crop Sci.* **2019**, *59*, 833–842. [CrossRef]
88. Sadras, V.O.; Lawson, C. Genetic gain in yield and associated changes in phenotype, trait plasticity and competitive ability of South Australian wheat varieties released between 1958 and 2007. *Crop Pasture Sci.* **2011**, *62*, 533–549. [CrossRef]
89. Borger, C.P.; Hashem, A.; Pathan, S. Manipulating crop row orientation to suppress weeds and increase crop yield. *Weed Sci.* **2010**, *58*, 174–178. [CrossRef]
90. Maino, J.L.; Renton, M.; Hoffmann, A.A.; Umina, P.A. Field margins provide a refuge for pest genes beneficial to resistance management. *J. Pest Sci.* **2019**, *92*, 1017–1026. [CrossRef]

MDPI
St. Alban-Anlage 66
4052 Basel
Switzerland
Tel. +41 61 683 77 34
Fax +41 61 302 89 18
www.mdpi.com

Plants Editorial Office
E-mail: plants@mdpi.com
www.mdpi.com/journal/plants

www.ingramcontent.com/pod-product-compliance
Lightning Source LLC
Chambersburg PA
CBHW061522210926
51850CB00008B/2001
9783039360086